# Advances in Sustainability Science and Technology

The book series aims at bringing together valuable and novel scientific contributions that address the critical issues of renewable energy, sustainable building, sustainable manufacturing, and other sustainability science and technology topics that have an impact in this diverse and fast-changing research community in academia and industry.

The areas to be covered are

- Climate change and mitigation, atmospheric carbon reduction, global warming
- Sustainability science, sustainability technologies
- Sustainable building technologies
- Intelligent buildings
- Sustainable energy generation
- Combined heat and power and district heating systems
- Control and optimization of renewable energy systems
- Smart grids and micro grids, local energy markets
- Smart cities, smart buildings, smart districts, smart countryside
- Energy and environmental assessment in buildings and cities
- Sustainable design, innovation and services
- Sustainable manufacturing processes and technology
- Sustainable manufacturing systems and enterprises
- Decision support for sustainability
- Micro/nanomachining, microelectromechanical machines (MEMS)
- Sustainable transport, smart vehicles and smart roads
- Information technology and artificial intelligence applied to sustainability
- Big data and data analytics applied to sustainability
- Sustainable food production, sustainable horticulture and agriculture
- Sustainability of air, water and other natural resources
- Sustainability policy, shaping the future, the triple bottom line, the circular economy

High quality content is an essential feature for all book proposals accepted for the series. It is expected that editors of all accepted volumes will ensure that contributions are subjected to an appropriate level of reviewing process and adhere to KES quality principles.

The series will include monographs, edited volumes, and selected proceedings.

More information about this series at http://www.springer.com/series/16477

Amit Joshi · Atulya K. Nagar ·
Gabriela Marín-Raventós
Editors

# Sustainable Intelligent Systems

Springer

*Editors*
Amit Joshi
Global Knowledge Research Foundation
Ahmedabad, Gujarat, India

Atulya K. Nagar
School of Mathematics
Computer Science and Engineering
Liverpool Hope University
Liverpool, UK

Gabriela Marín-Raventós
University of Costa Rica
San José, Costa Rica

ISSN 2662-6829 ISSN 2662-6837 (electronic)
Advances in Sustainability Science and Technology
ISBN 978-981-33-4900-1 ISBN 978-981-33-4901-8 (eBook)
https://doi.org/10.1007/978-981-33-4901-8



This Springer imprint is published by the registered company Springer Nature Singapore Pte Ltd.
The registered company address is: 152 Beach Road, #21-01/04 Gateway East, Singapore 189721, Singapore

# Preface

Sustainable Intelligent Systems explains and aims at giving a gist to exigency of the technological solutions with a forward move towards sustainability. The book presents issues related to ICT, intelligent systems, data science, AI, machine learning, sustainable development and overall their impacts on sustainability and provides an overview of the technologies awaiting unveiling. The book also discusses novel intelligent algorithms and their applications to move from a data-centric world to sustainable world. The book also includes the research paradigms on SDGs and societal impacts. The book provides an overview of cutting-edge techniques towards sustainability and ideas to help researchers who want to understand the challenges and opportunities of using for Smart Management Prospective for Sustainable Society. It covers wide ranges of audience from computer science, data analysts, AI technocrats and management researchers.

A great pool of more than 100 papers were submitted for this volume by researchers related to the theme intelligent sustainable systems. This book appeals to a broad readership, including academics, researchers and industry professionals.

| | |
|---|---|
| Ahmedabad, India | Amit Joshi, Ph.D. |
| Liverpool, UK | Atulya K. Nagar, Ph.D. |
| San José, Costa Rica | Gabriela Marín-Raventós, Ph.D. |

# Contents

# Editors and Contributors

## About the Editors

**Amit Joshi** is currently Director of Global Knowledge Research Foundation and also Entrepreneur Researcher who has completed his masters and research in the areas of cloud computing and cryptography in medical imaging. Dr. Joshi has an experience of around 10 years in academic and industry in prestigious organizations. Dr. Joshi is an active member of ACM, IEEE, CSI, AMIE, IACSIT, Singapore, IDES, ACEEE, NPA and many other professional societies. Currently, Dr. Joshi is International Chair of InterYIT at International Federation of Information Processing (IFIP, Austria). He has presented and published more than 50 papers in national and international journals/conferences of IEEE and ACM. Dr. Joshi has also edited more than 40 books which are published by Springer, ACM and other reputed publishers. Dr. Joshi has also organized more than 50 national and international conferences and programs in association with ACM, Springer and IEEE to name a few across different countries including India, UK, Europe, USA, Canada, Thailand, Egypt and many more.

**Atulya K. Nagar** holds Foundation Chair as Professor of Mathematical Sciences and is Pro-Vice-Chancellor for Research and Dean of the Faculty of Science at Liverpool Hope University, UK. He is also Head of the School of Mathematics, Computer Science and Engineering, which he established at the University. He is an internationally respected scholar working at the cutting edge of theoretical computer science, applied mathematical analysis, operations research and systems engineering. He received a prestigious Commonwealth Fellowship to pursue his doctorate (D.Phil.) in Applied Nonlinear Mathematics, which he earned from the University of York (UK) in 1996. He holds a B.Sc. (Hons.), an M.Sc. and M.Phil. (with distinction) in Mathematical Physics from the MDS University of Ajmer, India. His research expertise spans both applied mathematics and computational methods for nonlinear, complex and intractable problems arising in science, engineering and industry.

**Gabriela Marín-Raventós** received a M.Sc. in Computer Science from Case Western Reserve University in 1985 and a Ph.D. in Business Analysis and Research from Texas A&M University, USA, in 1993. She has been a Computer Science faculty member at Universidad de Costa Rica (UCR) since 1980. She was Dean of Graduate Studies and Director of the Research Center for Communication and Information Technologies (CITIC), both at UCR. Currently, she is Director of the Graduate Program in Computer Science and Informatics. She has organized several international and national conferences, and has been, and actually is, Chair of several program and editorial committees. From 2012 to 2016, she was President of the Latin American Center for Computer Studies (CLEI), becoming the first woman to occupy such a distinguished position. Since September 2016, she is Vice-President of the International Federation for Information Processing (IFIP), in charge of the Digital Equity Committee. Her research interests include smart cities, human–computer interaction, decision support systems, gender in IT and digital equity.

## Contributors

**Roberto O. Andrade** Facultad de Ingeniería de Sistemas. Escuela Politécnica Nacional, Quito, Ecuador

**Viacheslav Belenko** LG Electronics Inc., Seoul, South Korea

**David Bird** The Institution of Engineering Technology, London, UK

**Jose A. Brenes** Research Center for Communication and Information Technologies (CITIC), University of Costa Rica, San José, Costa Rica

**María Cazares** IDEIAGEOCA Research Group, Universidad Politécnica Salesiana, Quito, Ecuador

**Valery Chernenko** LG Electronics Inc., Seoul, South Korea

**Haejin Cho** LG Electronics Inc., Seoul, South Korea

**Franco Cicirelli** CNR—National Research Council of Italy, Institute for High Performance Computing and Networking—(ICAR), Rende, Italy

**César A. Collazos** Universidad del Cauca, Popayán, Colombia

**Paweł Dymora** Faculty of Electrical and Computer Engineering, Rzeszów University of Technology, Rzeszów, Poland

**Markus Eger** Graduate Program (PPCI), University of Costa Rica, San José, Costa Rica

**Mezzour Ghita** National and High School of Electricity and Mechanic (ENSEM), HASSAN II University, Casablanca, Morocco;
Research Foundation for Development and Innovation in Science and Engineering, Casablanca, Morocco;
Innovation Lab for Operations (ILO), Mohammed VI Polytechnic University (UM6P), Ben Guerir, Morocco

**Hafid Griguer** Innovation Lab for Operations (ILO), Mohammed VI Polytechnic University (UM6P), Ben Guerir, Morocco

**César D. Guerrero** Universidad Autónoma de Bucaramanga, Bucaramanga, Colombia

**Medromi Hicham** National and High School of Electricity and Mechanic (ENSEM), HASSAN II University, Casablanca, Morocco;
Research Foundation for Development and Innovation in Science and Engineering, Casablanca, Morocco

**Mihaiela Iliescu** Department of Mechatronics and Robotics, Institute of Solid Mechanics of the Romanian Academy, Bucharest, Romania

**Ashok Jetawat** Pacific Academy of Higher Education & Research Society, Udaipur, Rajasthan, India

**Manuj Joshi** Pacific Academy of Higher Education & Research Society, Udaipur, Rajasthan, India

**Mustapha Kabil** Hassan II University, Casablanca, Morocco

**Maxim Kalinin** Peter the Great St.Petersburg Polytechnic University, St. Petersburg, Russia

**Vasiliy Krundyshev** Peter the Great St.Petersburg Polytechnic University, St. Petersburg, Russia

**My Driss Laanaoui** Cadi Ayyad University, Marrakesh, Morocco

**Sangwon Lim** LG Electronics Inc., Seoul, South Korea

**Gina Maestre-Gongora** Universidad Cooperativa de Colombia, Medellín, Colombia

**V. V. Mahesh** Division of Electronics Engineering, School of Engineering, Cochin University of Science and Technology, Kochi, India

**Marvy Badr Monir Mansour** Department of Electrical Engineering, Faculty of Engineering, The British University in Egypt, Cairo, Egypt

**Gabriela Marín-Raventós** Graduate Program (PPCI), University of Costa Rica, San José, Costa Rica

**Mirosław Mazurek** Faculty of Electrical and Computer Engineering, Rzeszów University of Technology, Rzeszów, Poland

**Yurley Medina-Cárdenas** Universidad Francisco de Paula Santander Ocaña, Ocaña, Colombia

**Libero Nigro** DIMES—Department of Informatics Modelling Electronics and Systems Science, University of Calabria, Rende, Italy

**Iván Ortiz** Universidad de Las Américas, Quito, Ecuador

**Marius Pandelea** Department of Mechatronics and Robotics, Institute of Solid Mechanics of the Romanian Academy, Bucharest, Romania

**Shrikant Pawar** Yale University, New Haven, CT, USA

**Julio Proaño** IDEIAGEOCA Research Group, Universidad Politécnica Salesiana, Quito, Ecuador

**Corina Radu Frenţ** Department of Mechatronics and Robotics, Institute of Solid Mechanics of the Romanian Academy, Bucharest, Romania

**Alexander Raikov** Institute of Control Sciences RAS, Moscow, Russia

**Dewar Rico-Bautista** Universidad Francisco de Paula Santander Ocaña, Ocaña, Colombia

**Abdelrahman Said** Department of Electrical Engineering, Faculty of Engineering, The British University in Egypt, Cairo, Egypt

**T. K. Shahana** Division of Electronics Engineering, School of Engineering, Cochin University of Science and Technology, Kochi, India

**Benhadou Siham** National and High School of Electricity and Mechanic (ENSEM), HASSAN II University, Casablanca, Morocco;
Research Foundation for Development and Innovation in Science and Engineering, Casablanca, Morocco

**Mouad Tantaoui** Hassan II University, Casablanca, Morocco

**Isabela Todiriţe** Department of Mechatronics and Robotics, Institute of Solid Mechanics of the Romanian Academy, Bucharest, Romania

**Luige Vlădăreanu** Department of Mechatronics and Robotics, Institute of Solid Mechanics of the Romanian Academy, Bucharest, Romania

# Performance Analysis of Classification Algorithms Used for Software Defect Prediction

**Manuj Joshi and Ashok Jetawat**

**Abstract** The software quality and reliability are being the major issues while developing software, and as in the software project management, the major cost overruns are due to the deployment of bug prone software at the customer side. Mainly, warehouses are used for maintaining the databases regarding fixing the defects. In this research work, six different classification techniques were being selected based on previous researches which showed better results while developing efficient predictive model. The classification techniques being considered were Naive Bayes, multilayer perceptron, random forest, support vector machine, PART and simple CART, and these various techniques were being measured on various performance measure included as correctly classified instances or accuracy, recall, precision, ROC, F-measure, etc. The experimental analysis used the WEKA environmental tool with calibration setting of tenfolds cross-validation which was being applied on PROMISE repository datasets AR1, AR3, AR4, AR5 and AR6 so as to measure the performance of various algorithms on parameters like accuracy or correctly classified instances, recall, F-measure, precision and ROC measure values, and finally, the overall performance was being evaluated. For the all over performance evaluation, the method of average and correlation was being used. The findings suggest that support vector machine classification algorithm was the best and most promising when applied on software defect prediction supported by the highest average accuracy measure of 87.22%. Similarly, other measures like recall, F-measure, precision have higher values of 0.872, 0.847 and 0.857, respectively.

**Keywords** PART · Support vector machine · Naive Bayes

## 1 Introduction

Classification approach used for software defect prediction is basically based on finding an appropriate predictive model which divides the dataset into different

---

M. Joshi (✉) · A. Jetawat
Pacific Academy of Higher Education & Research Society, Udaipur, Rajasthan, India
e-mail: manujjoshi@gmail.com

 
A. Joshi et al. (eds.), *Sustainable Intelligent Systems*, Advances in Sustainability Science and Technology, https://doi.org/10.1007/978-981-33-4901-8_1

classes. In this approach, the unclassified data elements are being classified by providing a particular class label based on the developed model. Applying the classification algorithms in software defect prediction helps to identify defect in the code and find whether the code is defect free or not so that it can further be reused for software development. The test dataset is being classified based on the existing trained model. In this paper, various classification algorithms are being compared and analyzed on different software defect datasets.

The efficient prediction models used for software defect identification are required to find the defects in the code so that by proper approach the latent defects can be rectified before the deployment of the software after post delivery the cost of bug correction is quite high as compared to pre-delivery. Almost every year billion of dollars are being spent on bug fixing or correction. Here, the efficiency of the predictive models comes in play if the models are having high accuracy then these can save large amount of money for the software companies as in case small bug identification and correction can save huge amount of cost.

In the research work, basically six data mining classification algorithms were being chosen, the selection was based on the previous research works being conducted for building an effective predictive model, so the algorithms have better performances were being considered. The paper included the classification algorithms like multilayer perceptron, Naive Bayes, random forest, support vector machine, PART and simple CART which were analyzed on various performance measures or parameters like precision, accuracy, recall, ROC, F-measure, etc. For evaluating the accuracy, the total number of correctly classified instance was being taken in consideration, and for error value, the incorrectly classified instances were taken in consideration. The total performance score was being calculated using the average and correlation methods. According to correlation method, the measures showing either positive or negative values were being used for further analysis, and the remaining were not being taken under consideration.

## 2 Background

Software products are today used in every functional area of the business organizations. According to the present scenario, software quality plays an important role, and it can be defined as the parameter to evaluate design and implementation aspects. The previous researchers suggest many measures based on attributes for evaluating software quality, and the various factors or attributes are product quality, accuracy (correctness), scalability, error free, etc. Although the quality measure adopted by one organization may differ from the other organization, so software metrics standard measures should be used for quality assessment. The software defect predictor can use the software metrics parameters as input to obtain the software quality. The study covers the application of machine learning classifier algorithms to predict the software quality [1].

Quality and reliability are being considered as the major issues while developing software. According to software project management, the major cost overruns are due to the deployment of bug prone software at the customer side. Mainly, software warehouses are used for maintaining the databases regarding fixing the defects. The research work focused on identifying the causes of software failures in recent past with detailed literature review. The timely identification of these bugs requires the use of data mining techniques and also emphasized on the development of efficient bug tracking system. The author suggests that bug repositories play important role in bug tracking system which contains the history of particular codes success or failure, the prediction models related to bug fix time are also useful for marinating the quality and reliability of the software products [2].

According to the Peng experimental study which focused on the software bug prediction using the software measures for the evaluation of the classification algorithm. The approach adopted by author mainly aimed at performance measure evaluation in the study experiment was conducted on ten open projects (that were 34 releases) of the PROMISE repository. The major finding in the study suggested that top-k parameters or measure gives better outcomes as evaluated to standard predictors [3].

Software defect prediction is one of the important aspects in decreasing the costs of the software design, development & maintaining the standards of quality assurance for building quality software solutions. Mainly, classification approaches including the extraction by features and categorization in classes are basically applied to identify bugs or defects which are present in software code at the time of testing conducted in each phase. To improve the accuracy of software bug prediction model, various algorithms and techniques which are based on statistical and mathematical models are being used particularly in the feature extraction phase. Lastly, it can be suggested that the process actually reduces the cost by applying the prediction model as compared to the previous models having lower accuracy measure values and also helps in finding the most promising algorithm for software defect prediction [4].

Software defect prediction is an approach in which an appropriate prediction model is being created so as to predict the future outcomes based on the software faults previous dataset as obtained from software warehouse repository. There are many approaches for developing the prediction model, the author proposed NB, DT and ANNs techniques which were being used or applied on various experimental datasets, and also, various performance measures were being used for evaluating the classifiers. The study basically evaluated the above-mentioned three machine learning algorithms in software defect prediction problem or case. The evaluation process was conducted using three real-time datasets. The performance measures like accuracy, mean absolute error, precision, recall, F-measure, ROC and RMSE were being evaluated for all the three datasets, and the results revealed that machine learning classification algorithms are the best approaches for developing a predictive model so as to maintain the software quality. When comparative analysis was done between the three algorithms, it was found that the decision tree (DT) classifier was the best or appropriate over the others algorithms. The outcomes suggested that machine learning approaches like the classifier are far more better than the AR linear approach and POWM model [5].

## 3 Methodology

The proposed work includes the analysis of six different classification algorithms which are mainly applied for the software defect prediction, and according to the previous studies, the most appropriate algorithms are Naïve Bayes, multilayer perceptron, random forest, support vector machine, PART & simple CART. Selection of the classification techniques was based on the concept that they are built on different mathematical models along with different properties as the Naive Bayes approach is based on the concept of combined probability in which dependent variable being related to various categories of independent variables, and the algorithm requires both type variables to be categorial. The algorithm PART is basically adopted from CART approach which is classification & regression trees, and in this approach, particularly a decision tree is being formed based on the measure information entropy by subdividing the datasets as training based on various independent variables. When considering the SVM classifier, the model develops a hyper-plane that basically categorize the training datasets into two different class labels. Random forest is basically a collaborative technique which combines various CARTs based on various features.

The proposed methodology includes the following steps:

Firstly, the identification & data collection related to software defect prediction datasets which consists of attributes Blank_LOC, Total_LOC, Unique_operands, Comment_LOC, Code_and_comment_LOC, etc., related to software measures.

The second step includes data pre-processing consisting of cleaning, transformation, etc. The data cleaning includes the removal of missing or inconsistent data from the dataset, while transformation is used to convert datasets in different formats to the desired format which is being further considered as input to the WEKA tool.

Thirdly, the experimental environment is basically used for applying the six suggested classification algorithms on five software defect prediction datasets using tenfold cross-validation testing configuration as the algorithm calibration for measuring, analyzing & comparing the algorithms on six different parameters.

In the fourth step, various performance analysis measures such as correctly classified instances or accuracy, incorrectly classified instances, recall, F-measure, ROC curve values were being obtained for further evaluation. Mainly, the observations were taken at tenfold cross-validation configuration settings that is datasets were portioned using random sample partition method by using WEKA tool.

Finally, the best algorithms would be selected based on various performance measure comparison, averages and correlation methods.

The experiment was being conducted on PROMISE repository datasets AR1 [6], AR3 [7], AR4 [8], AR5 [9] and AR6 [10] with configuration settings tenfold cross-validation on the WEKA platform. The observations were taken separately for all the five datasets from software units. All the different algorithms were being individually compared based on different performance measures. Lastly, the overall score of performance is being obtained to rank the algorithms.

## 4 Results

According to the experimental analysis, the outcomes as shown in the table below suggest that when applying the configuration setting with cross-validation of tenfolds on the dataset or database AR1, it was found that support vector machine was the best data mining classification algorithm with the highest accuracy or correctly classified instances with value of 91.73%. The second most promising classification algorithm was found to be the rule-based PART which was having the accuracy measure value of 90.90%. Further applying the other measure like mean absolute error on the classification algorithms, it was obtained that the accuracy measure and mean absolute error value justify each other as support vector machine has the lowest mean absolute value of error that was calculated as 0.0826 (Table 1).

The table below for the dataset AR3 suggests that accuracy measure of multilayer perceptron was found to be 93.65% which is higher than the other algorithms. Similarly, the algorithms random forest and Naive Bayes have the accuracy values 92.063 and 90.48%, respectively. Here, for this particular dataset, the support vector machine algorithm has shown lower accuracy with value 88.89% as compared to the dataset AR1. According to the mean absolute error (MAE) measure, the least value was calculated to be of Naive Bayes algorithm with 0.1085, and similarly, the multilayer perceptron algorithm has also shown lower value of 0.1101. The performance evaluation results for AR3 dataset reveal that Naive Bayes algorithm &

**Table 1** Classifiers performance analysis with cross-validation folds (10) on AR1 dataset

| S. No. | Classifier | Accuracy (%) | Mean absolute error | Precision | Recall | F-measure | ROC area |
|---|---|---|---|---|---|---|---|
| 1 | Naïve Bayes | 85.12 | 0.1519 | 0.899 | 0.851 | 0.871 | 0.688 |
| 2 | Multilayer perceptron | 90.08 | 0.1037 | 0.89 | 0.901 | 0.895 | 0.701 |
| 3 | Random forest | 90.08 | 0.1289 | 0.855 | 0.901 | 0.877 | 0.641 |
| 4 | Support vector machine | 91.73 | 0.0826 | 0.856 | 0.917 | 0.886 | 0.496 |
| 5 | PART | 90.90 | 0.1048 | 0.914 | 0.909 | 0.911 | 0.756 |
| 6 | Simple CART | 90.08 | 0.1583 | 0.855 | 0.901 | 0.877 | 0.417 |

multilayer perceptron algorithms were the most appropriate algorithms in software defect prediction (Table 2).

According to the experimental outcomes generated for the dataset AR4, it is observed that correctly classified instances or accuracy value for support vector machine was obtained to be the most appropriate and best having value 85.05% and also has the lowest mean absolute value of 0.1495 as compared to other classification algorithms. Similarly, the second best algorithm was found to be Naive Bayes having the accuracy and MAE values 84.11% and 0.153, respectively (Table 3).

**Table 2** Classifiers performance analysis with cross-validation folds (10) on AR3 dataset

| S. No. | Classifier | Accuracy (%) | Mean absolute error | Precision | Recall | F-measure | ROC area |
|---|---|---|---|---|---|---|---|
| 1 | Naïve Bayes | 90.4762 | 0.1085 | 0.916 | 0.905 | 0.909 | 0.824 |
| 2 | Multilayer perceptron | 93.6508 | 0.1101 | 0.933 | 0.937 | 0.933 | 0.748 |
| 3 | Random forest | 92.0635 | 0.1651 | 0.914 | 0.921 | 0.913 | 0.718 |
| 4 | Support vector machine | 88.8889 | 0.1111 | 0.87 | 0.889 | 0.866 | 0.616 |
| 5 | PART | 88.8889 | 0.142 | 0.883 | 0.889 | 0.886 | 0.567 |
| 6 | Simple CART | 88.8889 | 0.1568 | 0.87 | 0.889 | 0.866 | 0.639 |

**Table 3** Classifiers performance analysis with cross-validation folds (10) on AR4 dataset

| S. No. | Classifier | Accuracy (%) | Mean absolute error | Precision | Recall | F-measure | ROC area |
|---|---|---|---|---|---|---|---|
| 1 | Naïve Bayes | 84.1121 | 0.153 | 0.828 | 0.841 | 0.832 | 0.809 |
| 2 | Multilayer perceptron | 81.3084 | 0.2081 | 0.799 | 0.813 | 0.805 | 0.71 |
| 3 | Random forest | 82.243 | 0.2187 | 0.792 | 0.822 | 0.794 | 0.794 |
| 4 | Support vector machine | 85.0467 | 0.1495 | 0.838 | 0.85 | 0.823 | 0.639 |
| 5 | PART | 81.3084 | 0.2046 | 0.793 | 0.813 | 0.8 | 0.603 |
| 6 | Simple CART | 81.3084 | 0.26 | 0.787 | 0.813 | 0.794 | 0.69 |

**Table 4** Classifiers performance analysis with cross-validation folds (10) on AR5 dataset

| S. No. | Classifier | Accuracy (%) | Mean absolute error | Precision | Recall | F-measure | ROC area |
|---|---|---|---|---|---|---|---|
| 1 | Naïve Bayes | 83.3333 | 0.1666 | 0.851 | 0.833 | 0.84 | 0.907 |
| 2 | Multilayer perceptron | 69.4444 | 0.2665 | 0.708 | 0.694 | 0.701 | 0.799 |
| 3 | Random Forest | 83.3333 | 0.2 | 0.833 | 0.833 | 0.833 | 0.893 |
| 4 | Support vector machine | 83.3333 | 0.1667 | 0.833 | 0.833 | 0.833 | 0.759 |
| 5 | PART | 83.3333 | 0.2018 | 0.833 | 0.833 | 0.833 | 0.717 |
| 6 | Simple CART | 77.7778 | 0.2642 | 0.778 | 0.778 | 0.778 | 0.676 |

The accuracy measure values for support vector machine, random forest, PART & Naive Bayes were found to be highest when applied on dataset AR5 with values 83.33% each, but support vector machine and Naive Bayes were only having the lowest mean absolute error values of 0.167 each (Table 4).

The table below for the dataset AR6 suggests that accuracy value of support vector machine was obtained to be 87.13% which is higher than the other algorithms. Similarly, the algorithm random has the accuracy values 84.15%. According to the mean absolute error measure, the lowest value was obtained to be of the support vector machine (SVM) algorithm having value 0.1287, and similarly, the Naïve Bayes algorithm has also shown lower value of 0.1744. The performance evaluation results for AR6 dataset reveal that SVM algorithm was observed to be the most appropriate algorithm in software defect prediction (Table 5).

The overall performance evaluation based on average and correlations suggests that support vector machine classification algorithm is the most appropriate algorithm for the software defect prediction as it has the highest average accuracy value of 87.22%, lowest MAE error value of 0.1277 and high precision, F-measure & recall values of 0.857, 0.847 and 0.872, respectively, as compared with other algorithms (Table 6).

## 5 Conclusion

The results conclude that for the above-mentioned datasets of software defect prediction the accuracy and mean absolute error performance measures for all classification algorithms have shown variation. The accuracy and mean absolute error follow the negative correlation between them. In all the cases except AR3 dataset, it was found

**Table 5** Classifiers performance analysis with cross-validation folds (10) on AR6 dataset

| S. No. | Classifier | Accuracy (%) | Mean absolute error | Precision | Recall | F-measure | ROC area |
|---|---|---|---|---|---|---|---|
| 1 | Naïve Bayes | 82.1782 | 0.1744 | 0.812 | 0.822 | 0.816 | 0.652 |
| 2 | Multilayer perceptron | 82.1782 | 0.2074 | 0.812 | 0.822 | 0.816 | 0.751 |
| 3 | Random forest | 84.1584 | 0.2198 | 0.806 | 0.842 | 0.816 | 0.663 |
| 4 | Support vector machine | 87.1287 | 0.1287 | 0.888 | 0.871 | 0.827 | 0.567 |
| 5 | PART | 80.198 | 0.235 | 0.764 | 0.802 | 0.781 | 0.526 |
| 6 | Simple CART | 84.1584 | 0.2573 | 0.724 | 0.842 | 0.778 | 0.397 |

**Table 6** Overall performance based on average measure including all datasets

| S. No. | Classifier | Accuracy (%) | Mean absolute error | Precision | Recall | F-measure | ROC area |
|---|---|---|---|---|---|---|---|
| 1 | Naive Bayes | 85.04 | 0.1509 | 0.8612 | 0.8504 | 0.8536 | 0.776 |
| 2 | Multilayer perceptron | 83.33 | 0.1792 | 0.8284 | 0.8334 | 0.83 | 0.7418 |
| 3 | Random forest | 86.37 | 0.1865 | 0.84 | 0.8638 | 0.8466 | 0.7418 |
| 4 | Support vector machine | 87.22 | 0.1277 | 0.857 | 0.872 | 0.847 | 0.6154 |
| 5 | PART | 84.92 | 0.1776 | 0.8374 | 0.8492 | 0.8422 | 0.6338 |
| 6 | Simple CART | 84.44 | 0.2193 | 0.8028 | 0.8446 | 0.8186 | 0.5638 |

that support vector machine classification algorithm is the most appropriate algorithm for the software defect prediction. The results were being further justified by comparing the algorithms based on average overall performance which were in accordance with support vector machine classification algorithm being the best algorithm with highest average accuracy value of 87.22% also taking in consideration the measure mean absolute error value which was found to be lowest with 0.1277. Similarly, the other measures recall, F-measure and precision were found to be on higher side with values 0.872, 0.847 and 0.857, respectively.

## References

1. F. Akmel, E. Birihanu, B. Siraj, A literature review study of software defect prediction using machine learning techniques. Int. J. Emerg. Res. Manage. Technol. **6**, 300 (2018). https://doi.org/10.23956/ijermt.v6i6.286
2. T. Hall, S. Beecham, D. Bowes, D. Gray, S. Counsell, A systematic literature review on fault prediction performance in software engineering. IEEE Trans. Softw. Eng. **38**(6), 1276–1304 (2012)
3. P. He et al. An empirical study on software defect prediction with a simplified metric set. Inf. Softw. Technol **59**, 170–190 (2015)
4. A. Parameswari, Comparing data mining techniques for software defect prediction. Int. J. Sci. Eng. Res. (IJ0SER) **3**(5), 3221–5687 (2015).
5. A. Hammouri, M. Hammad, M. Alnabhan, F. Alsarayrah, Software bug prediction using machine learning approach. Int. J. Adv. Comput. Sci. Appl. **9**(2), 78–83 (2018)
6. G. Boetticher, T. Menzies, T. Ostrand, AR1/Software Defect Prediction Dataset (Classification), PROMISE Repository of Empirical Software Engineering Data https://promisedata.org/repository (West Virginia University, Department of Computer Science, 2007)
7. G. Boetticher, T. Menzies, T. Ostrand, AR3/Software Defect Prediction Dataset (Classification), PROMISE Repository of Empirical Software Engineering Data https://promisedata.org/repository (West Virginia University, Department of Computer Science, 2007)
8. G. Boetticher, T. Menzies, T. Ostrand, AR4/Software Defect Prediction Dataset (Classification), PROMISE Repository of Empirical Software Engineering Data https://promisedata.org/repository (West Virginia University, Department of Computer Science, 2007)
9. G. Boetticher, T. Menzies, T. Ostrand, AR5/Software Defect Prediction Dataset (Classification), PROMISE Repository of Empirical Software Engineering Data https://promisedata.org/repository (West Virginia University, Department of Computer Science, 2007)
10. G. Boetticher, T. Menzies, T. Ostrand, AR6/Software Defect Prediction Dataset (Classification), PROMISE Repository of Empirical Software Engineering Data https://promisedata.org/repository (West Virginia University, Department of Computer Science, 2007)

# Early Detection of Diseases in Precision Agriculture Processes Supported by Technology

**Jose A. Brenes, Markus Eger, and Gabriela Marín-Raventós**

**Abstract** One of the biggest challenges for farmers is the prevention of disease appearance on crops. Governments around the world control border product entry to reduce the number of foreign diseases affecting local producers. Evenmore, it is also important to reduce the spread of crop diseases as quickly as possible and in early stages of propagation, to enable farmers to attack them on time, or to remove the affected plants. In this research, we propose the use of convolutional neural networks to detect diseases in horticultural crops. We compare the results of disease classification in images of plant leaves, in terms of performance, time execution, and classifier size. In the analysis, we implement two distinct classifiers, a densenet-161 pre-trained model and a custom created model. We concluded that for disease detection in tomato crops, our custom model has better execution time and size, and the classification performance is acceptable. Therefore, the custom model could be useful to use to create a solution that helps small farmers in rural areas in resource-limited mobile devices.

**Keywords** Diseases detection · Precision agriculture · Machine learning · Feature selection · Convolutional neural networks

## 1 Introduction

Currently, the world economy is facing big challenges. Today, most countries are involved in a socioeconomic slowdown, augmented by the COVID-19 pandemic,

---

J. A. Brenes (✉)
Research Center for Communication and Information Technologies (CITIC), University of Costa Rica, 11501 San José, Costa Rica
e-mail: joseantonio.brenes@ucr.ac.cr

M. Eger · G. Marín-Raventós
Graduate Program (PPCI), University of Costa Rica, 11501 San José, Costa Rica
e-mail: markus.eger@ucr.ac.cr

G. Marín-Raventós
e-mail: gabriela.marin@ucr.ac.cr

 
A. Joshi et al. (eds.), *Sustainable Intelligent Systems*, Advances in Sustainability Science and Technology, https://doi.org/10.1007/978-981-33-4901-8_2

requiring all sectors to innovate. In Costa Rica, for example, the agricultural sector had a growth equivalent to the 3.7% of GDP, in 2017. Meanwhile, in 2018, the growth of the sector was only 2.4% of GDP. That represents a decrease in the contribution to the country's economy compared to the previous year [1]. Governments are expecting bigger decrements due to the pandemic recession. It is important to consider that with the reported decrease in production, a lot of families saw, in the last years, and will soon see their income reduced. According to the Costa Rican Performance Report of the Agricultural Sector [1], the reported GDP growth in 2018 was possible due to improvements in agricultural activities related to pineapple crops. In that area, farmers applied actions to control pests and diseases affecting pineapple crops fields.

Tomato and bell pepper crops are two of the most produced crops in the world [2]. In 2018, both crops reached high production rates in the order of 182 and 36 million tons of tomatoes and bell peppers, respectively.[1] In Costa Rica, tomatoes, and bell peppers are the two most consumed vegetables, reaching an intake of about 18.2 and 3.3 kg per capita of tomato and bell pepper, respectively [3, 4].

Nevertheless, even though horticulture is particularly important in Costa Rica, Ramírez and Nienhuis [5] emphasize on pests and diseases suffered by these types of crops. Pests and diseases force farmers to apply chemical pesticides with harmful consequences for the environment. Thus, detecting them as early as possible is especially important in order to minimize the need for these chemicals.

In addition, Abarca [6] performed an analysis of climate change and plagues in crops cultivated in the tropics. He mentioned that it is crucial to create solutions that enable early alarms against disruptive plagues. In this way, it can be possible to react fast to the plague and avoid the appearance of the disease with a high impact in crops.

Dimitrios et al. [7] state that tools implementing emergent technologies to support farmers in pest and disease management are imperative now more than ever. The high incidence of pests and plagues on horticultural crops forces us to join efforts to create tools focused on this type of crops. These solutions must enable farmers to act immediately. For practical applications, we must also consider that farms may not have ready access to powerful hardware. An important consideration is therefore also to provide solutions that can run with very little resources, such as on a standard smartphone device.

There are some IT research efforts around the world focused on contributing to the agricultural productive sector, but mainly on automatic fertigation [8], while pest and disease management have been less studied. We center our research on this area. We propose the use of machine learning techniques to detect disease appearance on crops early to enable farmers to attack them on time or to remove the affected plants avoiding its spread. To achieve it, we keep in mind all the time, that the resulting solution must have to be runnable on mobile devices with very low resources to make it accessible to small-scale farmers in rural areas.

This paper is organized as follows. In the next section, we define some concepts needed to understand our research. In the following section, we mention different studies made by other researchers that are related to our research. Afterward, we

[1] https://www.fao.org/faostat/en/#data/QC.

present statistical analysis to select the features and artificial intelligence techniques to use in the tool to be provided to small farmers. Next, we present a study case in which we evaluate the feasibility of the proposal by comparing two different convolution neural networks in terms of performance, the training execution time, and the model's resultant size. Finally, we present the conclusions of this research process.

## 2 Theoretical Framework

In this section, we introduce some technical terms that must be clear to achieve a comprehension of our research.

### *2.1 Precision Agriculture*

Dwivedi et al. [9] define precision agriculture as a farming managing method based on observing, measuring, and responding to variability in crops. The goal of precision agriculture is to define a decision support system (DSS) that can be used to manage the production optimizing its processes and increasing its yield.

Bongiovanni and Lowenberg-DeBoer [10] mention that precision agriculture is based on the idea of automation of site-specific management (SSM). This is achieved by using information technology to tailor input use to reach desired outcomes or to monitor those outcomes. Precision agriculture merges technologies with agricultural processes to achieve an integrated crop management [11].

### *2.2 Horticultural Crops*

The horticultural crops are defined as herbaceous plants cultivated with the goal of self-consumption as well as for commercialization in internal and external markets [12]. Vegetables are particularly important crops for mankind because they are an extraordinarily rich source of vitamins and nutrients. They cover crops cultivated directly in the land as lettuce, celery, and parsley; tubers including potatoes and onions; fruits such as tomatoes and bell peppers; and flowers such as broccolis and artichokes [12]. Crops that will be used in our research are described next.

**Bell Pepper**

Bell peppers belong to the family of Solanaceae. *Capsicum annuum* is the most cultivated species in countries located in the tropics [13]. According to [2], bell peppers are one of the most frequently produced vegetables in Costa Rica. It is cultivated in an opencast way, but it is also produced inside greenhouses. Currently, producers can get access to a hybrid version known as "Dulcitico" developed by the researcher Carlos R. Echandi at the Fabio Baudrit Moreno Agricultural Experimental Station (EEAFBM) of the University of Costa Rica in 2013 [3]. This permits producers to get access to high quality, cheap seeds locally. "Dulcitico" is a highly productive hybrid which is also adapted climatologically to the region [14].

**Tomato**

Tomato crops also belong to the family of Solanaceae. The more cultivated species is *Solanum lycopersicum,* which has its origins in the Andes region [15]. The tomato plant is a perennial herb that can be cultivated annually [16]. Its fruits are harvested for consumption. According to [15], tomato is the most cultivated and consumed vegetable in Costa Rica.

Under adverse climatic conditions (high temperatures and humidity), tomato crops suffer many problems related to the presence of diseases which affect their production [16]. Diseases cause low-yield and quality and can even cause total crop losses. Authors point out that due to diseases' aggressiveness, it is important to prevent their appearance.

## *2.3 Diseases in Crops*

Diseases are the main limiting factor in the production of vegetables in the tropics, according to [17]. In some cases, such as in tomato crops, there are pathogens resistant to existing treatments. Since plants do not respond to agrochemicals, farmers must remove affected plants from crops to avoid a greater affectation.

The high incidence of disease appearance is due to the use of seeds created for regions distinct to the one in which the crops are cultivated [17]. For example, farmers sometimes use seeds created for temperate zones in tropic regions. This limits the plants' resistance to diseases since crops are not adapted to the tropical climatic conditions.

The creation of hybrid plants, like "Dulcitico" in Costa Rica, enables farmers to fight some diseases. Nevertheless, even though we have seeds prepared for the prevailing climatic conditions, the best way to deal with pests and diseases is the prevention and know-how acquired over time [18].

It is important to provide farmers with tools for the early detection and control of diseases, so that they can reduce the uncontrolled propagation of diseases and can avoid loss of production. Disease detection is regularly done by examining the plants, and especially their leaves. This can be done manually but requires time and expertise. It can also be automated.

## 2.4 Computer Vision

Artificial intelligence (AI) techniques focused on computer vision and machine vision have been used in agriculture [19]. These techniques represent an alternative for the identification of diseases in crops achieving good results. The techniques used are based on the extraction of information (features) from images (photographs). Image information is used to create classifiers that permit categorizing new samples of images [19].

Tools that use computer vision techniques enable farmers to make a better use of pesticides [20]. When farmers detect a possible specific disease in the crop, they can apply the correct pesticide and control the quantity to apply according to the infection. Both parameters can be obtained automatically or semi-automatically through trained computer tools.

## 2.5 Classification Algorithms

In addition to computer vision, there are other techniques in the field of artificial intelligence that can be useful to classify images based on crop data. Many techniques are used to create classifiers which can be used to detect diseases in plant leaves images [21].

In this paper, we propose the use of techniques such as Naïve Bayes, linear discriminant analysis, and neural networks to classify images of plant leaves affected by diseases.

## 2.6 Image Classification

To classify images based on computer vision and machine learning techniques, it is required to extract information from the images to be analyzed. Features extracted from images represent characteristics associated with color, shape, and texture. Those features are used by algorithms in the learning process to carry out the classification.

It is possible to extract the following features from images:

- Color: It can be obtained by using different color models:
  - Color (RGB): The commonly used three channels representing the combination (additive) of colors Red, Green, and Blue. Each color is represented by a decimal number in the range 0–255. The color combinations can be analyzed as a cube in which the three dimensions are the three colors (RGB) [22].
  - Color (CYMK): This is a representation by using complementary colors (Cyan, Magenta, Yellow, and Black). This representation is mostly used for output devices like printers [22].

- Color (HSV): This is a representation of colors in a way that is more closely aligned with the way human vision perceives color making attributes [22]. The H represents the hue and takes values from 0 to 179; the S corresponds to the saturation which takes values from 0 to 255, and the V is the value in the range 0–255. Another representation uses lightness instead of value (in the same range). That last representation is referred to as HSL.
- Color Grayscale: This corresponds to a model that represents the amount of light in an image. The value for pixels can be in the range 0–255.

- Texture: texture is recognizable in tactile and optical ways. Texture analysis in computer vision is complex. According to [23], texture in images can be defined as a function of spatial variation of the brightness intensity of the pixels. To calculate the texture of the images, it is possible to use the Haralick method [24] with which it can be obtained 13 values representing textural features.
- Shape: for the calculation of the shape in the images, we use the Hu-moments invariants which are seven numbers, known as moment invariants, which are calculated by using weighted average of image pixel intensities. According to [25], moment invariants store all information about shapes in an image, so they can be used for shape recognition in images.

In the next section, we describe some of the existing techniques to help farmers detect diseases.

## 3 Related Work

The use of technological tools in agriculture is an active research field. Precision agriculture seeks to maximize agricultural production using technological solutions. This research area has been growing in the last years. Researchers around the world have been dedicated to the creation of systems and applications which enable farmers to process data collected from crops to transform it into valuable information for decision making.

Studies seeking the automation of irrigation in crops using data sensed from the cropfields in real time can be found in the literature. Meanwhile, other researchers focus on pest and disease prevention, control, and avoidance [8]. It is possible to find studies which propose automatic detection of diseases in tomato crops [26, 27], wheat [28, 29], grapes [28], among others. All these studies have similarities. They all propose solutions where real-time image processing is used to detect diseases affecting crops. Some of the mentioned studies use automatic analysis of plant leaves using machine learning and computer vision techniques, but in these cases, authors are more concerned on the precision and do not focus on the complexity of solutions and their computational cost.

The research presented in [27] shows a completely automated solution that includes from the image capturing process to the classification results. Nevertheless, even though the solution does not require farmers' participation, the system performance is quite similar to the one we propose in this research. So, we consider it is not necessary to automate all the process. Farmers can identify affected plants and take some pictures to feed the disease detection system. Additionally, putting static cameras in the crop field may be very expensive, and if few are used, it reduces the possibility to capture pictures of infected plants out of camera range.

Authors in [30, 31] analyze distinct techniques to recognize species through leaves color images analysis. A combination of techniques with which the researchers process color, shape, and texture features extracted from images to classify them is presented in [30]. Meanwhile, in [31], authors present a novel method to recognize plant species even with fragmented images, i.e.,, when the complete plant leaf is not present in the image. Both studies result relevant to this work. In both cases, the authors keep in mind a low computational cost requirement, a goal we also are seeking. Our research is somehow different. We are working with the classification of plant leaf images but we focused on disease detection present in leaves and not trying to determine the identity of the plant leaves belong to. We propose to do so analyzing only the color feature and by using convolution neural networks.

Other authors [32, 33] focus their studies on the creation of mobile applications to execute the identification of diseases. They depict that processing power in mobile devices results in a limitation. For that reason, authors recommend considering it in the creation of the solutions. A strategy to mitigate the lack of processing power is the use of cloud computing. However, we consider that solutions including data analysis in the cloud limit their use in rural areas in developing countries. In this setting, there is commonly no Internet connectivity or it is very poor.

It is also possible to find studies focusing on specific crops. First, a literature review of published proposals related to disease detection in bell pepper crops is presented in [34]. They conclude that in case of fungal diseases, automatic detection enables farmers to avoid high affectation by removing the affected plants (fungal diseases spread through spores that travel in the air).

In our research, we evaluate and propose an efficient algorithm for early detection of diseases in horticultural crops. The value of our proposal is related to our focus on two main requirements: mobile devices with very limited resources and local processing. These two requirements can enable farmers to use the solution without Internet access. Our focus also lies on the automation of the identification process, not in the control process. Our goal is to provide suggestions to farmers to help them act quickly.

## 4 Feature Selection for Automatic Diseases Detection: An Experiment

In 2019, a bell pepper crop (*capsicum annuum*) inside a greenhouse at EEAFBM was suffering the presence of a disease with high affectation and an extremely fast

**Fig. 1** Example of leaves from the analyzed crop

propagation. The disease was causing plants death and a lot of problems for the researchers.

Leaves from healthy and infected plants are different. It is possible to see that the leaves of disease affected plants present degradation of colors in some rounded formations across the leaf. That affectation can also be present in other parts of the plant like the stem and the fruits. Additionally, it was possible to note that rounded formations in leaves had a different texture in comparison with the healthy sections of the leaves. In Fig. 1, we show an example of a healthy leaf (Fig. 1a) and an affected leaf (Fig. 1b), both taken from the mentioned crop. In Fig. 1b, it is possible to see the described affectation, with the rounded formations.

Different diseases exhibit different visual markers on the leaves. Thus, computer vision techniques could be used for image classification to identify the presence of a disease, as well as to distinguish between different diseases. To do that we started exploring the different features that can be extracted from images to describe them. We decided to use color (RGB), texture, and shape for the analysis, considering that the disease affects leaf color, creates circles (rounded formations), and changes the texture of the leaves. We designed an experiment to identify which feature (or combination of features) provided the best results. Different artificial intelligence techniques were used for the classification task. We also wanted to identify if there is a significant difference in which technique to use for the classification.

### *4.1 Objectives*

The proposed three specific objectives for the experiment were:

- To determine if the features extracted from an image can affect the detection of a disease in an affected crop leaf in terms of the classification precision.
- To identify if the used classification technique can affect the precision in the detection of a disease in an image of an affected leaf.
- To determine if there exists an interaction between the features extracted from an image and the techniques used for the classification task.

The first objective allows identifying the features to use in the classification task. It is necessary to consider that we want to create a mobile solution for disease detection. In mobile solutions, it is important to take into consideration that processor and memory resources are limited. For that reason, we needed to identify the minimum number of features required in the process.

Through the second objective, we wanted to explore if there is a significant difference between the use of a classification technique instead of the use of another. That is useful to select the best technique or the most efficient technique to be run on a mobile device.

The third objective tries to determine the relation between classification techniques and features. Our goal was to identify the best combination of features and techniques to be used.

### *4.2 Methodology*

We defined as an experimental unit the different arrays formed by the combinations of features extracted from images: color (RGB), texture, and shape. The precision of the classification was the response variable. The precision[2] is obtained as the ratio between true positives and the sum of true positives and false positives, i.e., the percentage of leaves classified as infected by the algorithm that were actually infected.

As design factors we identify:

- The classification techniques, in which case we consider:
  - Neural Networks (NN): a MultiLayer Perceptron classifier implementing ReLu activation function, Adam optimizer, alpha = 0.0001, 100 neurons in the hidden layer, initial learning rate = 0.001.
  - Naïve Bayes (NB): a Gaussian Naive Bayes classifier.
  - Linear Discriminant Analysis (LDA): a classifier with a linear decision boundary implementing singular value decomposition solver and a tolerance of 0.0001.
- The features extracted from the images. We had seven combinations of features to be analyzed:
  - C: Color
  - T: Texture
  - S: Shape
  - CT: Color + Texture
  - CS: Color + Shape
  - TS: Texture + Shape
  - CTS: Color + Texture + Shape

[2]https://scikit-learn.org/stable/auto_examples/model_selection/plot_precision_recall.html.

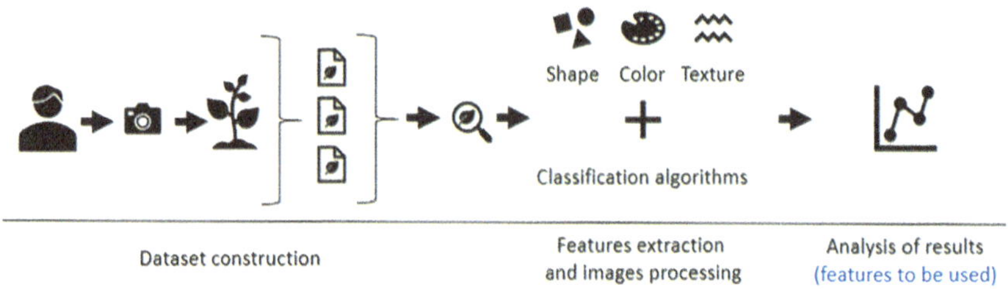


**Fig. 2** General overview of the process to be followed in the experiment

We decided to fix some factors to specific values for the experiment. Those factors were:

- Disease present in the images: The crop was infected with a leaf spot disease known as "ojo de gallo." All the leaves images were healthy or affected by that disease.
- Crop and plant species: The crop was a plantation of sweet pepper, a genetically modified hybrid known as "Dulcitico" (*capsicum annumm*) which was created by the Fabio Baudrit Moreno Agricultural Experimental Station[3] at the University of Costa Rica.
- Time of day during which photographs of the leaf were taken: Another fixed factor was the time of day during which the plant leaf images were produced. To produce good quality photographs of leaves, it is important to preserve the amount of light affecting the images. For that reason, we decided to take the photographs from 11:00 to 13:00 h Central American Time on a sunny day.

We defined as nuisance factors the presence of other diseases affecting the plants and the angle in which the images were taken. The first nuisance factor was minimized with the help of a professional in Agronomy (the person in charge of the original experiment using the crop), who checked that plants did not have another disease at the moment of image creation. The second nuisance factor was related to the amount of light present in the images since, if the angle of the camera changes, the colors in the image can vary significantly. To minimize the second nuisance factor, all the photographs were taken by a specific person, approximating the same angle and in a fixed setup with an uniform background.

We have two factors with distinct levels for each factor. For that reason, we designed a factorial experiment to analyze the previously mentioned factors and the interaction between them.

Taking into account the experimental factors mentioned above, Fig. 2 shows a general overview of the process to be followed. Each of the steps presented in Fig. 2 will be described in detail in subsequent sections.

[3]https://www.eeafbm.ucr.ac.cr/.

## 4.3 Dataset Construction

For the execution of the experiment, we decided to create a dataset from the infected crop. We took 121 photographs of healthy and infected leaves, distributed in 83 healthy and 38 infected leaves from distinct plants in the same crop. A sample of the obtained images can be seen in Fig. 1.

To increase the size of the dataset, we made a sampling with repetition without replacement, in which we created sets of 30 images, healthy (15) and infected (15), to use in the training phase. Additionally, with the remaining images we created a second sample of 20 images of healthy (10) and infected (10) leaves. That second sample was used for the testing phase. An important aspect to consider is that both sets were disjoint; i.e., images in the training set were not present in the test set.

Additionally, we repeated the sampling process 100 times for each treatment. We eliminated repeated samples. At the end of the sampling process, we got a total amount of 2100 sets for the training phase and 2100 sets for the test phase.

## 4.4 Features Extraction and Images Processing

We used Python programming language to create a script with which to extract features from images. For that task, we used the packages *mahotas*[4] and *opencv*[5] to extract texture, color, and shape from images, respectively. We decided to store the data extracted from images in csv files. Each row in the file represents an image, and each entry starts with the label for each image indicating if the leaf in the image was healthy or infected.

We created another script for the sampling process, which loads the csv files and runs the process, creating csv files with the different datasets.

For the processing of images and the creation of classifiers, we used the package *scikit-learn*[6] which is a library for machine learning in Python. That library provides off-the-shelf implementations of various machine learning approaches. In our case, we created several scripts to train and test classifiers using the selected algorithms.

## 4.5 Analysis of Results

We conducted statistical analysis to identify the effect of the classification algorithms and the features extracted from images on the classification precision. We started by exploring the data. Due to the results showed in Table 1, we discard the possibility to conduct a normality analysis.

[4]https://mahotas.readthedocs.io/en/latest/.

[5]https://pypi.org/project/opencv-python/.

[6]https://scikit-learn.org/stable/.

**Table 1** Normality assumptions test results

| Assumption | $p$-value | Used test |
|---|---|---|
| Normality | 5.79e–08 | Shapiro–Willk normality test |
| Independence | 0.056 | Durbin–Watson test |
| Homoscedasticity | 0.047 | Levene's test for homogeneity of variances |

**Table 2** Scheirer–Ray–Hare test results

| Factors/interactions | $p$-value |
|---|---|
| Feature | P < 1e–6 |
| Algorithm | 0.000001 |
| Feature: algorithm | 0.134886 |

Consequently, we ran the nonparametric Scheirer–Ray–Hare test in combination with a Dunn test for the post-hoc analysis [35]. The Scheirer–Ray–Hare test results are shown in Table 2. In that table, it can be observed that each of the chosen factors had a significant impact on the result, but that there is no interaction between factors.

The results of the Dunn test pointed out that treatments including Color (C) and Color + Shape (CS) are significantly different from the others. As a consequence, we prefer the use of the feature "Color" individually for a classifier or the combination of "Color" and "Shape." The results for the algorithms show that there are no significant differences for different classification algorithms. Thus, we will prefer to select the one with the best results for images classification tasks.

Through the described experiment we could determine that the most significant image feature to use in classification is color (RGB). Also, the experiment results show that it could be useful to include the shape feature in the analysis. Both features can be used in the creation of a classifier to automate the process of disease detection. Additionally, we concluded that it is possible to use at least Naïve Bayes, linear discriminant analysis or neural networks techniques to build an image classifier.

## 5 Disease Detection in Tomato Plant Leaves: A Study Case

We propose to execute a study case to determine the effectiveness of a machine learning technique for the classification of tomato diseases in plant leaves images. We also want to determine if the tested models can be used to run in a mobile scenario in which the available resources are very limited.

Dey et al. [36] analyze distinct scenarios in which machine learning technique can be used to get the best results. They state that artificial neural networks (ANN) can be used when the analysis is based on pixels from images, known as hard classification. ANNs may also be useful in scenarios where the analysis is nonparametric, i.e., not based on statistical parameters; and there is a high degree of variation in the spatial distribution of classes across the data [36].

In a literature review of distinct classification techniques, [37] argue that artificial neural networks have some advantages against other techniques. According to them, ANNs have high computational power to deal with noisy input efficiently, and they can learn from training data. That characteristic is extremely useful in contexts where the computational resources are limited, and the input data consists in photographs of plant leaves.

Albawi et al. [38] analyze the use of convolutional neural networks (CNN) for image classification. All the components of this type of ANN are described, including the most important component, the convolutional layer, an element that provides the possibility to detect and recognize features regardless of their position in the image. They report the convolutional layer gives CNNs good results in classification tasks.

We decided to use CNNs to create the classifier due to their particular suitability for the task. Before we describe the models, we used, we will first define the objectives for our case study.

## 5.1 *Objectives*

We want to identify which CNN implementation offers the best results while consuming the least computational resources. For that, we plan to compare a pre-trained CNN model and a custom CNN model regarding their relative performance in classification, the time consumed during the learning stage, and the size of the model, which is directly related with how computationally expensive it will be to use the model for classifications of new images.

## 5.2 *Methodology*

To achieve the proposed objective, we selected the densenet-161[7] pre-trained model, a dense convolutional network that connects each layer to the others in a forward way. According to the documentation, this CNN model is designed especially for image classification in which data correspond to 3-channel RGB images. It achieves a score of 22.35 for top-1 error (percentage representing the times when the classifier not assign the highest probability to the correct class), and a score of 22.35 for top-5 error (percentage representing the times when the classifiers not include the correct class in the first five guesses). These scores are the lowest rates from distinct pre-trained CNN models.

To use the pre-trained model, the following assumptions must be taken into consideration:

[7] https://pytorch.org/hub/pytorch_vision_densenet/.

**Table 3** Defined hyper-parameters for the models

| Hyper-parameter | Pre-trained CNN | Custom CNN |
|---|---|---|
| Input images' size | 224 × 224 | 64 × 64 |
| Batch size | 48 | 48 |
| Number of epochs | 100 | 1000 |
| Initial learning rate | 0.001 | 0.001 |
| Optimizer | Adam | Adam |
| Loss function | Negative log-likelihood | Cross-entropy |
| Activation function | ReLu | ReLu |

- Data must be arranged in shape $3 \times H \times W$, where $H$ = Height and $W$ = Width must have a size of at least 224 pixels.
- The data must be loaded in a range [0, 2] and then normalized using mean = [0.485, 0.456, 0.406] and standard deviation = [0.229, 0.224, 0.225]. It is important to consider that those values are fixed and cannot be changed.

We used Python as our programming language and the *pytorch*[8] package to load the data, create the models and run the training and test of the classifiers. For the pre-trained version of the classifier, we created a reference model, loaded the pre-trained one and overwrote the reference model with the pre-trained one. We then used this pre-trained version as a starting point and performed additional training particular to our dataset. For the custom CNN model, we created a custom network with two convolutional layers and three fully connected layers.

We defined the hyper-parameters in Table 3, for the two models.

We used a scientific workstation to run the training and test phases. We executed the learning process of the models on a Nvidia Quadro P400 GPU. To calculate the analysis metrics, we used the classification report[9] provided by the package scikit-learn. By using this package, we can directly obtain accuracy, precision, recall, and F1-score metrics from the test process. To calculate the time consumed in training and test stages, we used the computer clock. To estimate the size of the model, we exported the weights and the whole model to a file after the training and test stages.

Figure 3 shows the overall process to be followed. The three process steps will be described in the next sub-sections.

## 5.3 Dataset

The dataset we selected is provided by the project PlantVillage[10] which consists of 54,303 healthy and unhealthy leaf images divided in 38 classes by species and

[8] https://pytorch.org/.

[9] https://scikit-learn.org/stable/modules/generated/sklearn.metrics.classification_report.html.

[10] https://www.tensorflow.org/datasets/catalog/plant_village.

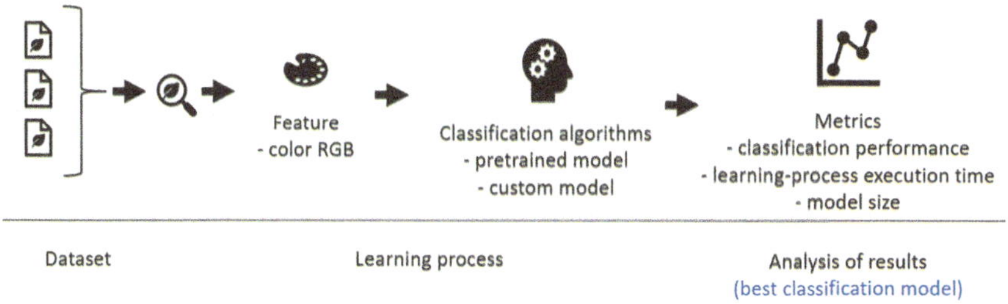


**Fig. 3** General overview of the process to be followed in the study case

diseases. The dataset contains images for 18 crops: apple, blueberry, cherry, corn, grape, orange, peach, pepper, potato, raspberry, soybean, squash, strawberry, and tomato.

We decided to run the study case for the tomato samples only, as the dataset contained a variety of different diseases for this crop. In the future, it is possible to create a model for each crop (and for its diseases). Solutions that support all crops and diseases can be harder to execute, and farmers can usually run the classification task only for a specific known crop.

We selected the tomato samples which consist of ten classes (one healthy and nine with different diseases). A total amount of 18,160 images of tomato leaves were considered. We divided the dataset into three parts: 70% (12,712 images) for training, 15% (2724 images) for test, and 15% (2724 images) for validation.

**Fig. 4** Tomato plant leaves healthy and affected by distinct diseases

In Fig. 4, we show several samples of photographs from this dataset. The (a) image represents a healthy leaf, while the others correspond to leaves affected by distinct diseases: *bacterial spot* (b), *leaf mold* (c) and *septoria leaf spot* (d).

## 5.4 Learning Process

The training and validation stages were run over the entire dataset. Since the learning process converges well within 100 epochs for the pre-trained model, based on empirical observation, we decided to set the number of epochs in 100 repetitions. Regarding the custom model, we decided to increase the number of epochs by 10-times based on the behavior of the training loss. In each epoch, we loaded the dataset in batches and proceeded to extract the RGB from those images. Next, we performed gradient descent for the total of iterations. At the end of the process, we calculated the performance metrics with the classification report based on the test dataset.

We used the test dataset to compare the results obtained from the training and validation stages. Test dataset provided an unbiased evaluation of the trained model, which was fitted with training and validation datasets. By loading the saved model, we tested the classification with the validation dataset and calculated the same metrics. Next, we compared the performance metrics for training and test stages to guarantee that they did not differ a lot, assuring the correctness of the classifiers.

We registered the execution time of the learning process. It is important to keep in mind that we use the same workstation to train both models. During the execution of the learning process, no other tasks were executed on the workstation.

As we stated above, we then analyzed the resulting size of each of the models and their weights. To do that, when the learning process finished, we proceeded to export the whole model (definition and weights), as well as only the weights.

## 5.5 Analysis of Results

After training, the pre-trained model densenet-161 was more accurate and precise than the custom model. Figure 5 shows the performance metrics for both models.

Additionally, Fig. 6 shows the results of the classification for the pre-trained densenet-161 model. In the figure, the confusion matrix shows the high classification rate obtained, in which the class "Target spot" is the one with most images classified incorrectly. However, in general, the classification produced by the pre-trained densenet-161 model results in a correct prediction.

Figure 7 shows the confusion matrix for the custom model. As seen in the matrix cells, most results lie in the diagonal, thus the corresponding classifications are good. "Early blight" and "Late blight" are the classes with more incorrect classifications, but they are not very numerous.

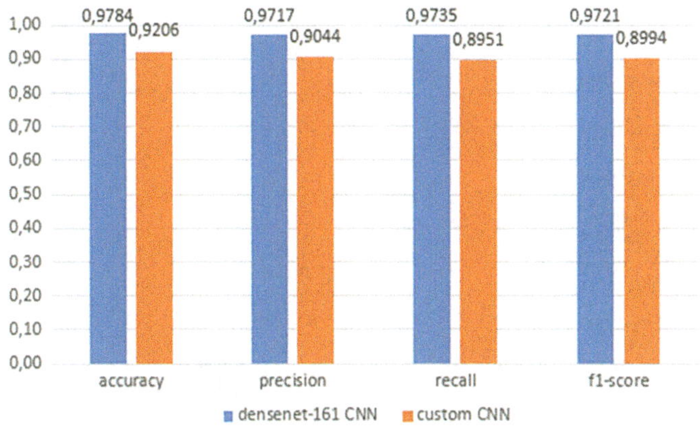


**Fig. 5** Performance metrics for densenet-161 and custom models

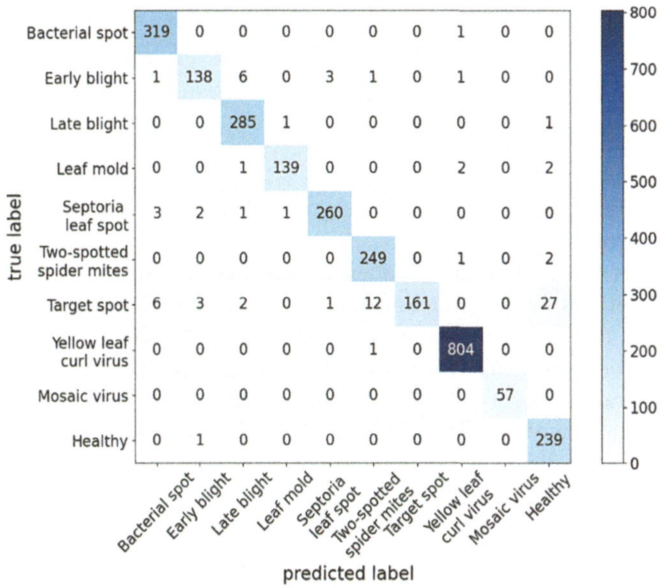


**Fig. 6** Confusion matrix for pre-trained densenet-161 model

Regarding the training time, we registered a total of 28 h, 31 min, and 10 s for the pre-trained model, and a total of 04 h, 47 min, and 59 s for the custom model. That difference in time execution is given by the structure of the model. In the pre-trained model, the network is very dense, which means that neurons have more connections, and therefore, more weights must be calculated each time. In the custom model, even though we worked with fully connected layers, there were fewer connections

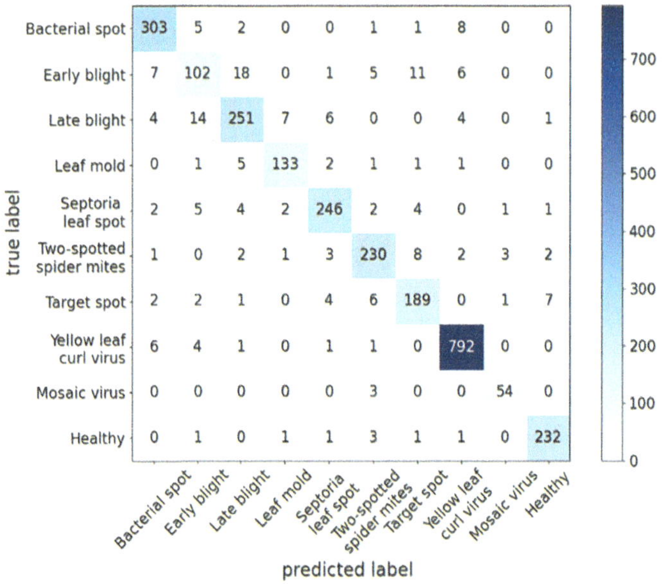


**Fig. 7** Confusion matrix for custom model

between neurons. This resulted in a training time that was much smaller than for the pre-trained network.

The size of the resulting model after training was also measured. We got a file size of 115 430 KB for the pre-trained model and 115 384 KB for only its weights. In the case of the custom model, we got a file size of 259 KB for the entire model and 244 KB for the weights. As can be seen, the size of the custom model is smaller than the pre-trained model in both cases (whole model and only-weights) by several orders of magnitude. When we try to load the model on a mobile device, it is easiest to load the smaller file due to limited resources. For this reason, the custom model is likely a better fit for our application.

According to the obtained results, while there are differences in accuracy and precision, they are negligible compared to the difference in model size. We may think that we can always prefer the model that has a greater accuracy and precision, but in the scenarios in which we work, we must also consider other parameters. If we consider the training time and the resulting file size of the two models, it is advisable to select the custom model.

We would like to reiterate that these results are only for tomatoes. In order to determine the suitability of our custom model, we intend to also run the process for other crops when we want to create a solution for multiple crops. In this case, having a low resource consumption algorithm becomes even more important, since different models for each crop must be included in the application, which further increases the file size and computational complexity of the overall solution.

## 6 Discussion

In the experiment presented in Sect. 4 of this paper, we aimed to select the best combination of features extracted from images and the most suitable artificial intelligence algorithm to be used in the construction of a low resource comsumption classifier. The features extracted from images were color, shape, and texture, while the algorithms we evaluated for the classifier were linear discriminant analysis, Naïve Bayes, and neural networks.

In [30] and [31], researchers work with the three mentioned features (color, shape, and texture) and with two of the considered algorithms: linear discriminant analysis and neural networks, respectively. Nevertheless, in both of these cases, the objective of the solution was the identification of plant species. As it was mentioned above, our objective is focused on disease detection where plant leaves are affected by color degradation, distinct formations in the leaves, and texture changes. For that reason, we decided to run first an experiment to select the features and algorithms to be used, getting as a result that we can use the color features individually for the analysis. Related to the algorithm, the results show that we can use any of the three considered algorithms, since they field similar classification performance.

Since convolution neural networks (CNN) have become popular specialized algorithms, we decided to compare two distinct CNN models and use the color (RGB) feature. Remember that our goal is to reduce the computational complexity of the final solution. Keeping this in mind, we planned in a study case to identify the model with an acceptable classification performance but with the best execution time and size.

Regarding the study case results, our custom model can achieve acceptable classification performance keeping the learning process execution time low and the size of the model extremely tiny. The latter helps us run the model in devices with very low resources, accomplishing the final goal we were pursuing.

As stated in [39], even if the machine intelligence cannot be defined, it could be measured. It is important to measure the intelligence of solutions that use artificial intelligence to make decisions and to do tasks in which it is necessary to act like the human brain. The problem that we propose to solve by using machine learning techniques represents a part of more complex decision support systems that can be developed to help farmers in their daily activities. We consider that measuring the intelligence of our solution can be useful, focusing on the task of recognizing diseases in plant leaves by using supervised learning. In this scenario, our solution tries to help un-experienced farmers to detect diseases in crops, an activity that can be done by experienced farmers or specialists easily.

## 7 Conclusions

After conducting the experiment and the study case, we can conclude that for the scenario of disease detection in tomato crops, it is possible to create a solution by using the PlantVillage dataset and by implementing convolution neural networks for the classifier. We demonstrate that the densenet-161 model presents better results in terms of classification performance in comparison with our custom convolution neural network model. Nevertheless, it also requires significantly more training time and the size of the resulting model is several orders of magnitude larger, limiting its applicability in the field.

In the scenario of disease detection in crops, by using resource-limited mobile devices, it could therefore be better to use a custom model instead of a pre-trained model. Custom model performance is already close to the full densenet-161 model, but could be improved in the future; however, we consider that the benefits in terms of execution time and size are already enough to justify its usage.

Our proposed classifier is currently designed to recognize diseases in only one type of crop. However, due to the low size of the model, we can create several models for disease detection for each crop type and load them on mobile devices. In addition, the results obtained for tomato crop diseases reveal that it could be better to focus our efforts on one crop at a time, adjusting the solution for each crop. The main effort to do this is to create datasets for each type of crop that include enough healthy and sick plant leaves.

As future work, we propose to improve our custom model implemented, by changing the CNN structure and hyper-parameters. Additionally, we propose to divide the disease detection into two stages: discriminating between healthy and infected, and when detecting that there is a disease present, running the classification to determine the type of disease.

We also plan to build a mobile application to carry out the detection in real time with the support of farmers. However, we consider that all solutions created for farmers must also be analyzed from the perspective of user experience to guarantee their usability and accessibility. Intelligent systems as the one we intend to create, having the farmer and its conditions in mind, can be part of our academic contribution to agriculture sustainability.

**Acknowledgements** This work was partially supported by the Research Center for Communication and Information Technologies (CITIC), the School of Computer Science and Informatics (ECCI), and the Fabio Baudrit Moreno Agricultural Experimental Station (EEAFBM) at the University of Costa Rica. Research Project No. 834-B9-189. We are grateful to G.B.S., M.A.C., and R.Z.M. for their contributions in this research.

## References


1. Secretaría Ejecutiva de Planificación Sectorial Agropecuaria (SEPSA), Desempeño del Sector Agropecuario. **3**, 1–10 (2018) Available at: https://www.sepsa.go.cr/docs/2019-003_Desempeno_Sector_Agropecuario_2018.pdf
2. J.E. Monge Pérez, M. Loría Coto, Producción de chile dulce (*Capsicum annuum*) en invernadero: efecto de densidad de siembra y poda". Posgrado Y Sociedad. Revista Electrónica Del Sistema De Estudios De Posgrado, **16**(2), 19–38 (2018). https://doi.org/10.22458/rpys.v16i2.2269
3. R. Martarrita, J. Aguilar, L. Meza, *Adaptabilidad de seis cultivares de chile dulce bajo invernadero en Guanacaste*. Alcances Tecnológicos. **12**(1), 13–23 (2018). ISSN-1659–0538
4. J. Rojas, M. Castillo, *Planeamiento de la agro-cadena del tomate en la región central sur de Costa Rica* (Puriscal, Costa Rica 2017). Disponible en https://www.mag.go.cr/bibliotecavirtual/E70-4158.pdf
5. C. Ramírez, J. Nienhuis, Cultivo protegido de hortalizas en Costa Rica. Tecnología En Marcha. **25**(2), 10–20 (2012)
6. S. Abarca, Análisis y comentario: Cambio climático y plagas en el trópico. Alcances Tecnológicos. **12**(1), 59–65 (2018). ISSN-1659–0538.
7. D.I. Tsitsigiannis, P. P. Antoniou, S. Tjamos, E. J. Paplomatas, Major diseases of tomato, pepper and eggplant in greenhouses. in *The Fruiting Species of the Solanaceae*, ed by H. Passam. European J. Plant Sci. Biotechnol. **2**(Special Issue 1), 106–124 (2008)
8. J.A. Brenes, A. Martínez, C.Y. Quesada-López, M. Jenkins, Sistemas de apoyo a la toma de decisiones que usan inteligencia artificial en la agricultura de precisión: un mapeo sistemático de literatura (Revista Ibérica De Sistemas y Tecnologías de Información RISTI 2020). https://doi.org/10.17013/risti.n.pi-pf
9. A. Dwivedi, R.K. Naresh, R. Kumar, R. Yadav, R. Kumar, Precision agriculture (2017)
10. R. Bongiovanni, J. Lowenberg-DeBoer, Precision agriculture and sustainability. Precis Agricul **5**, 359–387 (2004). https://doi.org/10.1023/B:PRAG.0000040806.39604.aa
11. G. Davis, W. Casady, R. Massey, *Precision Agriculture: An Introduction.* Available at https://extension2.missouri.edu/wq450#:~:text=Precision%20agriculture%20merges%20the%20new,areas%20within%20a%20farm%20field (1998)
12. Oficina de las Naciones Unidas contra la Droga y el Delito (UNODC), *Manual para el productor: El cultivo de hortalizas en Proyecto manejo integral de los recursos naturales en el trópico de Cachabamba y las Yungas de la Paz.* Primera edición (La paz, Bolivia 2017)
13. Ministerio de Agricultura y Ganadería (MAG) de Costa Rica, *Agrocadena Regional Cultivo CHILE DULCE.* (Grecia, Costa Rica 2007). Disponible en https://www.mag.go.cr/bibliotecavirtual/E70-4281.pdf
14 J. Bolaños, L. Barrantes C.Y. Echandi, K. Bonilla, *Manual téc-ni-co basado en experiencias con el híbrido 'Dulcitico' (Capsicum annuum).* (San José, Costa Rica Instituto Nacional de Innovación y Transferencia en Tecnología Agropecua-ria (INTA) 2018). 978–9968–586–34–4
15. Instituto Nacional de Innovación y Transferencia Tecnológica Agrope-cuaria INTA, *Manual técnico del cultivo de tomate (Solanum Lyco-persicum)* (San José, Costa Rica 2017). ISBN 978–9968–586–27–6. Available at https://www.mag.go.cr/bibliotecavirtual/F01-10921.pdf
16. Organización de las Naciones Unidad para la Alimentación y la Agricultura, *El cultivo de tomate con buenas prácticas agrícolas en la agricultura urbana y periurbana* (2013). Available at https://www.fao.org/3/a-i3359s.pdf
17. H.Y. Thurston, J. Galindo, *Enfermedades de cultivos en el trópico". Centro Agronómico Tropical de Investigación y Enseñanza (CATIE). Programa de Mejoramiento de Cultivos Tropicales, con permiso de la American Phytopathological Society* (Turrialba, Costa Rica 1989). ISBN 9977–57–072–8.
18. Australian Goverment, *Sweet Pepper IPM Factsheet* (2017). Disponible en https://ahr.com.au/wp-content/uploads/2017/11/Sweet-pepper-IPM-factsheet-20th-August-2017.pdf

19. B.S. Ghyar, G.K. Birajdar, *Computer Vision-Based Approach to Detect Rice Leaf Diseases Using Texture and Color Descriptors.* in 2017 International Conference on Inventive Computing and Informatics (ICICI) (Coimbatore, 2017), pp. 1074–1078. https://doi.org/10.1109/ICICI.2017.8365305
20. S. Phadikar, J. Sil, *Rice Disease Identification Using Pattern Recognition Techniques.* in 2008 11th International Conference on Computer and In-formation Technology (Khulna, 2018), pp. 420–423. https://doi.org/10.1109/ICCITECHN.2008.4803079
21. J.N. Reddy, K. Vinod, A.S. Remya Ajai, *Analysis of Classification Algorithms for Plant Leaf Disease Detection.* in 2019 IEEE International Conference on Electrical, Computer and Communication Technologies (ICECCT) (Coimbatore, India, 2018), pp. 1–6. https://doi.org/10.1109/ICECCT.2019.8869090
22. N. Ibraheem, M. Hasan, R. Khan, P. Mishra, Understanding color models: a review. ARPN J. Sci. Technol. **2** (2012)
23. L. Armi, S. Fekri Ershad. Texture image analysis and texture classification methods—a review. **2**, 1–29 (2019)
24. R. Haralick, K. Shanmugam, I. Dinstein, Textutal features for image classification. IEEE Trans. Syst. Man and Cybern. SMC **3**(6), 610–621 (1973). United States
25. M. Rhouma, S. Zafer, R. Khan, M. Hussain, Mohammad, Improving the performance of hu moments for shape recognition. Int. J. Appl. Environ. Sci (2014)
26. H. Sabrol, K. Satish, *Tomato Plant Disease Classification in Digital Images Using Classification Tree*, in 2016 International Conference on Communication and Signal Processing (ICCSP) (Melmaruvathur, 2016), pp. 1242–1246. https://doi.org/10.1109/ICCSP.2016.7754351
27. R.G. de Luna, E.P. Dadios, A.A. Bandala, *Automated Image Capturing System for Deep Learning-based Tomato Plant Leaf Disease Detection and Recognition.* in TENCON 2018—2018 IEEE Region 10 Conference (Jeju, Korea (South), 2018), pp. 1414–1419. https://doi.org/10.1109/TENCON.2018.8650088
28. H. Wang, G. Li, Z. Ma, X. Li,*Image Recognition of Plant Diseases Based on Backpropagation Networks.* in 2012 5th International Congress on Image and Signal Processing (Chongqing, 2012), pp. 894–900. https://doi.org/10.1109/CISP.2012.6469998
29. H. Wang, G. Li, Z. Ma, X. Li, Application of Neural Networks to Image Recognition of Plant Diseases. in *2012 International Conference on Systems and Informatics (ICSAI2012)* (Yantai, 2012), pp. 2159–2164. https://doi.org/10.1109/ICSAI.2012.6223479
30. A. Tharwat, T. Gaber, T.M. Awad, N. Dey, A.E. Hassanien, *Plants Identification Using Feature Fusion Technique And Bagging Classifier.* in The 1st International Conference on Advanced Intelligent System and Informatics (AISI2015), November 28–30, 2015, Beni Suef, Egypt (Springer, Cham, 2015), pp. 461–471
31. J. Chaki, N. Dey, L. Moraru, F. Shi, Fragmented plant leaf recognition: Bag-of-features, fuzzy-color and edge-texture histogram descriptors with multi-layer perceptron. Optik **181**, 639–650 (2019)
32. A.L.P.D. Ocampo, E.P. Dadios, *Mobile Platform Implementation of Lightweight Neural Network Model for Plant Disease Detection and Recognition.* in 2018 IEEE 10th International Conference on Humanoid, Nanotechnology, Information Technology, Communication and Control, Environment and Management (HNICEM) (Baguio City, Philippines, 2018), pp. 1–4. https://doi.org/10.1109/HNICEM.2018.8666365
33. N. Petrellis, *Plant Lesion Characterization For Disease Recognition A Windows Phone Application.* in 2016 2nd International Conference on Frontiers of Signal Processing (ICFSP) (Warsaw, 2016), pp. 10–14. https://doi.org/10.1109/ICFSP.2016.7802948
34. J. Francis, S. Anto, D. Dhas, B.K. Anoop, *Identification of Leaf Diseases in Pepper Plants Using Soft Computing Techniques.* in 2016 Conference on Emerging Devices and Smart Systems (ICEDSS) (Namakkal, 2016), pp. 168–173. https://doi.org/10.1109/ICEDSS.2016.7587787
35. S. Mangiafico, *Scheirer–Ray–Hare Test.* Available at https://rcompanion.org/handbook/F_14.html

36. N. Dey, G. Mishra, K. Jajnyaseni, S. Chakraborty, S Nath, *A Survey Of Image Classification Methods and Techniques*. in 2014 International Conference on Control, Instrumentation, Communication and Computational Technologies (ICCICCT, 2014). https://doi.org/10.1109/ICCICCT.2014.6993023
37. N. Thakur, D. Maheshwart, A review of image classification techniques. IRJET Int. Res. J. Eng.Technol. **4**(11) (2017)
38. S. Albawi, T. Abed Mohammed, S. Alzawi, *Understanding of a Convolutional Neural Network* (2017). https://doi.org/10.1109/ICEngTechnol.2017.8308186
39. L.B. Iantovics, A. Gligor, M.A. Niazi, A.I. Biro, S.M. Szilagyi, D. Tokody, Review of recent trends in measuring the computing systems intelligence. BRAIN. Broad Res. Artif. Intell. Neurosci. **9**(2), 77–94 (2018)

# Artificial Intelligence Sustainability Ensured by Convergent Cognitive Semantics

**Alexander Raikov**

**Abstract** The work addresses ensuring the sustainability of group decision-making processes supported by artificial intelligence (AI) with cognitive and denotative semantics of its models. The latter can be represented by applying the traditional digital and symbolic means like big data and images. Cognitive semantics cannot be represented by such means. This semantics reflects human consciousness, experiences, feelings, thoughts, etc. The interpretation of cognitive semantics is beyond the possibility of digital computers, logical predicates, knowledgebase, neuron networks, big data, and images. Cognitive semantics is a source of divergence and unsustainability in group decision-making, especially during brainstorming. This work shows that the problem of creating cognitive semantics can be resolved indirectly. Some methods from the electrodynamic, quantum field theory, chaos control, category theories are suggested for this aim. Combining such methods is carried out with the author's convergent approach, which creates a synergy. This approach is based on an inverse problem-solving method, genetic algorithms, cognitive modeling with its big data verification. Some practical results of the implementation are demonstrated.

**Keywords** Artificial intelligence · Chaos · Cognitive modeling · Convergent approach · Sustainability

## 1 Introduction

Many years of author's experiences in moderating the collective decision-making processes, including brainstorming and strategic meetings in situational centers and analysis of the scientific papers in this field, have shown that these processes could be separated on divergent and convergent [1]. The former aims to generate ideas as many as possible. It makes the processes unsustainable, i.e., a small deviation in supplying the participants with new information can lead the process aside. The latter and generating ideas aim to ensure the participants achieve a consensus concerning goals

---

A. Raikov (✉)
Institute of Control Sciences RAS, Moscow, Russia
e-mail: Alexander.N.Raikov@gmail.com



A. Joshi et al. (eds.), *Sustainable Intelligent Systems*, Advances in Sustainability Science and Technology, https://doi.org/10.1007/978-981-33-4901-8_3

and paths to achieve these goals. The divergent tendencies of collective discussion and ill-defined goals also make this process unsustainable.

These processes are becoming significantly more unsustainable in a network environment. In these conditions, the participants can see each other's only through the computer or smartphone monitor; they may not know each other at all. Due to it, they have limited opportunities to understand each other. Network brainstorming or as it called electronic brainstorming (EB) especially has been developing for the last decade.

During EB, which has a divergent character, the moderator's task is mainly to stimulate the participants' creative activity. In contrast, the convergent conversation requires the moderator to apply special technologies to structure the process to make the discussion more sustainable and purposeful. Apparently, nowadays, it is required to apply special AI methods to ensure these processes more convergent. The motivation of this work is to suggest the approach to support the sustainable convergent of network decision-making processes with taking into account different kinds of AI semantics.

The semantics of AI models can be divided into denotative and cognitive. The former has a formalizable, logical, or visualized representation; the latter does not. The mathematical spaces representing the denotative semantics can have many dimensions, and, in these spaces, some metrics can be defined. In comparison, cognitive semantics has a phenomenological character due to expresses non-formalizable human feelings, emotions, free will, experiences, etc. Once these human abilities anybody tries to represent verbally, they become formalized, and as a consequence, the semantics becomes denotative. Metrics cannot describe the cognitive semantics, and it must be represented somehow in an indirect way. The novelty of the work is the justification of the possibility to take into consideration the cognitive semantics by some local and non-local physical effects and by introducing the human participation in the decision-making process in a special way.

Both semantics are unsustainable sources, but cognitive semantics, as it turns out, is less controllable by modern AI tools due to its phenomenological nature and can be a reason for great unsustainability. This chapter is more concerned with cognitive semantics and contributes a special convergent approach that makes the group decision-making processes more sustainable and purposeful. The approach is tested during a long term of practical application, and some of its components are the subject for investigating and developing in the future.

The chapter's text consists of a related works analysis section, and the sections introducing the ideas of AI cognitive semantics sustainability, quantum sustainability of AI, and wave sustainability of AI. The author's convergent sustainability approach is suggested for integrating these ideas to ensure the decision-making processes in the network environment more sustainable. The practical implementation of the approach is shown at the end of the chapter.

## 2 Related Works

The study of AI's models' semantics covers themes of cognitive architectures, electronic brainstorming, and big data analysis. The cognitive architecture's direction tries to model the human mind including its flexible behavior, real-time operation, rationality, learning, linguistic abilities, awareness, etc. Scientific works in this field include aspects of evolutionary realism, adaptation, modularity of cognition, the structure of memory [2].

The core cognitive abilities are discussed in [3]. More than 2500 publications were analyzed. There are about three hundred cognitive architectures that describe these abilities. Paper [4] addresses the issues about broad areas of cognitive architectures such as perception, memory, attention, actuation, and emotion.

The analysis of these works shows that all methods have the logic and formalizable bases, and AI models are represented by symbols (labels, strings of characters, frames), production rules, probabilistic and non-probabilistic logical inference.

Numerous studies have devoted to electronic brainstorming (EB). EB can be divergent and convergent [5]. The former generates new ideas as many as possible; the latter ensures consent achievement in decision-making. In-network environment, participants of EB have limited opportunity in an understanding of each other's.

The results of BE study in different works are controversial. So, work [6] shows that EB did not become a widely used technology for idea generation and is not as effective as face-to-face brainstorming, but paper [7] noted the advantage of this process over face-to-face meeting due to the use of computer special application that structures the creative process in a special way.

In any case, it is useful to find and apply special technology that will reduce or eliminate the networked EB's restrictions. Quite obvious, this technology has to take into account the cognitive semantics of AI models that cannot be represented in a direct logical way.

Many works are devoted to big data technology. Undoubtedly, these technologies should be used to construct denotative and cognitive semantics of AI models. Now there are convenient tools to accumulate data and access to data from different fields of economy. For example, in [8] authors suggested an approach to improve the service metadata to maintain consistency without much compromising performance and scalability of metadata.

The tremendous growth of big data analytics in the past few years is shown in [9]. The book highlights state-of-the-art research on big data and the Internet of things (IoT). Modern information technology allows uploading, retrieving, storing, and collecting information, which forms big data. The IoT has entered the public consciousness, sparking people's imaginations. The work [10] justifies the important for our creating conclusion that the entire universe is moving toward digitalization and the key framework is the internet and connecting devices.

All these systems can fill any AI model with content in real time, thereby helping to create cognitive and denotative semantics of AI models.

## 3 AI Cognitive Semantics Sustainability

AI models' semantics, as usual, are represented in logical, image-like, or artificial neuronal network ways. The digital computer has to transform continuous signals about different events into the binary form. The continuous signals must be divided into pieces and replaced with values in single points, e.g., based on the well-known Nyquist–Shannon sampling theorem. This sampling operation truncates the signal spectrum and distorts the information about the events. With the raising of the data's volume, the mistakes are cumulating.

However, cognitive semantics cannot be recorded as a digital signal. This semantics has continuous nature and must be mediated through human consciousness inclusion or, e.g., indirectly using non-local physical effects. The difference between cognitive and denotative semantics is illustrated in Fig. 1. The cognitive semantics is located in the upper part of the figure. This semantics is interpreted by people with their thoughts and feelings or can be tried to be represented in different indirect ways and means. Every variant has its own peculiarity.

Talking about the role of people in cognitive semantics, let us suppose that it is required to solve one of the following problems:

- Find an optimal decision for a megapolis tourism strategic development;
- Create an effective architecture of an agricultural national digital platform;
- Generate optimal supply chains in the condition of ill-defined market behavior.

Creating cognitive semantics by human participation can be supplied using the author's method of conducting convergent strategic conversations with cognitive

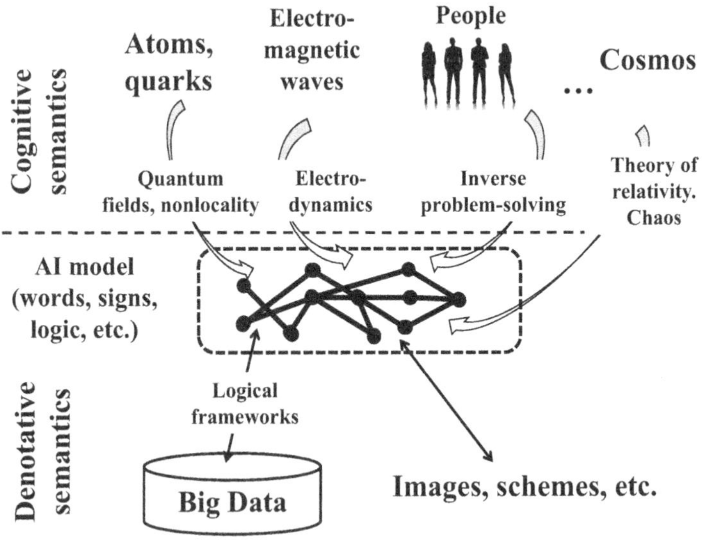


**Fig. 1** Difference between cognitive and denotative semantics

modeling [1]. During these conversations, people bring a phenomenological potential into the process, which constitutes the cognitive semantics. This process has an important peculiarity—every participant feels different things and may have contradictory inner desires. This part of the process cannot be represented verbally due to its latent character. It is a source of the unsustainability of the decision-making process, which leads to wasting time. Apparently, this process's latent part due to its non-metric nature cannot be modeled as on the bases of extrapolation of the tendencies got from the analysis of retrospective big data, as by applying deep neural networks.

Cognitive modeling helps to accelerate group decision making [11]. This modeling consists of revealing by experts the factors that affect the problem's solution and assessing directed connecting between them. The big data analysis is then used to verify or even automate the construction of the cognitive model by directly mapping the model's components on the relevant subsets of the data. To do this, each component of the model is supplied by the search query to the big data used to search for information. Based on the results of the search, a recommendation to refine the cognitive model is making.

But the verification and creating of the cognitive model by mapping it on the big data ensures to use only denotative semantics regardless of whether the logical mappings or neural networks are using. This mapping, in any case, can be described by algorithms or "if–then" rules. The denotative semantics can be represented by these tools, which puts things in order in the system. Let us denote this order by the symbol $P$.

Both types of semantics contain hidden sources of divergency and turbulence that introduce chaotic disorder into the system. Let us denote this mess by the symbol $S$. It increases the AI system's unsustainability. The question arises: in what ratio and in what dynamics should there be ordered and disordered for the AI system's behavior remains sustainable?

In our work [12], we tried to find the answer to the stated question, and then have got some convincing practical proves of the formula:

$$P * P' + (S_{\text{int}} - S_{\text{exch}}) * (S'_{\text{int}} - S'_{\text{exch}}) < 0, \tag{1}$$

where $P$ and $P'$, respectively, the level and speed of changing of the order in the system, $S_{\text{int}}$ and $S'_{\text{int}}$, respectively, the level and speed of changing of internal disorder (chaos), $S_{\text{exch}}$ and $S'_{\text{exch}}$, respectively, level and speed of exchanging of chaos between internal and external of the system. Formula (1) shows that the AI system's unsustainability can become even more with the accelerated establishment of order in the system, and with a decrease in the order—sustainability can increase. The system's sustainability can be increased with improving communication.

In addition to introducing into the cognitive semantics a component related to the regulation of people's participation in the decision-making process, Fig. 1 shows that the unsustainability sources can also be located in atomic, waves, and cosmos structures, which may take part in the cognitive semantics creating. This study looks shortly at how quantum and wave may influence the cognitive semantics, and respectively, on AI sustainability.

## 4 Quantum AI Sustainability

Formula (1) helps to understand how the AI system's sustainability depends on the relationship between chaos and order. But it does not give a clue about the interdependence between cognitive and denotative semantics but gives a hint of finding a path of definition connection cognitive semantics with the environment. For this, the deeper levels of the brain construction than neuronal have to be taken into consideration.

An artificial neuron is a natural neuron model, and their networks can be described by logic and put into a digital computer. Apparently, it is not enough for the representation of cognitive semantics. It has to take into account different physical fields and atomic behavior effects. For this, quantum field theory can be used [13, 14]. It, e.g., allows introducing into the consideration of the semantic interpretation of the complex number with its imaginary part $i$ (square root of $-1$), Euler's number e, ensure working with wave and particle together, consider a relativistic context, take into account hidden and different non-local effects, use the advantages of the possibilities of various physical and mathematical spaces and tools.

The quantum aspects of consciousness are much more complex than logical ones. Every brain's neuron consists of about $10^{10}$ atoms, as the human brain consists of about $10^{10}$ neurons. Due to it, the power of AI can rise many times. Using the quantum paradigm will support to make a holistic and polyphonic semantic representation of events and their interconnections, which are modeled with AI. For example, $e$ ties together $i$ and $\pi$, which was being used in many calculations—from of compound interest, to the power of cannons. If something varies over time, this approach will be useful for calculating the exact value at a certain time.

The quantum approach in AI gives the ability to describe events and objects' behavior, taking into account their internal structure by literally duplicating them, making the atomic doubles. The semantics interpretations of AI models become more holistic due to interpreting the modeled phenomenon's non-formalized and uncertain characteristics [15]. The dynamics of a quantum system represent such characteristics.

The quantum calculation model is based on a closed or isolated quantum mechanical system [16, 17]. This is a system without interactions with its surrounding, which is the source of unsustainability. But the system in real practice cannot be closed and exits a signal outside decoherence develops. This interaction follows to decay information and generates a quantum error, but it can be taken into account during calculation.

Quantum semantics means mapping the AI model on quantum computation, which transforms a state of the quantum system into another state. A unitary operator can represent mathematically quantum computation $U$, which is a matrix form of a linear operator, if $U^{\dagger}U = I$, where $U^{\dagger} = (U^T)^*$ is the complex conjugate transpose of the $U$ matrix and $I$ is the identity operator. Unitary transformations preserve inner products between vectors, the vectors' lengths, and the angles between vectors. These transformations sometimes are called *quantum gates*.

Quantum semantics can be useful for raising the quality of predicting and pattern recognition [18]. For example, in [19], it was used to predict educational services' long-term needs. The first step was the cognitive modeling of the situation by experts. The second step consisted of making the model's denotative semantic through its mapping on the relevant big data's subsets. It helped to verified and correct the model. On the third step, a quantum operator could be used for forecasting. It may be the quantum genetic algorithm, the entanglement operator, or the Hadamard operator. The latter, e.g., makes it possible to create a superposition of all randomly chosen solutions in the form of a generalized solution space. But in the referenced paper, only genetic algorithms on the cognitive (conceptual) model was applied—it occurs enough for this case.

Some quantum operators introduce into the cognitive semantics interpretations and calculations imaginary number $i$, which expands semantics interpretations. For example, some important quantum gates are so-called *Pauli matrices*:

$$X = \begin{pmatrix} 0 & 1 \\ 1 & 0 \end{pmatrix}, \quad Y = \begin{pmatrix} 0 & -i \\ i & 0 \end{pmatrix}, \quad Z = \begin{pmatrix} 1 & 0 \\ 0 & -1 \end{pmatrix} \tag{2}$$

These matrices can be exponentiated that give more complex further unitary matrices for calculating the rotation of the modeled object in the system coordinate about $x$-, $y$-, $z$-axes.

The use of quantum semantics methods seems to be quite promising for the construction of cognitive semantic interpretations of AI models. Of particular interest is the effect of the collapse of quantum states, as well as the entanglement effect of quantum particles, which reflects the instantaneous interconnection of the states of particles located at large distances, albeit without instantaneous transmission of information.

The collapse of the quantum state of quantum particles resembles the "Eureka effect" [20] when a person finds a solution to a problem instantly after lengthy reflections and due to an unexpected external impulse. True, in the case of quantum physics, it is believed that information from one participant in events to another is not transmitted, just both participants at the same time will see the predefined quantum state of the system from some of quantum superposition.

The nonlocality effect of quantum states or the Einstein–Podolsky–Rosen paradox has various explanations. Here is one of the interpretations of this phenomenon in constructing cognitive semantics [21]. Let there be two quantum particles, one of which is in one of the human neurons, and the other is anywhere in the Universe. Let the first particle can take two different states $|u_1\rangle$ and $|u_2\rangle$, and the second one—$|v_1\rangle$ and $|v_2\rangle$. Then, the quantum system in its most general form can be represented as a tensor product (superposition) of the sums of quantum states, and each state of each particle can correspond to a certain coefficient.

At the same time, another state is also possible, represented in the form of the sum of the superposition of particles' states. Moreover, the tensor product in the second case is non-commutative, and it is non-factorizable; that is, it cannot be reduced to the form of the product of the states of two particles, since in the more general first case, it is impossible to select the necessary values of the coefficients for this. That is, various observers, each of whom can observe one of the particles, cannot control another particle's behavior. These states are, as it were, fixed at a certain point in time. This moment comes when one of the observers makes a look at the state of his particle.

In practice, it might look like this. Let there be two observers separated by a large distance, each observing its own particle. When one of them fixes the state of its particle, a quantum collapse occurs—the second observer at the same absolute moment will be presented with the fact of seeing only one of the states of his particle, and no other. As above noted, the information is not transmitted from one observer to another observer instantaneously, since this would contradict the principles of the special theory of relativity.

At the same time, the given explanation may seem not very correct. For example, the observation time is considered the same. Even if we admit the synchronization of two observations, which can now be done with an accuracy of up to $10^{-18}$ s, it turns out that the collapse occurs with some accuracy in time. If such synchronization is admitted, then the idea about the impossibility of instantaneous transmission of information is not correct too. Although this aspect can be omitted, because before the moment of observation, the quantum (mental) system is in a state of superposition, it enters a state of collapse only after the moment when one of the observers pays attention to it, the second observer may measure the state of his/her particle later.

This subtle temporal aspect of an instantaneous event may not be taken into account, and then the canons of the relativistic theory are not violated. Although the idea that the properties of a quantum particle observed during measurement actually existed even before the measurement may be incorrect [22]. However, we will not go into these details. Giving this example, our task is only to indicate such issues, giving additional opportunities for constructing semantic interpretations of AI models while answering them.

Studying quantum aspects of cognitive semantics allows us to fall risks of AI's decision-making processes' unsustainability. The progress in this direction will be developing with solving such issues, as follows:

- The quantum particles behavior is represented in infinite-dimensional space, which can help to represent the non-formalizable cognitive semantics;
- Changing the location of a particle means changing this particle on another, which looks like changing the meanings of the words in another situation;
- Particle is represented as both a particle and a wave, which leads to the different interference effect and give synergetic effects;
- Any attempt to detect the state of the particle leads to collapse, which is associated with the trying to look into the inside of an object by the destruction one, etc.

At the same time, the quantum representation of cognitive semantics has a problem. The Q-bit construction relies on the superposition of binary representation of signals (see the sampling operation above) that destroys the continuous nature of consciousness and thinking. Perhaps, the representations of these phenomena by qubits need to be added by other tools that are more continuous in their base. For example, it may be tools that work with electromagnetic fields.

## 5 Wave AI Sustainability

Human consciousness can be compared to a physical field, e.g., an electromagnetic one. Apparently, the low-frequency signals that the brain sends outward are only communication signals. These signals are more reminiscent of the speech signals of a person who verbally transmits his thoughts to another person. Thought itself is a much more complex field-like phenomenon.

In addition to the quantum-relativistic, this complexity can also be expressed through the biochemical, wave, and acoustic nature of consciousness. Being under the influence of an external wave field, the human body and his/her brain as a physical object can interact with the field, be a kind of resonant receiver, translator, and generator.

The fundamental basis for the description of fields and the wave nature of signal propagation is d'Alembert and Helmholtz's well-known equations, formalisms of electrodynamics [23]. At the same time, "mental events" occur in some infinite-dimensional space, in which some points can be a source of radiation of a signal (wave, quantum, sound), and some are the result of radiation. The radiation source and the radiation receiver can be located on both sides of the participants' physical bodies of the thought process as inside the bodies. It means that one side is the body or at least the brain of a person, and the other is the external environment.

The field equations show the possibility of expanding the field of semantic interpretation of thought processes. They allow the inhomogeneity of the medium, that is, the dependence of the signal propagation on various factors, including the coordinates and the observer (participants). The behavior of the field can be quantized, continuous, and discrete. Thoughts and events take place in time. Impacts can be carried out in a forward and reverse relationship; direct and inverse problems can be solved.

The scalar representation of a field can have different physical meanings. It can describe the electromagnetic field's magnitude in its classical and quantum manifestations, sound pressure, and other oscillatory processes in the medium. Such processes can influence thinking on the subatomic level.

The above wave formalisms concern only the initial aspects of representing the thought process in a physical field. They only demonstrate that each point of mental space, distributed between the human body and the environment, fixing the fact of the presence of a radiation source and the result of radiation, determines an event reflecting a phenomenon of a certain nature—this can be, for example, the state of a quantum particle, the behavior of which can be subject to non-local laws, as discussed above.

The use of the wave fields for representing cognitive semantics might compensate for the limitations of discrete computation. This is not entirely so. For example, replacing discrete representations of data with waves can be realized with optical tools. Such a substitution does not give an absolute continuous (analog) representation of cognitive semantics because optical transformations have a corpuscular-wave nature. The corpuscular properties of light can be described by using the quantization procedure. As well known, a quantum particle of light is the photon that has a zero-rest mass, spin 1, and does not carry an electric charge.

Simultaneously, the replacement of the digital form of representation of the consciousness by a wave has a fundamental difference. It lies in the fact that the wave representation of the thinking process allows one to consider the atomic and relativistic-spatial levels in this process, the power of the set of elements of which is many orders of magnitude higher than the logical-neural level of modern AI methods of knowledge representation. This makes it possible to embrace the cognitive semantics of models much more holistically and thereby increase the stability of the work of AI systems.

## 6 Convergent Sustainability

In the decision-making process, especially in hybrid systems, when a person or a group of people are the bearers of cognitive semantics, the process's unsustainability is immanently embedded in it from the very beginning. This unsustainability increases significantly in conditions of great uncertainty, when circumstances develop in an emergency way, or when participants are not clear about their goals. The events in these cases cannot be described only by logic means or immersed in the metric spaces.

To accelerate the collective decision-making under these conditions, the problem, which is solving, has to be divided into many parts. Then the results of the decision-making of every part must be assembled into a holistic solution. The author's convergent approach with cognitive modeling helps to accelerate this process and make it more sustainable.

Cognitive modeling consists of revealing the set of factors, which describe the problem, assess the interaction between factors, and track changes in goal factors overtime during variation control factors [11]. The tact of time is the parameter of modeling, for example, six months, a year, etc. Cognitive modeling allows tracking trends of changing the situation connected with the process of problem-solving, and thus evaluating the success of the solution from various influences on the control factors of the model. During the modeling, the best scenarios are selected. But this modeling does not guarantee fast convergence of the solution to the goals, especially in cases where the solution's goals are not precisely known. Cognitive modeling helps to answers to the questions like these: "What will happen if make this decision?", "What should be done to achieve the goal?"

Cognitive modeling differs from another kind of modeling by increasing attention to the semantics. Cognitive models have real semantic interpretations (denotative) that can be represented by formalized visual images, texts, and logic as mental (cognitive) interpretations without formalized representations. From the very beginning, the model is created by an expert. Then the model can be verified by its mapping to the relevant big data. This process can be used for deep learning of neural networks that helps then to generate a draft of the cognitive model automatically for different problems on the base of analysis information about these problems, which is required especially in an emergency.

The mapping of the model on the big data ensures only taking into account denotative semantics. Cognitive semantics cannot be created in this way. For example, for supplying models with cognitive semantics based on quantum effects, the quantum computer is needed. But this process can also be emulated by some quantum operators or gate, which was mentioned above. For this, the statistic distribution for every message about the new problem or event has to be calculated, and then the quantum operator or gate is applied to this distribution. The processes of creating and using a cognitive model are illustrated in Fig. 2.

The author's convergent methodology helps to ensure the decision-making and cognitive modeling processes more holistic, purposeful, and sustainable. The convergent approach is based on applying the inverse problem-solving method in the topological spaces, fundamental thermodynamics, and genetic algorithm. In [24], it was shown that to ensure the decision-making convergence, the space of problem has to be decomposed by taking into account the following axioms in terms of the category theory and the topology theory:

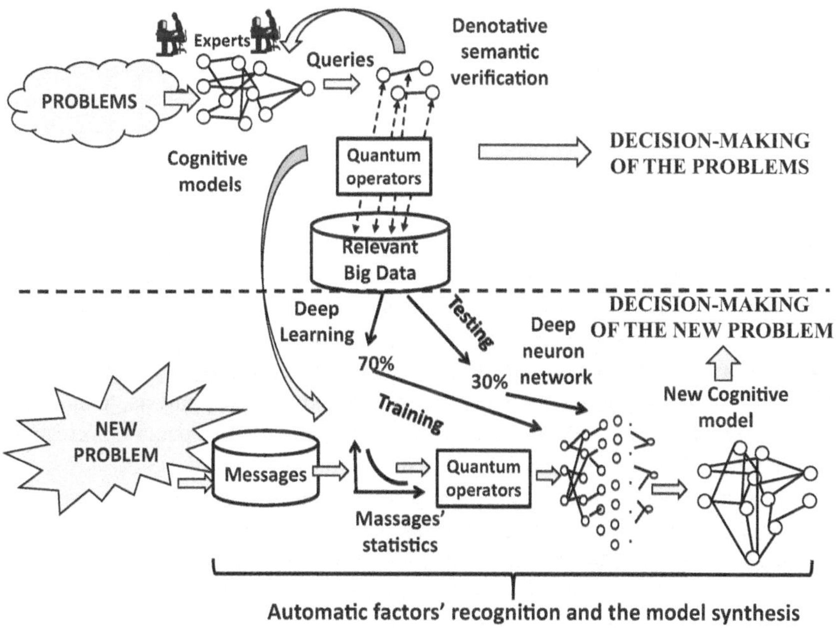


**Fig. 2** Cognitive modeling processes

- *D*: ***Set*** → ***Set***; the number of elements in the system of the sets ***Set*** is infinite, and the maps of the objects are maps with the *closed graph*;
- $\beta$ is a non-empty finite subcover of the ***Set*** (*Compactness*);
- For each pair of points in ***Set***, there are always exist disjoint neighborhoods, in the topological sense—open set contains this point (*Hausdorffness*).

To make the process of group decision-making more purposeful and sustainable, these axioms have to be transformed into the following rules that can be used during conducting collective conversations and meetings:

- Ensure quickly inclusion in the discussion all participants by asking them to assess the problem in one phrase;
- During discussion, identify in the problem such components as team, goals, resources, and actions;
- Goals have to be arranged hierarchically and assessed by important: main goal, internal and external;
- The resources, including institutional, financial, and material, must be separated into a finite and visible number of parts;
- The AI model has to cover all aspects of the problem, and all model's components have to be connected;
- Never underestimate small factors, etc.

These are only the necessary conditions for ensuring the purposefulness and sustainability of the decision-making process. A genetic algorithm can be used to provide sufficient conditions. This algorithm can provide several different solutions to achieve the stated goal. From these decisions, the team leader, according to a certain procedure, selects the most adequate for the current situation.

## 7 Practical Implementation

The list of three examples of practice implementations of the suggested approach to ensure the purposefulness and sustainability of AI applications was given above. These examples are connected with a hybrid version of AI implementations, i.e., the decision-making system includes peoples. A special place in these realizations occupied accelerating decision-making processes based on exploiting the convergent methodology, including cognitive modeling and big data analysis, to automate models creating. Some experimental attempts at using quantum operators were made to enrich the cognitive semantics of these models.

For example, let us take the first point from the list connected with the megapolis tourism development [25]. Figure 3 illustrates the situation of 2018.

Figure 3 shows that the situation development's statistical extrapolation does not lead to the strategic goals set by the power of the metropolis. From strategic decisions were required to transform the tourist situation strongly. Such decisions are of the

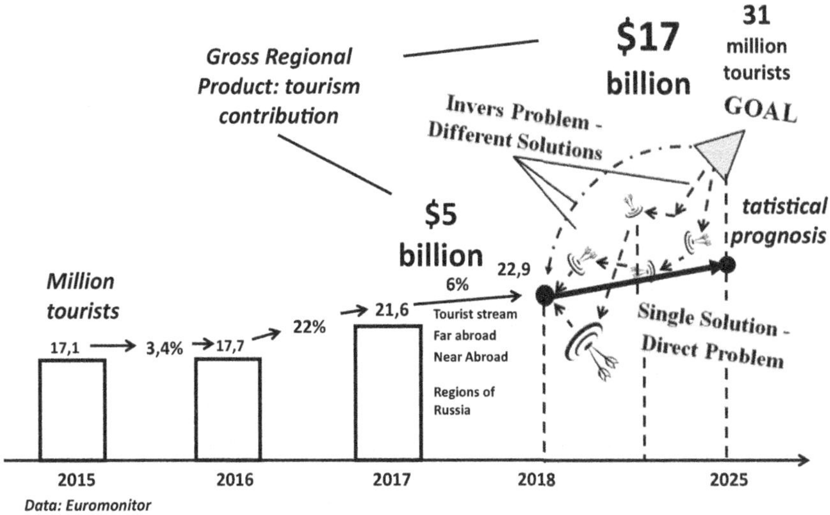


**Fig. 3** Megapolis tourism development and its strategic goals

inverse nature, which means they are unsustainable, that is, small changes in the situation may result in other decisions on the ways to achieve goals.

For revealing factors that were influencing the situation, 35 brainstormings were run simultaneously. Due to using the convergent methodology and cognitive modelling, this process took only four hours. During brainstorming with the SWOT analysis method, which helps identify factors of strengths, weaknesses, opportunities, and threats for the situation, about 70 factors were revealed. Then by using special procedures, they were convoluted into the substantial 15 factors, the weights of importance of their mutual influences were evaluated, and the cognitive model was made. Then the big data analysis helped to verify this model.

To answer the question "What should be done to achieve the goal?" (the inverse problem solving), the genetic algorithm to the cognitive model was applied. With its help, the optimal ratio of controlling factors to ensure the megapolis strategic tourism development's optimistic scenario was calculated. This result was used to substantiate the strategy for the development of tourism in the megapolis.

The quantum approach for creating cognitive semantics had an experimental character. It showed the possibility of raising the quality of verifying a cognitive model by mapping it to the relevant big data. So far, this approach can be classified as a discussional for future research, including using quantum calculations.

## 8 Conclusion

The sustainability of the systems that include AI components has different peculiarities. The sustainability of the system's behavior is difficult to ensure and take under control when AI has a hybrid nature, i.e., including the people as its inner component. The sustainability of these systems can be raised by taking into consideration the cognitive semantics, which cannot be formalized in a direct way.

Cognitive semantics can be embraced in an indirect way and with using some methods from quantum and field theories. The indirect way can be realized by the inverse problem solving on topology spaces method, cognitive modeling with their subsequent verification by mapping models on relevant big data, and using genetic algorithms. The quantum and field theories are at the stage of forming an idea and the beginning of theoretical and experimental research.

Two decades of author's practical implementations of the proposed convergent approach using AI in the field of the state and corporate strategic planning have proved its fruitfulness. The author's convergent approach creates the necessary and sufficient conditions for achieving the required goals, which can be ill-defined. Group decision-making processes with AI due to its use become more sustainable and greatly reduced in time.

The suggested approach has limitations. It can be used, e.g., for creating corporate strategy by organizing networked brainstorming with 35 groups simultaneously. The number of participants can be 5–7 members in every group; one brainstorming duration is 4-5 h.

## References

1. D. Gubanov, N. Korgin, D. Novikov, A. Raikov, *E-Expertise: Modern Collective Intelligence*.in Series: Studies in Computational Intelligence, vol. 558, XVIII, (Springer, 2014), 112 p. https://doi.org/10.1007/978-3-319-06770-4
2. R. Sun, Desiderata for cognitive architectures. Philos. Psychol. **17**(3), 341–373 (2004)
3. I. Kotseruba, J.K. Tsotsos, 40years of cognitive architectures: core cognitive abilities and practical applications. Artif. Intell. Rev. **53**(1), 7–94 (2020). https://doi.org/10.1007/s10462-018-9646-y
4. S. Adams et al., Mapping the landscape of human-level artificial general intelligence. AI Mag. **33**(1), 25–42 (2012)
5. S. Klimenko, A. Raikov, *Virtual Brainstorming*. in Proceedings of the International Scientific-Practical Conference on Expert Community Organization in the Field of Education, Science and Technologies (Triest, Italy, 2003), pp. 181–185
6. A.R. Dennis, B.A. Reinicke, Beta versus VHS and the acceptance of electronic brainstorming technology. MIS Q. **28**(1), 1–20 (2004). https://doi.org/10.2307/25148622
7. L.A. Liikkanen, K., Kuikkaniemi, P. Lievonen, P. Ojala, *Next Step in Electronic Brainstorming: Collaborative Creativity With the Web*. in Proceedings of the International Conference on Human Factors in Computing Systems, CHI 2011, Extended Abstracts on Human Factors in Computing Systems (Vancouver, BC, Canada, 2011). pp 2029–2034. https://doi.org/10.1145/1979742.1979929

8. H. Matallah, G. Belalem, K. Bouamrane, Towards a new model of storage and access to data in big data and cloud computing. Int J Ambient Comput Intell (IJACI) **8**(4), 31–44 (2017). https://doi.org/10.4018/IJACI.2017100103
9. N. Dey, A.E. Hassanien, C. Bhatt, A. Ashour, S.C. Satapathy (eds), Internet of Things and Big Data Analytics Toward Next-Generation Intelligence. (Springer, Berlin, 2018). https://doi.org/10.1007/978-3-319-60435-0
10. M.G. Sarowar, M.S. Kamal, N. Dey, *Internet of Things and its Impacts in Computing Intelligence: A Comprehensive Review–IOT Application for Big Data*. in Big Data Analytics for Smart and Connected Cities (IGI Global, 2019), pp. 103–136. https://doi.org/10.4018/978-1-5225-6207-8.ch005
11. A.N. Raikov, Z. Avdeeva, A. Ermakov, *Big Data Refining on the Base of Cognitive Modeling*. in Proceedings of the 1st IFAC Conference on Cyber-Physical&Human-Systems, Florianopolis, Brazil. 7–9 December (2016), pp. 147–152 https://doi.org/10.1016/j.ifacol.2016.12.205
12. S.V. Ulyanov, A.N. Raikov, Chaotic Factor in Intelligent Information Decision Support Systems, Eds. by Aliev R. end etc., Proc. 3d International Conference on Application of Fuzzy Systems and Soft Computing. Wiesbaden, Germany, pp. 240–245 (1998)
13. A. Dalela, *Quantum Meaning: A Semantic Interpretation of Quantum Theory,* Kindle Edition (Shabda Press, 2012), 196 p
14. H. Atmanspacher, *Quantum Approaches to Brain And Mind. An Overview with Representative Examples*. in The Blackwell Companion to Consciousness, ed by S. Susan Schneider, V. Max (Wiley, 2017), pp. 298–313. https://doi.org/10.1002/9781119132363.ch21
15. D. Aerts, M. Czachor, Quantum Aspects of Semantic Analysis and Symbolic Artificial Intelligence. In: J. Phys. A.: Math. Gen. **37**, L123–L132 (2004) in IEEE Computer Graphics and Application "Collaborative Visualization, pp. 58–68 (2009)
16. O.V. Ivancova, V.V. Korenkov, S.V. Ulyanov, *Quantum Software Engineering. Quantum Supremacy Modelling. Part I: Design it and Information Analysis of Quantum Algorithms: Educational and Methodical Textbook* (Dubna, Joint institute for Nuclear Researches, INESYS (EFCO Group), Moscow, KURS, 2020) 328 p
17. O.V. Ivancova, V.V. Korenkov, S.V. Ulyanov, *Quantum Software Engineering. Quantum Supremacy Modelling. Part II: Quantum Search Algorithms Simulator—Computational Intelligence Toolkit: Educational and Methodical Textbook* (Dubna, Joint institute for Nuclear Researches, INESYS (EFCO Group), Moscow, KURS, 2020), 344
18. S.V. Ulyanov, Quantum Fast Algorithm Computational Intelligence PT I: SW /HW smart toolkit. Artif. Intell. Adv. **1**(1) (2019). https://doi.org/10.30564/aia.v1i1.619
19. A. Raikov, Strategic Analysis of the Long-Term Future Needs of Educational Services. in *Proceedings 3rd World conference on smart trends in systems, security and sustainability (WorldS4)* (Roding Building, London Metropolitan University, London, UK. IEEE, 2019), pp. 29–36. https://doi.org/10.1109/WorldS4.2019.8903983
20. D. Perkins, *The Eureka Effect. The art and Logic of Breakthrough Thinking* (W.W.Norton & Company, NY, London, 2000) 293 p
21. B. Zwiebach, *Entanglement* [online], https://ocw.mit.edu/courses/physics/8-04-quantum-physics-i-spring-2016/video-lectures/part-1/entanglement/. Accessed 2 July 2020
22. T. Scheidl et al., Violation of Local Realism with Freedom of Choice. PNAS **107**(46), 19708–19713 (2010)
23. R.Pike, P.C. Sabatier, (eds.) Scattering. Scattering and Inverse Scattering in Pure and Applied Science (Academic Press, San Diego, San Francisko, New York, Boston, London, Sydney, Tokyo, 2002) 1831 p
24. A. Raikov, A. Ermakov, A. Merkulov, Self-organizing Cognitive Model Synthesis with Deep Learning Support. Int. J. Eng. Technol. (IJET). Special Issue Comput. Eng. Inform. Technol. **7**(2.8), 168–172. (2018). https://doi.org/10.14419/ijet.v7i2.28.12904
25. A. Raikov,*Megapolis Tourism Development Strategic Planning with Cognitive Modelling Support*. ed. by Fourth International Congress on Information and Communication Technology (London). Advances in Intelligent Systems and Computing, vol. 1041 (Springer, Singapore, 2020). https://doi.org/10.1007/978-981-15-0637-6_12

# Personal Data as a Critical Element of Sustainable Systems—Comparison of Selected Data Anonymization Techniques

**Paweł Dymora and Mirosław Mazurek**

**Abstract** In order to facilitate the sharing and dissemination of public data, the open data concept has been developed in recent years. Apart from its unquestionable positive effects, the whole process of opening data can also lead to negative ones. In most of the sectors processing vast amounts of data, such as medical, financial, or environmental, there are also legally protected data, the publication of which has far-reaching pejorative consequences. The article presents selected methods for sharing data containing confidential information in open data systems. The depersonalization of data subject to modifications makes it impossible to identify the subject that has been moved. Anonymization and pseudo-anonymization techniques were used. The most popular methods of data anonymization were compared: masking, permutations, adding noise, k-anonymization, l-diversification, and t-closeness. The risk of data extraction, correlation, or reasoning was determined based on the research.

**Keywords** Open data · Anonymization · Privacy · k-anonymization · l-diversification · t-closeness

## 1 Introduction

Sustainability is a broad discipline that provides insight into most aspects of the human world from business to technology, environment, and social sciences. Sustainability aims to protect the environment, human health, and the environment while driving innovation and not threatening our lifestyle. The core of any system is data. The security of information is the basis for the system to function correctly. A critical element in such a system is the security of data published on the Internet. For a given process to be considered sustainable, it should not cause irreversible changes in the

P. Dymora (✉) · M. Mazurek
Faculty of Electrical and Computer Engineering, Rzeszów University of Technology, 35959 Rzeszów, Poland
e-mail: Pawel.Dymora@prz.edu.pl

M. Mazurek
e-mail: Miroslaw.Mazurek@prz.edu.pl

 
A. Joshi et al. (eds.), *Sustainable Intelligent Systems*, Advances in Sustainability Science and Technology, https://doi.org/10.1007/978-981-33-4901-8_4

environment, should be economically viable, and should bring benefits to society. An example of such a system is called open data as also open government.

Personal data is a critical element in sustainable systems. Data security must take into account maintaining the integrity and reliability of the information and ensuring continuous access to authorized persons' information. The main purpose and motivation of the presented research are to answer the question of what tools should be used to ensure that the provided data does not endanger the subjects concerned while ensuring the integrity and protection of personal data. The article describes the primary way to protect data against disclosure, which is anonymization. It analyzes the latest free programs for anonymizing data, showing their real impact on security levels. The analysis of selected methods of anonymization allowed us to determine the optimal parameters of transformation, making the present research unique. Preliminary research of the analysis allowed to determine the highest levels of risk of extraction, risk of linking, and risk of reasoning for each of the analyzed methods.

The article consists of six chapters with an introduction (chapter first). In the second chapter, a literature review was conducted, and the main challenges in this field were discussed, as well as the research gap this article fills. The third chapter describes the most popular anonymization techniques: masking, permutations, adding noise, k-anonymization, l-diversification, and t-closeness. The next chapter introduces research methodology, in particular, free data anonymization tools such as ARX Data Anonymization Tool and Amnesia. Chapter five presents the results and analysis of the simulations. The last chapter presents a summary and future research.

## 2 Related Works

By definition, these are generally available and understandable data, intended for reuse and redistribution by any person. It is data containing useful information, stored in a convenient form that allows any person to use it freely and redistribute it regardless of the field of action. The correct use of this information results in open knowledge. To systematize the concept of open data, the open knowledge foundation established to disseminate open knowledge resources and has developed vital features that each open data set must meet [1]:

- Accessibility—data must be accessible in its entirety, in a convenient, open, and modifiable form.
- Reuse and redistribution—open data must be provided under conditions that approve its reuse, consolidation with other data sets, and redistribution. The data must also be machine-readable.
- Participation by all—data must be made available so that any person, regardless of their field of activity, or membership of a particular social group, can use and redistribute it. This means that no restrictions on reuse are allowed.

Personal data is all information by which an individual can be identified. It includes information that enables a person to be identified directly utilizing a unique identifier in the form of an identification number and, at the same time, information indirectly linked to the entity to which it relates. It includes, for example, information on physiological, mental, social, cultural, economic, or physical characteristics. More generally, personal data is information that makes it possible to identify the identity of a particular person, at a real cost, with real time, and effort. Examples of personal data include names, surnames, but also data such as cookie identifiers, applications, tools, IP addresses, protocols, and location data. [2, 3].

The identification of an individual can be made directly and indirectly. Direct identification occurs when an entity is identified directly, using the information in its possession, which is unambiguously identifiable as a data subject. This is possible through the use of data such as identity card number, employer identification number (EIN), social security (SSN), bank account, or other information linked to a specific person. Indirect identification is based on the relationship between the data at hand or using external information sources. Unlike direct identification, it does not occur immediately and does not indicate a straightforward subject. It requires different techniques, such as reasoning. The issue of data security and protection is described in detail in [4]. Medical records of at least 173 million people gathered since October 2009 have been breached and might have adversely influenced over half of the population in the USA. It takes a considerable amount of time to educate the public, and it takes substantial financial resources to prevent data breaches, why we need a lower cost solution like using data anonymization.

An average shared data file consists of attributes representing features and information about a particular entity represented by a single line called a record. To identify information that is considered to be sensitive, the attributes are subject to categorization. This determines how and what action must be taken before a data set is anonymized and then published. Three main categories of features are distinguished in literature [5–7]:

- **Unambiguous identifiers**—it also called explicit attributes that clearly and uniquely identify a particular person. They contain information such as identity card number, employer identification number (EIN), and social security (SSN). Usually, the only effective way to anonymize such attributes is to remove them or completely masking them.
- **Pseudo-identifiers (PID)**—it also known as quasi-identifiers and are attributes that ambiguously identify a particular person. However, by combining several of them or combining them with knowledge from an external source, it is possible to make a sufficient identification. The concept of a pseudo-identifier is often used to identify individual attributes but is usually used as a combination of a set of attributes of a given data file used to carry out privacy-intrusive attacks. An example of a pseudo-identifier can be a set of two attributes of one file: address and first name.
- **Sensitive attribute**—these are attributes containing information called private. An example is a medical information about the health of patients. Disclosure of

this information may cause harm to people connected with it so that it may be an interesting target for the attacker.

Additionally, the fourth category of attributes is considered to be insensitive attributes, representing attributes that are not anonymized in any way and thus remain in an explicit form [8].

In recent years, big data systems have become an essential trend in IT, especially in the context of IoT and Industry 4.0. These systems are characteristic for branches of industry and business, where a large amount of data with great diversity is a very characteristic feature. Another important feature of systems of this class is the fact that this data is generated by various sources to a large extent by sensors, including wireless sensors transmitting data in an open medium. That is why new tools are needed to secure the transmitted data and then retrieved it in the cloud. As the authors emphasize in [9], high data security, and in these, privacy protection, data anonymization, and security policies are critical challenges for the protection of big data and big data analytic techniques. In this context, our work fills an important gap in the systematization of current knowledge in data security concerns and privacy protection, especially in terms of analysis and availability of tools to implement a high level of these protections. Therefore, it is an essential contribution of the authors to assess the effectiveness of selected anonymization techniques and ensure an appropriate level of data privacy.

An exciting area for further research presented in [10] may be the aspect of improving the privacy and security of user data in location-based services (LBS). As the authors emphasize in [10], user data in LBS systems are vulnerable to interception and malicious attacks, and the process of improving security in LBS, which has been far from a complete solution. An important direction of further research that the authors would like to expand in the future is using caching techniques and space-efficient probabilistic data structure, which is Bloom filter.

In [11], authors described how to integrate the process of anonymization and data analysis. The appropriate noise distribution can be determined. The first stage uses the addition of random noise with known distributional properties. The second stage of the study consists of specifying the model of interest so that parameter estimation accounts for the added noise. The Bayesian MCMC algorithm was used to recovers consistent estimates of the correct model parameters.

## 3 The Most Popular Anonymization Techniques

There are many anonymization techniques. The choice of the appropriate method is closely linked to the type of information that will be anonymized. Each deperson-alization process should be approached individually, in order to achieve the highest possible level of privacy, with the least degradation of the information value contained in the original state of the anonymized data set. The most common techniques for anonymizing data include [12, 13]:

- **Masking**—the masking method involves hiding part of the characters of a specific value with a selected character, which is usually an asterisk: "*". The advantage of this method is that there is no need to modify actual identities. Thanks to this, there is also no need to create a generalization hierarchy using a dictionary containing all the provinces to which the anonymous cities belong. Still, each level is made based on a different number of masked characters [14].
- **Permutations**—the permutation technique is based on shuffling attribute values, as a result of which they are artificially correlated with random entities. This protects the privacy of the individuals concerned by the anonymized data file, without modifying the data, which may be crucial in case of reuse. By replacing the content of attributes between the rows, the original range and value distribution are maintained, so that the anonymized file remains a complete and complete source of information [14].
- **Adding noise**—the technique of adding noise is based on modifying the value of the attributes with small interferences, depending on the scale of its value. These are usually size disturbances, which will reduce their accuracy without degrading the usability of the data after it has been anonymized. Appropriate application of this technique will make it impossible for the attacker to identify a particular entity and prevent the attacker from diagnosing the magnitude of the interference that has been applied to specific attribute values [14].
- **k-anonymization**—the k-anonymization model is one of the most popular privacy models. It prevents the identification of the subjects of anonymized data file records. It is seen as a basic anonymization technique that meets the basic requirements that an anonymized data file should offer. According to the model, once a data file has been anonymized, each record should not be able to be distinguished from at least ($k$-1) other records. This means that records are subject to a kind of grouping, whereby each entity can be mistakenly identified with k other records, making it impossible to identify a specific person directly. Table 1 shows an anonymized data set that meets the assumptions of the k-anonymization model with a $k$-value of 3 [7, 15].

Generalization makes it possible to achieve the goals of the k-anonymization model. To obtain a data file that meets the assumptions of this privacy model, one

**Table 1** Example of k-anonymized data set

| Row index | Age | ZIP code | Disease |
|---|---|---|---|
| 1 | [20, 30] | Northeastern USA | HIV |
| 2 | [30, 40] | Western USA | Hepatitis C |
| 3 | [20, 30] | Northeastern USA | HIV |
| 4 | [30, 40] | Western USA | Hepatitis C |
| 5 | [30, 40] | Western USA | Diabetes |
| 6 | [20, 30] | Northeastern USA | HIV |

can use Samarati's algorithm, which was first proposed in the context of the k-anonymization model. If a domain is generalized, a grid is created for the combination of attributes. It consists of nodes, each of which determines the level of generalization for the entire attribute combination. The resulting nodes are also called transformations. Anonymizing a data file with one of them may result in meeting the requirements of the k-anonymization model, with a different value of *k*, depending on the node used.

- **l-diversification**—the l-diversification model is an extension of k-anonymization. It eliminates the occurrence of situations in which a sensitive attribute is revealed. This type of case takes place in Table 1, where the HIV virus has been exposed to people who are referred to the zip code attributes with the value "northeastern USA" and age with the value "[20, 30]". It aims to maintain a minimum level of diversity in a given record group. This means that a single pseudo-identifier, being a group consisting of several records, must contain l different values in the sensitive attribute [6, 14].
- **t-closeness**—this method does not guarantee full protection against privacy intrusion. Still, it does impose limitations on the probability distribution of the value that occurs in sensitive attributes in specific pseudo-identifiers and the entire data set. This method improves the quality of the results of the anonymization process but will not achieve the same results for each data set. The general definition says that the task of the t-closeness model is to keep the distance between the distribution of a sensitive attribute in a given group (PID) and the distribution in the whole data set below the set threshold t [15, 16].

To ensure adequate security of confidential information, it is necessary to obtain a recommended level of privacy and determined by technical documentation describing the security standard. By using various anonymization or pseudonymization techniques, it is possible to protect the data file against the three most common privacy-intrusive vulnerabilities listed in the data.gov.pl documentation [17]:

- *Extraction*—this involves isolating individual or all records that are able to identify some or all persons in a given data set.
- *Linking*—it consists of linking at least two records that relate to the same entity. They may be in the same database or in separate databases.
- *Reasoning*—it allows deducting the value of an attribute from the value of other attributes together with a high probability.
- Different depersonalization techniques have different levels of resistance to these privacy-intrusive vulnerabilities. Therefore, anonymization or pseudonymization techniques should be chosen in such a way as to achieve the highest possible level of security together with resistance to all listed attacks, while maintaining a high level of use of the open file. Table 2 shows the vulnerabilities to privacy risks of the most popular, recommended, and most frequently used privacy protection methods, according to data.gov.pl [17, 18].

**Table 2** Vulnerabilities to privacy risks of depersonalization methods recommended by data.gov.pl [18, 19]

| Data depersonalization technique | Risk of extraction | Risk of linking | Risk of reasoning |
|---|---|---|---|
| Adding noise | Yes | Maybe not | Maybe not |
| Permutations | Yes | Yes | Maybe not |
| Aggregation or k-anonymization | No | Yes | Yes |
| l-diversification | No | Yes | Maybe not |
| Differential privacy | Maybe not | Maybe not | Maybe not |
| Tokenization | Yes | Yes | Maybe not |
| Pseudonymization | Yes | Yes | Yes |

## 4 Research Methodology

Anonymization of data is a process that aims to break the links between individuals and personal data by which a person to whom the data relate can be identified. Anonymization transforms personal data so that it is no longer possible to identify the data or to link it to an identified natural person [5, 17]. Once anonymized, new categories of data are no longer covered by the protection of personal data. However, it should be checked and specified for all cases whether the new classes qualify as "anonymized data" [18, 19].

Many tools can be used in the anonymization process. The article analyzes the most popular free applications, which include ARX Data Anonymization Tool and Amnesia.

- **ARX Data Anonymization Tool**—this is an open-source program designed to anonymize files containing sensitive personal data. It offers many privacy models together with techniques such as masking, generalizing by creating compartments, and a generalization hierarchy based on them. It also has several risk models that can be used to assess the security level of an anonymized file. The program supports many privacy models, including the most popular ones, such as k-anonymization, l-diversification, and t-closeness. It allows us to verify the risk of reidentification, according to the prosecutor attacker model, the journalist attacker model, and the marketer attacker model [20].
- **Amnesia**—it is a free anonymization tool based on the k-anonymity method. The data is removed from the classified information employing generalization and suppression procedures. Thus, it allows us to transpose the values of individual attributes from the exact form to the generalized form. An example of this can be the conversion of the value of an attribute containing names of cities, to names of countries or at a higher level of generalization hierarchy, to names of continents. The program also allows exploring a grid of solutions containing a combination of levels of the generalization hierarchy. The user can select any node from the entire grid, view anonymized data, and view statistics of anonymized attributes [20].

In the anonymization research, a CSV extension data file was created using a free "Online test data generator" available at the address: https://www.onlinedatagenerator.com. The generated data file contains information about social benefits collected by fictional corporate employees. It contains 1000 records and the following attributes:

- **PIN**—Personal identification number is a public identifier,
- **First Name**—the attribute that is part of the pseudo-identifier,
- **Last Name**—the attribute that is part of the pseudo-identifier,
- **City**—the city of residence of the employee and the attribute that is part of the pseudo-identifier,
- **Address**—the precise residential address of the employee and the attribute that is part of the pseudo-identifier,
- **Social Benefits**—information on the social security benefits collected by the employee. An attribute to be reused after depersonalization of the data file.
- **SUM**—the number of social security benefits received by the employee. An attribute to be reused after depersonalization of the data file.

The data file was depersonalized in programs: ARX Data Anonymization Tool and Amnesia. The depersonalization took place according to data.gov.pl portal guidelines [20, 21]. The initial assumption is to achieve the highest possible level of privacy, ensuring the protection of the data file against the three most successful attacks: extraction, linking, and reasoning.

The first step to being taken before starting the anonymization process is to categorize the attributes according to their impact on the data subject's identification risks:

- **First Name, Last Name, City, Address**: these attributes form a pseudo-identifier (PID) and have therefore been categorized as quasi-identifying.
- **PIN**: This is a non-confidential identifier, which can lead to direct reidentification and has therefore been classified as "identifying", which will result in its complete deletion from the data set.
- **Social Benefits, SUM**: These attributes have been classified as non-sensitive, as they are the main ones that can be relied upon to create a compilation and statistics file.

## 5 Results and Analysis of the Simulations

Presented research and analysis are based on data generalization. The most important activity is to create hierarchy steps. The higher the level, the more intensive the generalization is, which in the case of, e.g., character string masking method, means that the lowest level contains the openly presented data. As the hierarchy level increases, the data is masked, which at the highest level leads to the total hiding of the value of a given attribute.

Mainly due to generalization, it is possible to achieve the assumptions of the k-anonymization model. To obtain a file that meets the assumptions of the aforementioned privacy model, you can use Samarati's algorithm, which was first proposed in the context of the k-anonymization model concept. When you generalize a domain, a grid is created for the combination of attributes. It consists of nodes, each of which determines the level of generalization for the entire attribute combination. The resulting nodes are also called transformations. Anonymizing a file with the power of one of them may result in meeting the requirements of the k-anonymization model, with different k-value, depending on the node used.

Table 3 shows the parameters of the six carried out experiments using the ARX Data Anonymization Tool.

Table 4 presents the detailed results of the experiments in the ARX Data Anonymization Tool. For the first experiment, the results of the transformation are presented, using the pseudo-identifier value masking method on individual attributes, as well as the implementation of the whole data set of the k-anonymization privacy model (k-anonymity), where $k = 2$. For experiment 2, the "Social Benefits" attribute

**Table 3** Selected anonymization techniques in experiments (ARX Data Anonymization Tool)

| Experiment no | Depersonalization techniques used |
|---|---|
| 1 | Generalization, k-anonymization ($k = 2$) |
| 2 | Generalization, k-anonymization ($k = 2$), l-diversification ($l = 2$) |
| 3 | Generalization, k-anonymization ($k = 4$), l-diversification ($l = 2$) |
| 4 | Generalization, k-anonymization ($k = 4$), l-diversification ($l = 2$), t-closeness ($t = 0.001$) |
| 5 | Generalization, k-anonymization ($k = 20$) |
| 6 | Generalization, k-anonymization ($k = 20$), l-diversification ($l = 2$) |

**Table 4** Detailed summary of the experimental results (ARX Data Anonymization Tool)

| Experiment No. | Deleted records | PID number | Min PID | Max PID | Maximum risk by prosecutor model (%) | Records at maximum risk (%) |
|---|---|---|---|---|---|---|
| 1 | 0 (0%) | 107 | 3 (0.3%) | 18 (1.8%) | ~33.3 | 0.3 |
| 2 | 341 (34.1%) | 98 | 2 (0.2%) | 43 (4.3%) | 50 | ~7.6 |
| 3 | 0 (0%) | 23 | 5 (0.5%) | 153 (15.3%) | 20 | 0.5 |
| 4 | 843 (84.3%) | 2 | 57 (5.7%) | 100 (10%) | ~1.6 | ~36.3 |
| 5 | 42 (4.2%) | 19 | 20 (2%) | 100 (10%) | 5 | ~2.09 |
| 6 | 42 (4.2%) | 19 | 20 (2%) | 100 (10%) | 5 | ~2.09 |

(classified as a "sensitive" attribute), a privacy model l-diversification with a default value of l of 2, has been imposed. It guarantees the occurrence in each pseudo-identifier of at least l different values of a given attribute, which in our examination takes values "true" or "false". Therefore, the only value that can be assigned to the parameter l is value 2. In order to improve the results of experiment 2, in experiment 3, the number $k$ of the k-anonymization model has been increased to value 4. In experiment 4, additionally, the t-closeness privacy model has been used with a default $t$-value of 0.001. This model provides the maximum distance of probability distribution of the variable values of the sensitive attribute between the distribution in the given pseudo-identifier and the distribution in the entire data file. In experiment 5, a k-anonymization model with a k-value of 20 was used. It is intended to check whether the use of a single k-anonymization model with a high $k$-value makes it possible to achieve data file anonymization at the level of the third experiment, in which an additional l-diversification model was used. Experiment 5 showed that with a significant increase in the k-anonymization model from 4 to 20, the degree of data file anonymization increased considerably. Therefore, in experiment 6, the configuration from experiment 3 was used, together with an increased $k$-value ($k = 20$), the l-diversification model ($l = 2$).

Table 5 presents the results of the protection level for each of the three most popular privacy breach attacks.

From the results summarized in Tables 4 and 5, it can be concluded that the highest results in terms of personal data security were achieved in the fourth experiment. The minimum number of pseudo-identifiers (PIDs) is 57, resulting in a probability of identifying the data subject at best 1/57, which is approximately 1,8%. It resulted in a data file reduced by more than 84% of the number of all records, making it completely useless.

A data file with a high level of security, as well as a high usability value, due to a record reduction of 0%, was obtained as a result of experiment three. It is also threatened with the risk of attack according to the prosecutor's model at the level of 20%, together with the number of vulnerable records at the level of 0.5% from the whole data file, which confirms the properly conducted depersonalization process. After increasing the number of $k$-values in the sixth experiment, the data file has achieved the risk of attack according to the prosecutor's model at the level of 5%, together with the number of vulnerable records in the amount of about 2.1%

**Table 5** Summary of security levels (ARX Data Anonymization Tool)

| Experiment No. | Risk of extraction | Risk of linking | Risk of reasoning | File usefulness |
|---|---|---|---|---|
| 1 | None | High | High | High |
| 2 | None | High | Low | Low |
| 3 | None | High | Low | High |
| 4 | None | High | Low | None |
| 5 | None | High | High | High |
| 6 | None | High | Low | High |

from the whole data file. This is the experiment that has achieved the highest level of anonymization of the six experiments, minimizing privacy risks, according to the three models implemented in the program, while gaining resistance to privacy-intrusive attacks such as reasoning and extraction, maintaining the usefulness of a data file that has been reduced as little as 42 records. The fifth experiment, despite the right results, according to the risk analysis implemented in the program, guarantees only resistance to one of the three most popular privacy violation attacks, which is the extraction risk, devalued by the k-anonymization model. At the same time, it is not resistant to reasoning attacks, unlike the results of experiment 6.

The impact of the implementation of the single l-diversification model on the data set has been omitted because, in the case of the attribute "Social Benefits", the number of its values ("false" and "true") determines the minimum number of different values of the sensitive attribute that are contained in one pseudo-identifier. This also translates into a maximum value of parameter l of 2 when using the model on this data set. Using only the l-diversification model to anonymize the used data file leads to a similar effect as when using the k-anonymization model with a $k$-value of 2 and l-diversification with an $l$-value of 2 at the same time.

Tables 6 and 7 show the vulnerabilities to the most common anonymized data privacy breaches, together with their level of usability obtained in Amnesia program. In all experiments, the k-anonymization model effectively excluded the risk of extraction by creating groups of pseudo-identifiers. In experiment 2, the use of high degrees from the generalization hierarchy led to a high generalization of attribute values, resulting in a decrease in the usability of the depersonalized data file.

However, this has led to obtaining a level of privacy according to the k-anonymization model with a k-value of 379, which excludes the risk of an extraction attack. Unfortunately, it cannot be clearly stated that the file is immune to linking or reasoning attacks, as the tool does not provide anonymization methods that guarantee protection against these privacy-intrusive methods. In experiment 3, lower levels of generalization have been used, while the suppression method eliminated

**Table 6** Detailed summary of the experimental results (Amnesia tool)

| Experiment No. | Transformation | k-anonymization privacy level | Deleted records (%) | PID number |
|---|---|---|---|---|
| 1 | [1] | 7 | 0 | 6 |
| 2 | [1, 3] | 379 | 0 | 2 |
| 3 | [1] | 65 | 1.8 | 4 |

**Table 7** Summary of security levels (Amnesia tool)

| Experiment No. | Risk of extraction | Risk of linking | Risk of reasoning | File usefulness |
|---|---|---|---|---|
| 1 | None | High | High | High |
| 2 | None | High | High | Low |
| 3 | None | High | High | High |

the smallest pseudo-identifiers, most vulnerable to privacy-intrusive attacks. This resulted in a higher level of privacy of k-anonymization with a k-value of 65, which, in total, increased the usability of the data file, maintaining a high level of security.

The risk of reasoning and linking has not been eliminated in any of the conducted three experiments. The reason for this is the limitation of the tool used, which does not allow the use of methods such as adding noise, permutations, or any privacy model ensuring the declaration and achievement of a given distribution of a sensitive attribute ("Social Benefits"). As a result, an anonymized data file may contain a pseudo-identifier, which will consist of "Social Benefits" and "SUM" attributes with identical ("Social Benefit") or similar ("SUM") values, allowing the attacker to assign a known person to a dangerous pseudo-identifier, with a high or even 100% probability, leading to a violation of the privacy of the personal data.

## 6 Conclusion

While there are many anonymization tools available, the only way to achieve the highest possible level of the personal data security to make data securely available in open data systems is to protect against as many privacy-intrusive attacks as possible by depersonalizing the data with several anonymization methods. All the techniques used during the anonymization processes, including free and publicly available solutions, allowed for correct anonymization of the depersonalized data file but did not provide all the security recommendations of the data.gov.pl portal. The choice of the application, and most importantly, the anonymization methods must be closely related to the type of data contained in the anonymized data file.

The use of specific anonymization methods also depends on the type of anonymized information. Each of the methods degrades the risk of a personal data security breach. However, none of them will autonomously protect the data file. Particularly noteworthy are the recommendations of the data.gov.pl portal, which indicates that no anonymization technique is able to entirely exclude the risk of privacy violation through linking or reasoning attacks. This means that many anonymization techniques should be used simultaneously in such a way as to reduce the possibility of these attacks as much as possible. In doing so, it is always essential to ensure that the anonymized data file is not deprived of useful information for reuse for security reasons. As technology develops, the capabilities of the attacker increase, so the attacker can use innovative methods of violating the privacy, for which no adequate protection policy has currently been developed. Therefore, it is necessary to continually monitor and increase the level of security of the data being shared by using the latest and most up-to-date anonymization techniques.

Fully synthesized data sets have a reasonably low risk of disclosure. In a partially incorporated data set, the risk of exposure is reduced by partially integrating the data, those with the highest risk. In the hybrid approach data sets, the risk should be lowest due to the mixing of original and synthetic records.

The methods presented have also limitations. The problem is the choice of appropriate depersonalization tactics, strictly dependent on an anonymous data set, and requires knowledge of the law. In the case of creating a hierarchy, each generalized value must have a specific amount at each level of generalization. This is a very demanding process, but it allows us to keep as much information as possible that can be used to reuse the anonymized file. Additionally, these tools do not contain implemented privacy models, so it is necessary to prepare the data properly using suppression methods in order to eliminate the risk of extraction attack, leading to meeting the requirements of the k-anonymization model. Unfortunately, such a process requires considerable knowledge and experience.

**Acknowledgements** We are thankful to the graduate student Marcin Żurek of Rzeszów University of Technology, for supporting us in the collection of useful information.

## References

1. B. Wessels, R. Finn, K. Wadhwa, *Thordis Sveinsdottir* (Open Data and the Knowledge Society, Wydawnictwo Amsterdam University Press, 2017).
2. Rozporządzenie Parlamentu Europejskiego z dnia 27 kwietnia 2016 r. o ochronie danych osobowych.
3. P. Dymora, M. Mazurek, B. Kowal, *Open Data—An Introduction to the Issue*. in Computing in Science and Technology (CST 2018), ITM Web Conference, vol. 21 (2018). https://doi.org/10.1051/itmconf/20182100017
4. W. Koczkodaj, J. Masiak, M. Mazurek, D. Strzałka, P.F. Zabrodskii, Massive health record breaches evidenced by the office for civil rights data. Iran J Public Health **48**(2), 278–288 (2019)
5. A. Liber, *Problemy anonimizacji dokumentów medycznych*. Część 1. (Wydawnictwo Państwowa Medyczna Wyższa Szkoła Zawodowa w Opolu, 2014)
6. C.C. Aggarwal, *Data Mining. Wydawnictwo* (Springer International Publishing, 2015)
7. P. Dymora, M. Mazurek, B. Kowal, *The effectiveness of the Use R-Language in Anonymizing Open Data*. in Computing in Science and Technology (CST 2019), pp. 29–40. ISBN 978–83–7583–930–2
8. https://arx.deidentifier.org/overview/privacy-criteria/.
9. S. Tamane, V. K. Solanki, N. Dey, (eds.) *Privacy and Security Policies in Big Data* (IGI Global, 2017)
10. M. Yamin, A. Abi Sen, Improving privacy and security of user data in location based services. Int. J. Ambient Comput. Intell. (IJACI), **9**(1), 19–42 (2018). https://doi.org/10.4018/IJACI.2018010102
11. H. Goldstein, N. Shlomo, A probabilistic procedure for anonymisation, for assessing the risk of re-identification and for the analysis of perturbed data sets. J. Official Statis. **36**(1), 89–115 (2020). https://doi.org/10.2478/JOS-2020-0005
12. J. Domingo-Ferrer, V. Torra, *A Quantitative Comparison of Disclosure Control Methods for Microdata*, in *Confidentiality Disclosure and Data Access: Theory and Practical Applications for Statistical Agencies*. ed. by P. Doyle, J.I. Lane, J.J.M. Theeuwes, L. Zayatz (North-Holland, Amsterdam, 2001), pp. 111–134
13. J.P. Daries, J. Reich, J. Waldo et al., privacy, anonymity, and big data in the social sciences. Commun. ACM **57**(9), 56–63 (2014)
14. https://gdpr.report/news/2017/09/28/data-masking-anonymization-pseudonymization/

15. J. Domingo-Ferrer, D.Sánchez, J.Soria-Comas, *Database Anonymization, Privacy Models, Data Utility, and Microaggregation-based Inter-model Connections* (Wydawnictwo Morgan & Claypool)
16. A. Liber, *Problemy anonimizacji dokumentów medycznych.* (Część 2. Wydawnictwo Państwowa Medyczna Wyższa Szkoła Zawodowa w Opolu, 2014)
17. https://www.gov.pl/web/cyfryzacja/otwarte-dane-publiczne.
18. https://dane.gov.pl/knowledgebase/useful-materials/preview/standardy-otwartosci%20danych
19. Ministerstwo Cyfryzacji, Standard techniczny. Opracowany w ramach projektu Otwarte dane—dostęp, standard, edukacja. https://dane.gov.pl/knowledgebase/useful-materials/preview/standardy-otwartosci-danych
20. https://arx.deidentifier.org/
21. https://amnesia.openaire.eu/amnesiaInfo.html

# Sustainable Solutions for Overcoming Transportation and Pollution Problems in Smart Cities

**Shrikant Pawar**

**Abstract** The urban transportation industry is changing rapidly. The most recognized example is cellphone-controlled shared cars or taxis. But other changes are suddenly noticeable including use of scooters, bikes, and walking. Many other changes based on cellphone reservations and billing are related to mass transit, road lanes with occupancy-related tolls, parking, and shared cars. One of the most recent additions to mobility as a service (MaaS), battery-powered scooters is the subject of this article. The objective is to have a scooter particularly suitable for more general use. More general use implies greater safety, comfort, and weather tolerance, particularly for adults over 25. The scooter to be commercialized should have 3 (or 4) wheels so balancing is not required, provide for better conditions in the rain or cold or hot, allow but not require seated use, be able to be stored in less land space, be very light based, have advanced location-based speed control for ease of use and safety, and be fun to use. Fun to use defines a secondary market; a scooter is not just for getting from *A* to *B*, but often can be sold for recreational use. This article focuses on production of such electric scooters for overcoming transportation and pollution problems in smart cities.

**Keywords** Sustainable solutions · Environment · Transportation

## 1 Introduction

A normal individual can stroll with a most extreme speed of 3.1 miles/h [1]. The quantity of individuals age 65 and more established in the USA on July 1, 2015, records for 14.9% of the complete populace [2]. Engine bicycles and vehicles that cause contamination are bulky and need permit, gas, upkeep, enlistment, and protection [3]. Besides, there is consistently an acquired issue of conveying additional load in movement (PC, cellphone, and so forth.). There appears to be a consistent need for upgrades in current travel techniques particularly in urban communities. These

S. Pawar (✉)
Yale University, 333 Cedar Street, New Haven, CT 06520, USA
e-mail: shrikant.pawar@yale.edu

 
A. Joshi et al. (eds.), *Sustainable Intelligent Systems*, Advances in Sustainability Science and Technology, https://doi.org/10.1007/978-981-33-4901-8_5

challenges motivated to work on a sustainable solution in the form of an electric bike which can be a potential answer for conquering the essential issues of city transportation swarming and contamination. These zippy (top helped speed is around 25–30 miles for every hour (mph)) bicycles can sneak around gridlocks and are a breeze to stop. The battery type utilized can be revived in a couple of hours so the bikes are consistently prepared to work. These points of interest have provoked numerous city inhabitants to purchase battery bicycles as a practical and hip transportation option in contrast to the car. The beta-client's suburbanites find that the engine gives enough additional capacity to dispense with perspiring. For the most part, such engines can create from 100 to 700 W [4]. It additionally makes it simpler to convey the additional load of business-related material with the client. This item supports adaptability and upgrades client voyaging experience. Since electric bicycles are named bikes, no permit, enlistment, and protection are required, sparing a pack (and there are no gas costs and just modest, essential upkeep needs) of client costs and with no exceptional prerequisites. The speed of accessible items can change from 20–30 km for each hour (km/h) (12.4–18.6 mph) with a charger time going from 1.5 to 6 h and a value scope of $399–999.95 barring tax assessment and transportation charges. Here, we present a novel approach of exploring a gigantic potential for a predominant quality and proficiency in a moderate electric bike.

## 2 Materials and Methods

### *2.1 Motor Using 150 a.H./6 h Discharge Time*

A superior quality scooter should have a powerful motor, speed, good battery discharge time, and stability. A 24 V (V) 500 W 28.5 A electric brushed engine with a 2800 RPM engine can work in forward or reverse by switching the extremity. A Duracell 12 V 1.3AH AGM SLA battery with F1 terminals can be completely energized in 1.5 h with a release of around 6 h; it comprises of an absorptive glass tangle framework; it has a support free activity and is without position and spillage free with stable quality and longer assistance life. An in-house bike brush speed regulator and bike choke turn grasps has an under-voltage assurance of 20 V with a choke quickening agent and force show with 5-LED battery marker. The model bike should think about numerous elective engines with an accentuation on high capacity to gauge. Such engines may all around run at a lot higher RPM and have a 5 to 10 kilowatt of yield for every kg.

## 2.2 *Choice of Section of Flooring and Skeleton Material*

The current models are completely made of plastic and are less steady, dependable, and inclined to breakage and mishaps. To limit weight and improve inflexibility different materials ought to be thought of. Steel, aluminum, and different aviation materials ought to be considered for the model. A side estimation demonstrates that the essential basic segments should to be around the 3.5 lb. weight.

## 2.3 *Technical Goal*

The technical objective can be summarized as accomplishing low weight with low assembling cost. The essential assembling parts give off an impression of being the plank of flooring, front directing cylinder, wheels, conceivable safeguard, front fork, back wheel axel with engine joining, and gadgets for advanced control and interface with the client's and vehicle's wireless. Also, an erectable seat and holder for an umbrella (and an umbrella as inherent part) and wireless holder ought to be thought of. An objective in general load of around 14 lb. gives off an impression of being reasonable with passenger(s) weight constrained to 300 lb. At exceptionally high traveler weight, the presentation regarding rate and slope climbing capacity may be constrained. It is planned than downhill speed should also be controlled by using the drive motors as generators as is common in electric drives for braking as well.

## 2.4 *Manufacturing a Scooter*

Research ought to be confined to building up a working model indicating practicality of protected and productive travel. Examination of large-scale manufacturing subtleties ought to be attempted in Phase II. Yet, large-scale manufacturing expenses should be a factor in Phase I. The Phase I work plan should incorporate the accompanying undertakings for accomplishing the following tasks.

### 2.4.1 Task I

Manufacturing front head tube, safeguard, and fork: Steel projects ought to be used to make head cylinders and fork. 20 1/4″ inch distance across chrome steel bearing balls G25 to be utilized for 360° head tube revolution.

### 2.4.2 Task 2

Manufacturing back wheel axel, with engine joining: A tempered steel round bar 1″ (A) × 12″ should be used as an axel. Two back wheels (250 mm each) with a 2500 RPM engine consolidation to be made in the axel component.

### 2.4.3 Task 3

Manufacturing bike deck: Permanent form projecting is a metal projecting procedure that utilizes reusable molds ("perpetual molds"), normally produced using metal. The most widely recognized procedure utilizes gravity to fill the form. Nonetheless, gas pressure or a vacuum is additionally utilized. A minor departure from the run of the mill gravity projecting procedure, called slush projecting, produces empty castings. Normal projecting metals are aluminum, magnesium, and copper amalgams. Different materials incorporate tin, zinc, and lead amalgams and iron and steel are likewise projected in graphite molds. Changeless molds, while enduring more than one projecting despite everything, have a restricted life before wearing out. The ingate is part of assembling plan that identifies with the progression of metal through the projecting's framework. Stream contemplations for metal giving production start a role as soon as the liquid metal enters the form. The fluid metal for the projecting goes from the pouring bowl through the down sprue, so keeping away from enormous warmth masses in areas far off to risers. At the point when cementing of a projecting starts, a slight skin of strong metal is first framed on a superficial level between the projecting and the shape divider. When choosing to make a section by projecting one must consider the material properties and potential imperfections that this assembling procedure produces. The essential method to control metal projecting imperfections is through acceptable shape plan contemplations in the formation of the projecting's mold and gating framework. The key is to structure a framework that advances directional cementing.

### 2.4.4 Task 4

Assembly stage: Assembly of front set, back wheels with engine, deck and electrical ought to be performed at this stage.

### 2.4.5 Task 5

Battery testing stage: A Duracell 12 V 1.3AH AGM SLA battery with F1 terminals can be completely energized in 1.5 h and a release of around 6 h, it comprises of an absorptive glass tangle framework, has an upkeep free activity, is sans position and spillage free with stable quality and longer help life. Packaging, security from or resistance to, power yield—load test, safeguard gadgets, mechanical and ecological tests ought to be performed. Variables like warming, temperature cycling, elevation, moistness, presentation to fire, squash tests, nail infiltration tests, stun test, vibration test, sway test, drop test, impede (time), voltage, over-release, voltage inversion, high temperature, low temperature, abuse, misuse, quality, unbending nature and combustibility, form pressure (temperature), venting, protection, electrolyte not under tension, no spillage and no blast or fire hazard ought to be tried.

### 2.4.6 Task 6

Motor speed testing stage: An in-house bike brush speed regulator and bike choke contort grasps has an under-voltage insurance of 20 V with a choke quickening agent and force show with 5-LED battery pointer. The speed of the engine is generally estimated in RPM (cycles every moment). It is one of more troublesome assignments; however, it is one of the most significant things to think about the engine. There are a few different ways to do it. A strobe light (stroboscope) might be utilized to gauge RPM. Some strobe lights have dynamic counters indicating the glimmer rate. Optical tachometers might be utilized to decide turn rate.

### 2.4.7 Task 7

Scooter load testing stage: Load testing gives trust in the framework and its unwavering quality and execution. Load testing recognizes the bottlenecks in the framework under overwhelming client stress situations before they occur in a creation domain. Burden testing gives phenomenal assurance against horrible scenarios and obliges reciprocal methodologies for checking of a creation domain. Testing reaction time for every exchange, execution of framework segments under different burdens, and execution of database parts under various burdens organize delay between the customer and the worker, programming configuration issues, and equipment restriction issues should be addressed. There are various procedures of burden testing, some of them are as per the following.

#### Manual Burden Testing

This is one of the procedures to execute load testing; however, it does not create repeatable outcomes and cannot give quantifiable degrees of weight on an application.

#### In House Created Load Testing Instruments

An association, which understands the significance of burden testing, may fabricate their own apparatuses to execute load tests.

Open-Source Load Testing Instruments

There are a few burden testing devices accessible as open source that are for nothing out of pocket. They may not be as modern as their paid partners, yet on the off chance with a careful spending plan, they are the most ideal decision.

Enterprise Class Load Testing Instruments

They as a rule accompany catch/playback office. They bolster countless conventions. They can reenact an uncommonly huge number of clients.

### 2.4.8 TASK 8—Testing on Beta-Clients

The completed items ought to be given to beta-clients to be tried for in any event 90 days for a normal use followed by assortment of client input on enhancements for plan and determinations.

## 3 Equipment's Needed

### *3.1 Metal Cutting, Employing, and Wood Cutting*

Fusion wet saw: M-D constructing items' wet tile saw for cutting an assortment of clay, porcelain and stone tiles ought to be used, a 7-inch consistent edge precious stone covered sharp edge plunged into a water bowl to abstain from warming is significant. It should have an unbending steel outline, slip safe feet, and an inclining table which permits miter cuts and the 3/4 HP movable tear direct for cutting exactness.

### *3.2 DEWALT D28710 14-Inch Abrasive Chop Saw*

The DEWALT D28710 14-inch rough cleave saw: This saw has a 15.0 Amp/4.0 HP engine with overburden assurance to build execution and strength. Brisk lock tight clamp innovation takes into consideration quick bracing on various size and the 45° rotating wall takes into consideration quick and exact edge cuts with a steel base that permits client to weld jigs or stops straightforwardly onto the base. A hard-core lock down pin permits the top of the saw to be secured in the conveying position without the utilization of a chain. The saw length required is 18–3/16 inch, a wheel arbor if 1 inch, with rectangular measurements 4″ × 7–5/8 inches.

### 3.3 *Flux 125 Welder*

An adaptable motion center circular segment welder (FCAW) intended to utilize self-protecting transition cored welding wire is perfect for this venture. It has a variable speed wire control and warm overburden assurance, to weld mellow steel from 18 checkup to 3/16 inches' thickness is perfect. A base yield up to 125 amps with a class obligation cycle 90A AC @17 V, 20% and a constant wire feed speed control for ideal welding execution is required.

### 3.4 *Skeleton Materials*

Testing research center ought to be prepared to gauge wear opposition to break down resultant wear and de-union zones. The testing ought to be actualized on skeleton created on a hot steel square cylinders A36 (1″ × 0.065″) made of cold shaped and flawlessly welded hot rolled steel. It offers high weldability and is utilized oftentimes in auxiliary applications in scaffolds and structures just as in both the car and apparatus industry. Cold moved steel sheet is basically hot moved steel that has been additionally handled to expand its quality and solidarity-to-weight proportion. It can hold tighter resilience's than hot rolled when machined or in any case manufactured and gives a superior in general surface completion. In cold rolling, steel sheet is cooled at room temperature (after hot rolling) and is then strengthened as well as tempers rolled. A 3/4″ thickness measure fortified steel with 36″ widths and a 96″ length ought to be utilized as a bike deck.

## 4 Results

Scooters may encounter an assortment of sidelong and longitudinal powers (dependability) and movements. On most bicycles, when the front wheel is gone aside or the other, the whole back edge pitches forward marginally, contingent upon the controlling pivot edge and the measure of trail. With suspensions, either front, back, or both, trim is utilized to depict the mathematical setup of the bicycle, particularly in light of powers of slowing down, quickening, turning, drive train, and streamlined drag [5]. Speed vibrations is another worry, notwithstanding pneumatic tires and customary bike suspensions, an assortment of strategies have been created to clammy vibrations before they arrive at the rider. These incorporate materials, for example, carbon fiber, either in the entire casing or simply key parts.

## 5 Discussion

Here, we propose a sustainable solution for overcoming transportation and pollution problems in smart cities. Keith Code constructed a bike with fixed handlebars to explore the impacts of rider movement and position on directing [6], Richard Klein additionally manufactured a "Force Wrench Bike" and a "Rocket Bike" to examine guiding forces and their belongings [7]. Parallel and longitudinal steadiness measures ought to be focal point of Phase II endeavors. Around 75% of carbon monoxide emanations originate from autos and the normal driver purchases around 11 gallons of gas seven days, which implies they will go through $1,400 per year, just on gas. The bike item ought to be used as a compensation and ride in real money or online installment for hourly rides. In the current market, 1 assembling unit can create least of 100 items in 2 months, with 5 representatives working 12 h/day for 5 days per week. Basic, executional, asset and action cost drivers ought to be allocated fittingly. Potential endeavors ought to be made to tie the plan of action with scholarly organizations and government transportation administrations.

## 6 Conclusion

The urban transportation industry is changing rapidly. The most recognized example is cellphone-controlled shared cars or taxis. But other changes are suddenly noticeable including use of scooters, bikes, and walking. Many other changes based on cellphone reservations and billing should be coming related to mass transit, road lanes with occupancy related tolls, parking, and shared cars. The modern name MaaS, mobility as a service, has become popular particularly in Europe. One of the most recent additions to MaaS, battery-powered scooters, is the subject of this article. The objective is to have a high-tech scooter, particularly suitable for more general use. Although the solution we presented is important and feasible, confounding factors with low-income countries like economy, population density, infrastructure resources, other transportation alternatives, gender biases, etc., are needed to be examined before its implementation. Here, we have presented scooters as one of the sustainable solutions for overcoming transportation and pollution problems in smart cities.

## 7 References

1. *Study Compares Older and Younger Pedestrian Walking Speeds*. TranSafety, Inc. 1997–10–01. Retrieved 2009–08–24
2. *Facts for Features: Older Americans Month:May 2017* (United States Census Bureau. 2017–10). Retrieved 2018–08–09

3. C. Johansson, B. Lövenheim, P. Schantz, Impacts on air pollution and health by changing commuting from car to bicycle. Sci. Total Environ. **4**, 584–585 (2017)
4. L. Cabral. *Introduction to Industrial Organization* (2001–11)
5. *Serotta Technology Glossary: Vibration Damping*. Archived from the original on April 23, 2008. Retrieved 2008–06–24
6. R.E. Klein et al., *Bicycle Science*. Archived from the original on 2008–02–13. Retrieved 2008–09–09
7. C. Gromer, *STEER GEAR So How Do You Actually Turn a Motorcycle?* Popular Mechanics. Archived from the original on 16 July 2006. Retrieved 2006–08–07

# DLT-Based $CO_2$ Emission Trading System: Verifiable Emission Intensities of Imports

**Julian Kakarott, Kai Hendrik Wöhnert, Jonas Schwarz, and Volker Skwarek**

**Abstract** The European Emission Trading System is the primary tool of the European Union to reach net-zero greenhouse gas emissions by 2050. However, it faces several challenges regarding its effectiveness. To address those issues, a new mechanism based on distributed ledger technology is proposed. Previous work has shown how consumption-based emissions of European consumers can be limited by a reshaped emission trading system. An upstream approach guarantees complete coverage of carbon dioxide emissions. In order to ensure that domestic firms stay competitive, both import and export border adjustments are proposed. This principle is further investigated for import border adjustments. It is shown that a decentralized bill of material could help foreign companies to prove their carbon footprint to European authorities when importing goods into the Union. To achieve verifiable proof, the models of verifiable object identity attributes and digital object memory storing carbon footprints of importing goods are used. It is shown how digital twins extend the concept to allow companies to increase supply chain efficiency.

**Keywords** Distributed ledger technology · Emission trading · Border adjustments · Digital twin · Digital object memory

## 1 Introduction

Climate change presents the world with new challenges. The increasing concentration of $CO_2$ and other greenhouse gases (GHG) due to the emissions of humanity, leads to an increased radiative forcing. As a consequence, an imbalance in the global radiation budget leads to new climate parameters to form a new equilibrium [29, 31]. The first worldwide commitment to the fight against global warming was the Kyoto Protocol in 1998 [35]. This was followed by a newer commitment in 2015 with the Paris Agreement [36]. Both played an important role in international emission trading and the enhancement of renewable technologies in developing countries with

J. Kakarott (✉) · K. Hendrik Wöhnert · J. Schwarz · V. Skwarek
RTC Digital Business Processes, Hamburg University of Applied Sciences, Hamburg, Germany
e-mail: Julian.Kakarott@HAW-Hamburg.de

 
A. Joshi et al. (eds.), *Sustainable Intelligent Systems*, Advances in Sustainability Science and Technology, https://doi.org/10.1007/978-981-33-4901-8_6

a clean development mechanism (CDM). However, both did not lead to decisive actions that seemed to stop or significantly slow down global warming. Due to the precarious situation, the European Parliament declared the climate and environmental emergency in 2019 [12]. Europes goal is to reach net-zero emissions by 2050 [10]. The primary instrument to lower emissions within the European Union is the so-called European Emission Trading System (EU ETS) [11]. Therefore, it is of interest to further examine its functionality.

## *1.1 The Technological Challenge*

To cover this process, two major challenges need to be solved: Consumption-based emissions represent the amount of emissions that are included in products and therefore consumed by individuals and companies. A consumption-based $CO_2$ footprint of products or production is not comprehensively and consistently measured yet. An effective system needs to be designed to determine the consumption-based emissions in products and energy.

The challenge is the political acceptance of this process. As numerous cross-national parties are involved and also laws need to be changed and created, such a process cannot simply be *introduced*. It ideally needs to implement an existing process with already existing political consensus and adapt it where necessary.

Additionally, this data is sensitive on one hand, as production numbers and technologies can be estimated by the knowledge of $CO_2$ consumption. Therefore, privacy needs to be granted. On the other hand, the information needs to be transparent and accountable for verification and also for prosecution as the $CO_2$ footprint and emission trade has a strong monetary component and may be subject to fraud. Consequently, a safe and secure, privacy-preserving system for $CO_2$ emission management and trade needs to be set up which is capable to process an extremely high number of transactions, if all $CO_2$ production and conversion steps shall be registered and monitored.

## *1.2 Research Questions*

These considerations lead to the research question: *How to determine the emission intensity of European imports?*

## *1.3 Structure*

To answer the research question, this contribution will be structured as follows. In Sect. 2.1 the current state of emission trading will be discussed. Including possible challenges of the current EU ETS. Sections 2.2 and 2.3 illustrate the current technical status quo regarding information security and distributed ledger technology. The European challenges to reach net-zero emission by 2050 will be addressed in Sect. 3. It will be shown how distributed ledger technology is able to improve climate policy. Following the renewed EU ETS concept, Sect. 4 presents how import border adjustments can be improved by documenting the emission intensities of products outside the EU member states. This is a highly complex technical process involving a combination of distributed ledger technology, a digital object memory and digital twins. Finally, the research question will be answered in Sect. 5 and an outlook and limitations are given in Sect. 6.

## 2 State of the Art

First, this section will summarize the state of the art of emission trading, information security and distributed ledger technology. Moreover, it present blockchain-based emission trading schemes previously designed by other authors.

### *2.1 Emission Trading Systems*

The underlying principle of any emission trading system is called cap & trade. In a cap & trade system, companies are only allowed to emit GHG, if they own an allowance for it. The allowances are distributed to companies by the primary legal authority. The sum of all allowances forms the cap, which is ideally lowered every year. If some companies don't need all the allowances they own, they can sell them to other companies on a secondary market. This trading mechanism ensures that the reduction of emissions is economically efficient. Emissions are reduced where it is most affordable to do so. The reduction of allowances over time leads to a price increase for allowances and it becomes profitable for market participants to adopt renewable technologies [6, 8]. Indifference to other policies that are focused on certain sectors or technologies, emission trading leads to a transformation of the whole economy. Viewing the atmosphere as a common good shared among all market participants, the cap and trade approach solves the so-called "the tragedy of the commons" problem described by Hardin [15]. Although climate change leads to a negative long-term effect for the vast majority of market participants, most have a short-term benefit from polluting the atmosphere. Emission trading individualizes emissions. Every market participant has to pay for their emissions and therefore has an immediate economic incentive to reduce GHG emissions.

Cap and trade systems are expanding all over the world. In 2018 almost 15% of worldwide GHG emissions were covered by emission trading systems. Those systems are hosted by countries that produce more than 50% of the global GDP and are inhabited by almost a third of the global population [21]. To fertilize the expansion of ETSs and their GHG coverage, it is of high interest to look at potential stumbling blocks and how to overcome them.

Regarding the European target of carbon neutrality by 2050, the EU ETS has three major weak spots. Like many emission trading systems, it only covers a minority of the GHG emissions. Roughly 45% of European emissions are regulated by the system. The rest is allocated between the EU member states and covered by their national climate goals [9].

Carbon Leakage is a severe problem for any kind of carbon policy. "Carbon leakage is the risk that increased costs due to climate policies in one jurisdiction, such as the EU, could lead companies to transfer their production to other countries that have laxer standards or measures to cut greenhouse gas (GHG) pollutant emissions" [8]. The EU reacts to this with free allocation of allowances to companies based on certain benchmarks. Giving free allowances to certain companies should incentivize them to stay at their current location. The following equation illustrates how free allowances are allocated to companies.

$$\begin{aligned}\text{Allocation} = \; & \text{Benchmark} * \text{Historical activity level (HAL)} \\ & * \text{carbon leakage exposure factor (CLEF)} \\ & * \text{cross-sectoral correction factor (CSCF)} \\ & \text{OR linear reduction factor (LRF)}\end{aligned} \tag{1}$$

The Benchmark reflects the kind of product that is produced, the HAL indicates the historic production per year regarding the benchmark. Depending on the sector and the year, companies receive more or fewer allowances per year. The CLEF indicates this share in percent. Finally, CSCF and LRF ensure that a maximum amount of free allocation per year is not exceeded [8].

The free allocation might reduce the risk of carbon leakage, but it also represents a direct subsidy for emission-intensive production. Companies with high energy consumption and high historic emissions are favored over others. This raises the question of whether there might be a better tool to protect domestic production.

A third problem of the EU ETS is its focus on production-based emissions. This means that only emission directly emitted within the jurisdiction of the participating countries are covered. However, the EU has higher environmental standards than most other jurisdictions and imports a lot of products that are consumed by European customers from overseas. Therefore, the EU creates a demand for GHG emissions elsewhere. This global responsibility is not reflected in the current instrument.

## 2.2 *Information Security*

Information security is a major requirement for an emission trading system, which leads to many design decisions within the system. Its importance is motivated by highly sensitive data such as production processes. The system has access to supply

**Table 1** Elements of the CIA-triad

| | |
|---|---|
| Confidentiality: | *information is not made available or disclosed to unauthorized individuals, entities, or processes* [17] |
| Integrity: | *data has not been altered or destroyed in an unauthorized manner* [20] |
| Availability: | *property of being accessible and useable upon demand by an authorized entity* [17] |

chain and production details, which companies do not want to share with business partners, competitors, or yet to the public. Also, the system handles billions of euros and therefore has to be secured so fraud is made uneconomically.

At its core, the security of a system has three dimensions, which are described by the CIA-triad. It consists of the elements shown in Table 1.

## 2.3 Distributed Ledger Technology

Choosing a fitting system architecture bases on several factors such as required security level, performance but also political factors. This makes the decision between a system with a trusted central authority or a distributed system more complex. Each has its strengths and weaknesses. Central systems require fewer resources and are faster than distributed systems. However, choosing a distributed system for a European emission trading system has a big advantage as it will be explained in the following.

In its basic principle, a DLT system has a ledger containing transaction data, which gains integrity and availability by sharing it across several participants and chaining the data using cryptographic functions [32]. The data is synchronized among the participants using a consensus mechanism.

A common approach to identifying whether a DLT system makes sense for a use case is a question method proposed by Wüst, Gervais [38]. Certainly, this approach is not sufficient to make the final decision as only four to six questions are provided but it gives direction. Since the emission trading system needs to store a state with multiple writers, which are not necessarily known nor trusted, a DLT system makes sense. However, an institution of the European Commission can function as an always-online trusted third party. Thus, the conclusion of this method is not to use DLT.

However, the mentioned methods do not take politics into account. In a central system, one entity such as an institution of the European Commission has full control over the system, which can strongly impact the local economy. Additionally, data, which may provide deep insights into the activities of European companies is not wanted to be stored in a central place. Ideally, no third party is involved in each process. A distributed system can provide this confidentiality of data [5] by sharing data only between parties being involved in a process. In conclusion, a DLT system shall be used.

## 2.4 Related Work

Climate change and distributed ledger technology have previously been associated with each other. Khaqqi et al. [23] propose one of the most detailed concepts for an emission trading system. The concept includes a reputation system that makes it possible to restrict market excess for companies that do not have a strategy for better environmental behavior. Therefore, it is more expensive for companies with a bad reputation to buy allowances, although they are the ones who probably need them the most. Without blockchain technology, this mechanism would suffer from high transaction costs due to the necessity of an intermediary who sorts offers and determines the matching of transaction partners.

In an earlier approach, Al Kawasmi et al. [1] proposed a Bitcoin-based system that includes smart meters of renewable power plants. Those smart meters produce carbon credit tokens equivalent to the amount of renewable energy. Owners of conventional power plants could buy those tokens to subsidize renewable energy. Although interesting from a technological standpoint, this concept lacks good use cases. The generated carbon credits do not equal carbon offsets and therefore do not reduce the emissions of the conventional power plant.

Yuan et al. [39] introduced a concept for an Emission Trading System that is based on the Hyperledger Fabric Blockchain framework. It targets to solve common issues of central data-keeping like the necessity for one trusted party and the associated threat of unreliability, inconvenient data sharing between different organizations with independent databases, and possible data tampering and data leakage. The authors designed a decentralized architecture and tested it regarding its performance. That being said, the concept does not include a solution to the political, economic, or ecological problems of climate change, but solely focuses on the decentralization of the emission trading system from a technological perspective.

This collection of previous research illustrates that the application of distributed ledger technology in the field of emission trading is not new. However, previous attempts have focused on the technological side instead of political and economic challenges. The concept of Khaqqi et al. [23] is very interesting, but does not tackle the current issues of the existing system. Our chapter targets this research gap and connects currently existing problems of the EU ETS with distributed ledger technology.

# 3 Distributed Emission Trading System

This section will take a deeper look into a possible solution to the current challenges described in Sect. 2.1. Kakarott and Skwarek [22] proposed a DLT-based emission trading system that is able to track emission information through European value chains. The goal of the concept is to cover all European consumption-based emissions while maintaining the competitiveness of domestic companies with import and export border adjustments.

The authors propose that instead of forcing companies to hold allowances when they emit carbon dioxide, they should be held accountable for the $CO_2$ emitting resources that they mine or import. This is called an upstream system, where the point of regulation affects the fuel producers rather than at the point of combustion (downstream) [16]. The overall cap would then be defined by the amount of $CO_2$ emitted over the life cycle of products consumed by Europeans. Boitier [4] defines consumption-based emissions as the following:

$$CO_2{}^{cons} = CO_2{}^{prod} + CO_2{}^{imp} - CO_2{}^{exp} \tag{2}$$

The consumption-based emissions (cons) equal the emissions resulting from production within the EU (prod) plus emissions included in net-imports (imp-exp). This consideration ensures that even if carbon leakage occurs, those emissions are still included in the EU climate policy.

As an alternative to free allocation, border adjustments can be used to prevent or weaken carbon leakage. Border adjustments try to equalize the competitive conditions for domestic and foreign companies at the border of the regulated jurisdiction. Firms that try to import products in the EU would have to pay for the amount of $CO_2$ that has already been emitted elsewhere for the production of it. This charge could be an import tariff on certain product groups or importers could be forced to purchase allowances on the secondary market of the EU ETS, just like regular European companies that emit their $CO_2$ within the Unions jurisdiction. Export border adjustments work the other way around. Domestic companies that paid for $CO_2$ allowances during production can be reimbursed when exporting their goods.

Despite a good theoretical concept, border adjustments face a practical issue. They only equalize the competition between domestic and foreign companies, if the underlying product emission intensity was determined correctly. Meaning that the $CO_2e$ emitted during the production and transport of a product is recorded accurately. Because a production process in a globalized value chain usually involves several companies and countries, the documentation of emissions has to take into account that those market participants do not necessarily trust each other and do not have the economic interest to self report their emissions accurately.

The proposed system by Kakarott and Skwarek [22] approaches this problem with the use of distributed ledger technology. $CO_2$ allowances are tokenized. Upstream, companies in the EU have to acquire tokens when they want to mine or import resources that have the potential to emit $CO_2$. According to Bajželj et al. [2] those are crude oil, natural gas, coal, calcium carbonate and emission from land-use changes. This first concept is limited to $CO_2$, because "carbon dioxide emissions account for 80% of the contribution to global warming of current greenhouse gas emissions" [27]. Companies also have to purchase tokens before importing products that have already emitted $CO_2$ during their production outside of the Union. When those products and resources are passed through the value chain towards the consumer, the tokens are always passed on to the next company together with the product. Every company receives a certain amount of tokens from its suppliers and energy provider and passes those tokens on to the next company together with the product. Therefore, the life cycle emissions of a given product sum up over time.

Internally, companies have to identify the material and energy streams of their production process with the use of material flow cost accounting (MFCA). That being said, the allocation of a company's tokens received for material and energy input, to the products it sells should be audited based on a standardized MFCA rule book. This concept leads to relatively accurate information about the emission intensity of products within the European Union. Despite benefits for the transparency for consumers, it enables the EU to enforce export border adjustments, because companies can prove the regulative burden that they are fraught with.

However, export border adjustments are only half of the necessary measures against carbon leakage. In order to enforce accurate import border adjustments, it is necessary to determine the emission intensity of products outside the European Union. This leads to new challenges because the EU cannot enforce its law on other jurisdictions.

## 4 Documentation of Emission Intensities in Non-member States

In contrast to the available MRV instruments of the EU ETS within the jurisdiction of the European member states, it is difficult to trace carbon emission in foreign countries. However, to ensure accurate import border adjustments, the EU has to be able to ascertain the emission intensities of foreign products.

A simple solution would be the categorization of products into their average emission intensity if produced in the EU. According to [26], this could work in several ways. For instance, the categories could be based on the estimated emissions in the exporting country or based on the predominant method of production. Latter is easier to determine and might be more practical in the context of global supply chains. However, there are several drawbacks. One, carbon emission for the production of the same product are likely higher in non-member states, due to lower environmental regulations. Therefore, foreign companies still have an advantage over domestic producers. Two, in a globalized market, some companies might want to produce in countries with access to renewable energy due to geographical advantages, like a hydroelectric or a solar power plant. Those companies are strongly disadvantaged in their striving to produce carbon-neutral products.

This leads to the necessity of a system that enables companies globally to prove their carbon footprint if they want to. Imports by companies not willing to adapt to EU rules could be judged based on standardized values for every product category. However, companies willing to improve their carbon footprint in their global value chain should be given a framework on how to prove their emissions. This is illustrated in Fig. 1. The upper value chain consists of companies that do not have a particularly good carbon footprint. Those are likely to not report their emissions.

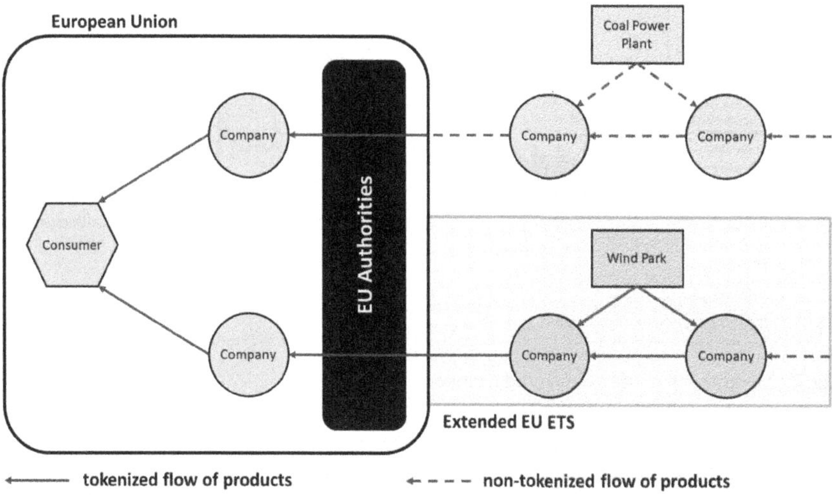


**Fig. 1** Extention of the EU ETS into the value chain of non-members of the EU ETS

Therefore, the EU assumes standardized values for their products when imported. These standard values could be high to incentives good behavior, as long as they are WTO compliant. The second value chain on the bottom uses renewable energies. Therefore, the assumed emissions are lower than average. These companies get the chance to extend the EU ETS into their value chain. Meaning that the companies involved voluntarily adopt EU norms, rules, and monitoring to be eligible for a lower import border adjustment. Due to a lack of trust in the value chain regarding both companies and foreign regulators, a decentralized approach seems reasonable. The goal of this approach is to document information about every component during the production process. In the following, a possible framework will be developed and evaluated.

## *4.1 Distributed Bill of Material*

In order to keep track of product emission information throughout production outside of the EU, a tokenized bill of material (BOM) is proposed. Such a bill of material contains needed materials. A specific carbon footprint of a newly produced product is represented by the sum of emissions of its components, namely the used materials and the energy consumed by manufacturing. As described above, companies operating outside of the EU have the choice of whether they want to participate in a system that helps them prove their carbon footprint. This system is an extension of the proposed EU ETS into non-member states. Dasaklis et al. [7] showed how to design a BOM-based system that merges tokens of materials into a new token that represents the

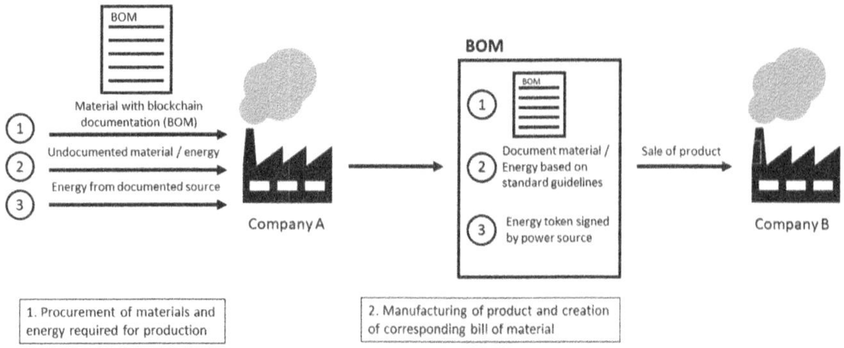


**Fig. 2** Merging input tokens into product token

finished product. As shown in Fig. 2, every company outside the EU, participating in the ETS extension has to consider two scenarios. The input materials and energy was already registered with the European authorities by the previous company or it is new to the system.

One, a production plant outside the EU receives materials or energy from a supplier that is not part of the extended EU ETS. Because this is the first touchpoint of those supplies with the EU ETS, the emission intensity of both materials and energy have to be estimated. To do so, the standardized values for each category that the EU usually would apply to imports should be used by the company. The company creates a BOM-token for materials and energy. Therefore, it is documented who valued the components with which emission intensity. Powerplants that use renewable energy have the opportunity to be audited by the EU. They can create energy tokens and sign those with their signature in order to make the source of energy transparent.

Two, a production plant outside the EU receives materials or energy from a supplier that already registered those with the EU and created a BOM-token. In this case, a verification instance, e.g. an involved production machine, checks if the tokens and their respective items match the BOM of the product. If matched successfully, manufacturing proceeds and tokens of input material and energy are merged into a product token, which is then passed along the supply chain. Inputs during production are allocated to the output according to MFCA. Therefore, proper tracking of a product's specific emission is guaranteed.

Each company participating in this extension of the main EU ETS needs to be certified by a European authority. The certification serves as a proof for the integrity of emission data. Therefore, a company does not have to disclose their BOM to the customer.

Companies that can prove that their value chain mainly consists of renewable energy sources and materials get a discount when importing into the EU and do not have to purchase as many $CO_2$ allowances on the secondary EU market. This incentive should attract more and more non-European companies. Moreover, other emission trading systems could already provide some extra information for the bill of material.

## 4.2 The Challenge of Identifying Physical Objects

A BOM token for the documentation of emission intensities only works if the digital token can be securely assigned to the physical product, which is a challenge. The secure identification is subject to the already mentioned CIA-triad as well as its common extension of accountability and authenticity [33].

Accountability is the *property that ensures that the actions of an entity can be traced uniquely to the entity* [17] and authenticity ensures *that an entity is what it claims to be* [19]. Both are identity-related attributes, which can be achieved with a secure identity management.

Identity management has been under ongoing development in the digital world for the last decades. Cryptographic functions have been improved so that algorithms such as SHA256 are regarded as secure with today's knowledge in the pre-quantum computer time [37]. Adding more factors besides knowledge such as a password or a private key, which can be leaked, increases the security of the authentication process so that all attributes of the extended CIA-triad can be achieved. However, security is impaired at the borders of the digital world. Problems arise when a system consists of digital and physical components, which is the case with the BOM-tokens. The connection between the physical and digital world still poses a challenge as it is possible to tamper it [28].

To cope with the challenge, the concept of multi-factor authentication can be applied, by using a biometric factor to identify a physical product [30]. Common approaches to connect products to the digital world are barcodes or RFID tags by physically attaching them to the products and therefore adding an artificial biometric attribute. Those tags can be tampered or removed [3] and therefore destroy the chain of trust [40]. The aim is to put the anchor of trust within the object itself. In order to achieve that, the actual physical attributes of the object shall form the identity. Similar to unique physical attributes of humans such as the fingerprint, products, and materials also have a unique fingerprint. Resources have unique chemical compounds, materials have a unique surface due to the production process, and technological products have processors with unique characteristics [3]. Those physical attributes can be digitalized into a byte string and used in the digital world as well as verified in the physical world. In sum, attributes compose the identity of an entity, which distinguishes the entity in its context [18].

For other entities, public getter functions as they are known in object-oriented programming languages are available. Depending on the circumstance a different case of getter functions can be used to identify an entity. The return value of a getter function represents one or more identity attributes. In a biometric system, it is a security risk, when attributes are made public since they cannot be changed like it is the case with compromised passwords. Sybil attacks in which an attacker uses these attributes are possible and the confidentiality of an object is no longer existent in this scenario. Therefore, the attributes may never be published in plain text. Rather, zero-knowledge-proofs (ZKP) shall be used, in which the attributes can be verified without revealing any information other than the correctness of the attribute [13].

An identity management system building on verifiable physical attributes of objects only needs a trusted authority once for verifying the correctness of the digitalized physical attributes. Apart from that, the system works fully distributed without a trusted authority. Thus, it can be used for the distributed emission trade system. Once the identity of an entity can be securely verified, the entity's specific carbon footprint can be related to it in a tamper-resistant way.

### *4.3 Digital Object Memory for Carbon Footprint Benchmarks*

Based on verifiable identities of objects, a system is needed that tracks the events of an object over its life cycle. Companies outside of the EU do not necessarily have access to the EU ETC to tokenize the BOM, therefore a system storing the BOM directly on the object can be necessary.

The digital object memory (DOM) is such a system and persists data of an object and makes it available for other entities using an interface. It allows to track the object during its whole life cycle starting with the creating and production of the object, continuing with the usage, and ending with its disposal [14]. Thus, the compositions of objects are stored in the object memory, as well as the completed production steps and transport routes. These serve as a basis for the calculation of the carbon footprint. Here, it was chosen to store this data decentrally on the object because the objects are already located in a distributed system and therefore no central authority is needed to manage this information as well.

The DOM can be passive or active. While the first mentioned just provides a memory unit, the latter comes in the form of an Internet-of-Things device enabling communication and computation. Tracking emissions of a product or material includes many different parties such as producer/miner, logistic companies, retailer, consumers, and recycling facilities. Therefore, a decentralized architecture for such a complex system is appropriate. Active DOM enables the system to operate more securely in such a scenario due to added cryptographic functions, which increases the integrity and authenticity of the stored data. Using DOM for tracking product- specific carbon footprints has already been demonstrated [25]. In this paper's emission trading system it is used to building the data foundation of the border adjustments.

### *4.4 Digital Twin*

On top of the DOM, the concept of digital twins allows data-based decision making when it comes to the processes within its supply chain. So companies can reduce the carbon footprint of a product.

Digital twins are used in many different areas, e.g. layout planning, product lifecycle, maintenance, and manufacturing. They are also used for different purposes and with different technologies. On account of this, several definitions of a digital twin exist, each varying in certain aspects. For an emission trading system, the following definition may fit best:

> Digital twin is an integrated multi-physics, multi-scale, probabilistic simulation of a complex product and uses the best available physical models, sensor updates, etc., to mirror the life of its corresponding twin. [34]

The digital twin model distinguishes between the physical and the digital object. Further categorization is based on the automation of exchanged data between the physical and digital objects. First, a system in which data is transferred only manually is called a digital model. Second, a unidirectional automated data flow towards the digital object where calculation can be made with that data characterizes a digital shadow. Third, the full digital twin allows bidirectional data flow, so that for instance the physical object can be controlled from the digital object [24].

For an emission trading system, each subcategory can add value and transparency to the supply chain of products, which is subject to the border adjustments. With a digital shadow, the border adjustments can be carried out automatically. The information regarding the carbon footprint, which is stored on the DOM of a product, can be automatically sent to the digital object within the European emission trading system. That physical object is a token in a DLT system and allows calculating the specific carbon footprint of that product. More advanced, however, also more investment intensive, is a digital twin with bidirectional automated data flow. In this scenario, the carbon footprint is constantly tracked and send to the digital object. There, it can be calculated, which next step of the supply chain a product should take based on the expected additional carbon emission. For instance, the mode of transportation can be dynamically changed so that costs due to transport and $CO_2$ token purchases at the EU border and time requirements can be weighed up to find an equilibrium.

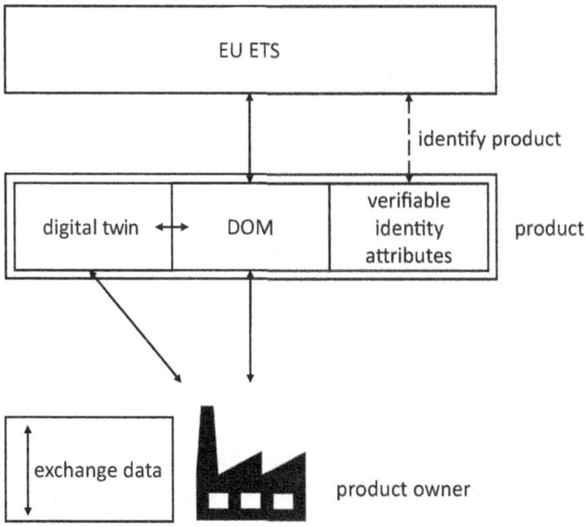


**Fig. 3** Extending models or securely documenting carbon footprints in non-member states are shown. The owner of a product can automatically exchange data with the DOM and the digital twin of the product. For the border adjustments, the product can verify its identity as well as its carbon footprint stored in the DOM

An efficient system can be created, in which companies can prove that their products have a smaller carbon footprint than a generic benchmark would have calculated. The connections between the models used in a system documenting emission intensities in non-member states can be seen in Fig. 3. However, it requires a highly digitalized supply chain so that extensive research and development is needed before such systems can be implemented.

## 5 Discussion

The exact and fair determination of border adjustments remains an unsolved problem for any cap and trade system. In a previous paper, it was shown that export border adjustments can be specified with the use of distributed ledger technology [22]. This work took the first leap forward in extending this concept to non-member states of the EU ETS. It was shown that companies should be provided with a framework that enables them to prove to European authorities how big the carbon footprint of their products is.

To do so, the ETS is enlarged by emission determination on the company level. The EU provides a fixed set of rules that allows companies to adapt their processes and documentation. This is voluntary for each company but could save them money in terms of import border adjustment if they can produce their good with renewable energy. To determine the carbon footprint of a product, a verifiable bill of material needs to be created, which bases on the secure identification of physical products. It was shown how a digital object memory can ensure the verifiable presentation of a product's carbon footprint to the EU ETS in a distributed way.

Initially, the research question was on how to determine the emission intensity of European imports? To summarize, it was shown that the EU ETS can be extended into non-member states of the European Union by allowing companies to voluntarily adopt the rules, norms, and monitoring of the EU. This is done by tokenizing product information and therefore, making the chain of emissions more transparent to European Authorities. However, this concept needs further research and is not yet functional. Nevertheless, this first approach underlined that a deeper investigation is reasonable and necessary. Also, implementing communication and computation units within products is not an applicable solution for every product due to physical and legal limitations. Even though the costs of IoT units are decreasing, their benefits still do not outweigh the costs for every product.

## 6 Conclusion

The research aimed to explore methods, which can be used to fairly and securely evaluate the carbon footprint of goods imported to the European Union. Basing the carbon footprint calculation on digital object memory and digital twins enable companies to prove its correctness and that it may be lower than the default value for the product set by the European Union. Digital object memory and digital twins can also add value through other usage scenarios. For example, it is valuable for companies within a supply chain to know what the history of a product looks like in order to identify sources of error. An ETS using these technologies can therefore benefit from synergy effects, making it worthwhile to equip the products with the necessary hardware and infrastructure for more products. Developers who already use such systems in production plants should design the systems modularly in order to be able to integrate them into the new application areas as well. One challenge is the current technological status of the manufacturing companies. Since production equipment is often used for decades, the diffusion of new technologies is a lengthy process. It is also uncertain whether the concept of border adjustments presented here will be accepted by the WTO. This is a necessary step, which is the more probable the more accurate and therefore fairer the border adjustments are calculated. The concept presented here could meet the requirements and be a chance for the extension of the European ETS to non-member states of the EU.

## References


1. E. Al Kawasmi, E. Arnautovic, D. Svetinovic, Bitcoin-based decentralized carbon emissions trading infrastructure model. Syst. Eng. 18(2), 115–130 (2015)
2. B. Bajželj, J.M. Allwood, J.M. Cullen, Designing climate change mitigation plans that add up. Environ. Sci. Technol. VII **47**(14), 8062–8069 (2013)
3. V. Balagurusamy, C. Cabral, S. Coomaraswami, E. Delamarche, D. Dillenberger, G. Dittmann, D. Friedman, N. Hinds, J. Jelitto, A. Kind, Crypto Anchors, IBM J. Res, Dev. (2019)
4. B. Boitier, $CO_2$ emissions production-based accounting vs consumption: insights from the WIOD databases in *WOID Conference* IV **23** (2012)
5. B. Carminati, C. Rondanini, E. Ferrari, Confidential business process execution on blockchain, in *2018 IEEE International Conference on Web Services (ICWS)*, pp. 58–65 (2018)
6. Climate Change 101 - Cap and Trade. Climate Change 101: Understanding and Responding to Global Climate Change (January 2011)
7. T.K. Dasaklis, F. Casino, C. Patsakis, C. Douligeris, A framework for supply chain traceability based on blockchain tokens, in *vol 362*, ed. by Business Process Management Workshops (Cham, Lecture Notes in Business Information Processing (Springer International Publishing, 2019), pp. 704–716
8. European Commission . EU ETS Handbook (2015)
9. European Commission. The EU Emissions Trading System (EU ETS). IX (2016)
10. European Commission, *Communication from the commission to the European Parliament, the European Council, The Council, The European Economic and Social Committee* (The Committee of the Regions and the European Investment Bank, XI, 2018)
11. European Parliament, Directive 2003/87/EC of the European Parliament and of the Council of 13 October 2003 Establishing a Scheme for Greenhouse Gas Emission Allowance Trading Within the Community and Amending Council Directive 96/61/EC. X 2005
12. European Parliament. Climate and Environmental Emergency I European Parliament Resolution of 28 November 2019 on the Climate and Environment Emergency (2019/2930(RSP)). XI 2019
13. O. Goldreich, Y. Oren, Definitions and properties of zero-knowledge proof systems. J. Cryptol. **7**(1), 1–32 (1994)

14. D. Gorecky, S. Weyer, F. Quint, M. Köster, Definition einer Systemarchitektur für Industrie 4.0-Produktionsanlagen (2016)
15. G. Hardin, The tragedy of the commons. Science **162**, 1243–1248 (1968)
16. T. Hargrave, US Carbon Emissions Trading: Description of an Upstream Approach. XII (1998)
17. ISO 7498-2, Information processing systems– Open Systems Interconnection – Basic Reference Model – Part 2: Security Architecture (1989)
18. ISO/IEC 24760-1, IT Security and Privacy – A framework for identity management (2019)
19. ISO/IEC 27000, Information technology - Security techniques - Information security management systems (2014)
20. ISO/TS 17573-2, Electronic fee collection – System architecture for vehicle related tolling – Part 2: Vocabulary (2020)
21. International Carbon Action Partnership (ICAP), Emissions Trading Worldwide - Status Report 2018 (2018)
22. J. Kakarott, V. Skwarek, An enhanced DLT-based $CO_2$ emission trading system, in *2020 Fourth World Conference on Smart Trends in Systems Security and Sustainablity (WorldS4)* (2020)
23. K.N. Khaqqi, J.J. Sikorski, K. Hadinoto, M. Kraft, Incorporating seller/buyer reputation-based system in blockchain-enabled emission trading application. Appl. Energy **209**, 8–19 (2018)
24. W. Kritzinger, M. Karner, G. Traar, J. Henjes, W. Sihn, Digital twin in manufacturing: A categorical literature review and classification, in 16th IFAC Symposium on Information Control Problems in Manufacturing INCOM 2018. IFAC-PapersOnLine **51**(11), 1016–1022 (2018)
25. A. Kröner, G. Kahl, L. Spassova, C. Magerkurth, T. Feld, D. Mayer, A. Dada, Demonstrating the application of digital product memories in a carbon footprint scenario, in *2010 Sixth International Conference on Intelligent Environments*, pp. 164–169 (2010)
26. O. Kuik, M. Hofkes, Border adjustment for European emissions trading: Competitiveness and carbon leakage. Energy Policy. IV **38**(4), 1741–1748 (2010)
27. A.L. Daniel, D.R. Ahuja, Relative contributions of greenhouse gas emissions to global warming. Nature IV **344**(6266), 529–531 (1990)
28. G.D. Martins, R.F. Gonçalves, B.C. Petroni, Blockchain in manufacturing revolution based on machine to machine transaction: a systematic review. Brazilian J. Oper. Prod. Manage. V **16**(2), 294–302 (2019)
29. 8 Anthropogenic and Natural Radiative Forcing, 82 (2013)
30. A. Ometov, S. Bezzateev, N. Mäkitalo, S. Andreev, T. Mikkonen, Y. Koucheryavy, Multi-factor authentication: A survey. Cryptography **2**(1), 1 (2018)
31. R. Ramaswamy, O. Boucher, J. Haigh, D. Hauglustaine, J. Haywood, G. Myhre, T. Nakajima, G.Y. Shi, S. Solomon, Radiative forcing of climate change **III**, 351–416 (2018)
32. M. Rauchs, A. Glidden, B. Gordon, G. Pieters, M. Recanatini, F. Rostand, K. Vagneur, B. Zhang, Distributed ledger technology system - A conceptual Framework, VIII (2018)
33. W. Stallings, *Computer Security: Principles and Practice*, 3rd edn. (Pearson, Boston, MA, 2015)
34. F. Tao, J. Cheng, Q. Qi, M. Zhang, H. Zhang, F. Sui, Digital twin-driven product design, manufacturing and service with big data. Int. J. Adv. Manuf. Technol. **94**(9–12), 3563–3576 (2018)
35. United Nations, Kyoto Protocol to the United Nations Framework Convention on Climate Change (1998)
36. United Nations, Paris Agreement. XII (2015)
37. M. Wang, M. Duan, J. Zhu, Research on the security criteria of hash functions in the blockchain, in *Proceedings of the 2nd ACM Workshop on Blockchains, Cryptocurrencies, and Contracts* (Association for Computing Machinery, New York, NY, USA, 2018), pp. 47–55 (BCC '18)
38. K. Wüst, A. Gervais, Do you need a blockchain? in *2018 Crypto Valley Conference on Blockchain Technology (CVCBT)*, pp. 45–54 (2018)
39. Y. Pu, X. Xiong, L. Lei, K. Zheng, Design and Implementation on Hyperledger-Based emission trading system. IEEE Access **7**, 6109–6116 (2018)
40. Y. Zheng, Y. Chunlin, F. Zhengyun, Z. Na, Trust chain model and credibility analysis in software systems, in *2020 5th International Conference on Computer and Communication Systems (ICCCS)*, pp. 153–156 (2020)

# Estimation of People Density to Reduce Coronavirus Propagation

**Mouad Tantaoui, My Driss Laanaoui, and Mustapha Kabil**

**Abstract** Today, the spread of coronavirus has become the number one concern of countries as it threatens human life and economics; therefore, the scientific community tries hard to discover the treatment to deal with this virus or at least find out a method to reduce its propagation. In this context, our concern is to estimate density of people being inside all different places of interest in the city in the purpose of distributing users of our application to different places by avoiding congestions of people. The use of big data is very important for data treatment for fast execution. In this time, oldest relational database technologies cannot anymore handle the enormous data created by various application sources, in this direction big data tools permit us to handle this enormous data in the purpose to mine important data from the voluminous data, if we rely on the support of big data tools, the treatment will be complicated to administer. In this chapter, firstly, we define a system that calculates number of people in diverse city's areas; it will help to inform users' best places to alleviate areas where there are congestions and redirect people to other place with low density. And secondly, we keep trace in a database all people contacts that have been near each other to prevent people who have been close to the positive cases.

**Keywords** Big data · Density management · Congestions handling · Spread of coronavirus

M. Tantaoui (✉) · M. Kabil
Hassan II University, Casablanca, Morocco
e-mail: tantaoui.mouad@gmail.com

M. Kabil
e-mail: kabilfstm@gmail.com

M. D. Laanaoui
Cadi Ayyad University, Marrakesh, Morocco
e-mail: d.laanaoui@uca.ma

 
A. Joshi et al. (eds.), *Sustainable Intelligent Systems*, Advances in Sustainability Science and Technology, https://doi.org/10.1007/978-981-33-4901-8_7

# 1 Introduction

Today, combatting covid-19 seems to be the first priority of scientific researchers, so here, we intend to develop management of people density all around the city; this domain of research catches the attention of researchers of different fields linked to either social and economic problems or citizen security. In another shutter, many generators of data generate enormous data characterized by its various genders and velocity that makes the process very complicated; big data technologies are a good alternative to process this voluminous data in real time. In our mechanism of computing people density, we rely on the help of big data tools advantages to process the huge data. Places or services desired in the whole city contain too many people with big waiting lines. The strong growth in waiting lines has engendered many environmental and social issues in towns mainly at present of the spread of covid-19, so fighting this viral disease is introducing the interest research field that concerns the researchers in diverse domains as a result of the problems found. In this chapter, we intend to discover a way for calculating number of users on different places through city, and a way of combatting covid-19 propagation; this permits us thereafter to engender a type of equilibrium on all services where there is a high congestion. To achieve this goal, we have defined a method of estimating density utilizing huge data gathered. In other words, our method intends to equilibrate people density in all locations, and it is a form of redirecting people to different places so that we obtain a medium density; this will reduce the risk of coronavirus propagation. After this introduction, the rest of our chapter is presented as following. Section 2 presents related works. Section 3 proposes the detailed proposed methodology with the experiment results. Finally, Sect. 4 concludes our chapter.

# 2 Related Works

Lambda architecture is an architecture that allows to process the huge data, and we are inspired on this architecture [1] characteristics for constructing our system; it consists of data processing that permits to process voluminous data. The architecture is composed of different elements: serving, batch, and speed layer.

In paper [2], authors review the evolution history of data controlling systems. After that, they point the state of the art of big data controlling mechanism concerning storage, modeling and request engines of enormous data managing systems. In addition to that, authors analyze the data and characteristics of the data mass controlling systems, big data controlling systems need to be more developed and analyzed because it progresses with a fast manner.

Demography is linked to statistic domain in citizen science. The rise of population is engendering many problems to monitor the huge mass of demographic data (with other words 'big data'). The help of science development progression and informatics progression, much architecture has been produced to control the mass of data. Almost architectures rely on old relational database, yet it cannot monitor big data management effectively. In paper [3], Bukhari et al. construct a system based on big data controlling system with the support of apache Hadoop platform to solve issues related to demography great increase data processing. The architecture is composed of Sqoop, HBase, and Hive. The system permits to export data from all RDBMS to apache Hadoop HBase. In addition to that, utilizing HBase and Hive seem to be a better alternative to store and request data to process accurately. The experience proves that the architecture monitors very well the mass data for future development. The system constructed will advance one step forward in monitoring complexes issues of big data and will reinforce processing in various domains such as economic or social domain. Big data is a domain which is becoming linked to our habits recently and engendering several data and solutions to develop and devoting on research field and new technologies of taking good solution in various fields like business and engineering. Data is no more such as supplementary element; big data gives the possibility to be an important piece of puzzle for taking good decision. In paper [4], authors utilized Guizhou as a solution to engender big data processing system depending on time and space relying GIS bussing technique, and after, authors discuss the results, and they concluded to construct technical assistance for the transformation of tourism by the specialists of Guizhou and efficient indicators.

Big data offers the opportunity of monitoring the huge datasets which were very complicated to process before. Actually, traditional relational database does not have in its power to administer the huge dataset which permits distributed database NoSQL to progress through the time. In paper [5], authors engendered and constructed a new distributed big data management system (DBDMS), which run under apache Hadoop technology and NoSQL mechanisms; it allows collecting the voluminous data, searching and infinite storage and showed by the results; DBDMS improves the treatment quality of huge dataset, and it is good for huge log backup and recovery, voluminous network packet capture and process, and other many different domains.

In [6], Matallah et al. establish a mechanism to well manage metadata of Hadoop technology to improve consistency and ameliorating performance and scalability of metadata by proposing a hybrid alternative on centering and apportionment of metadata to improve the quality and development of the mechanism.

Today, information technologies engender huge quantity of data every hour and every minute from various sources. Those quantities of data seem to be very complicated for the processing capability of traditional data treatment methods to monitor and manage in such a required limited time. This voluminous amount of dataset characterizes the big data. It is facing many problems in various processing on data such as capturing, treatment, searching, and filtering. Hadoop technology offers adaptive tools for various companies for big data monitoring mechanism. Hadoop concentrates on the concretization of various industry methods. Therefore, for handling Hadoop technology, it is obligatory to overcome how to master the operation of Hadoop ecosystem and Hadoop architecture. In this direction, the paper [6], Raj et al. discussed the Hadoop architecture and Hadoop ecosystem. All these methods lack to fast real-time data processing, because today, we need solution that handles and process the data immediately, and also we must adapt the solution to different situations.

In this age of high technology, innovation, production creativity and arranging have been mainly transformed toward open collaboration ecosystems. People organize and with an open manner their works concerning data treatment and collaborate on exploring new solutions or dataset transmission for lot of systems. Big data is an emerging tool to monitor the complexes systems of open collaboration, and community of research needs to work with open manner with collaboration researchers with new data that expose different aspects, and new numerical models and techniques. So in paper [7], authors focused on problems that is based on numerical collaboration and constructed the dataset analytical issues which have to be showed to remedy the essential research challenges.

In paper [8], a survey is elaborated to analyze the propagation of the coronavirus. They specify how the virus is spreading in cities; also authors clarify vision of the main indicators which affect the propagation of the virus, where most of indicators attached to the infection spread are taken into consideration. It is proved that the difference in terms of distance to epicentrum is greatly affecting indicator that impact the spread of covid-19. The propagation of the virus needs to be more investigated if we want to combat it because it still looks blurry on the way of its spread. In paper [9], several appearances of the covid-19 infection are showed; authors show a global vision of the spread of this virus, and exhibit some tools of data analytics on the virus. Firstly, they point a literature study on covid-19 specifying lot of factors like the origin, resemblance with old coronaviruses, the transfer power, symptoms, and more indicators. In the second part, they employed dataset diagnosis tools on a dataset of Hopkins University to know very well how this harmful virus spread.

In paper [11], authors proposed a case study of utilizing composite Monte Carlo which is supported by machine learning mechanism and fuzzy rule induction to obtain good stochastic vision concerning coronavirus progression. In paper [12], authors show some examples of technologies which are supported by artificial intelligence (AI) that enhances autonomous process, better technology, and good making decision, in the purpose of combatting the covid-19 and saving lives. These AI technologies are good but they need to be more developed because this virus is new that is why we need some novel tools to combat it.

## 3 Methodology and Experiment

In this chapter, firstly, we spot places where grouping of people are dense such as places of interest, we took different banks, all supermarkets, cashplus, and sales services; different places of interest should have equipment such as cellular detection sensor to log in a database all cellphones and compute number of phones in the location, and then we deduct number of people on all places within the city (sensor of cellphone detection is located at the entrance of the area to log the identifier number of the all entrances and exits of phones). When someone is entering to a particular location, he will connect to our system and he can select kind of area where he desire to attain, so the system will pick the place where the density is low; like this way, the system equilibrates load dispensation of all areas of interest in the sense of number of people, and so, reducing the covid-19 chance of propagation. Firstly, the methodology tends to build the database: every place of interest log incoming and outgoing of smart phones in the sensor of the area, and register them on a its own database for two purposes: we calculate smart phones number lying the place indoor, and in second hand, the system record the trace of cellphones which have been near to each other. This approach is described in Fig. 1, then the place system disseminates every 5 min the data and the number id to the application to append the data to the central base of all sites of services within the city, and the database is refreshed after each 5 min; therefore, the central system possess a total control on places densities located indoor in real time and cellphones which are near to each other. So, when an user search a wanted destination, the application selects the best place of service described by a lower number of people located there based on the base constructed previously, and therefore, transmits to the consumer the optimal

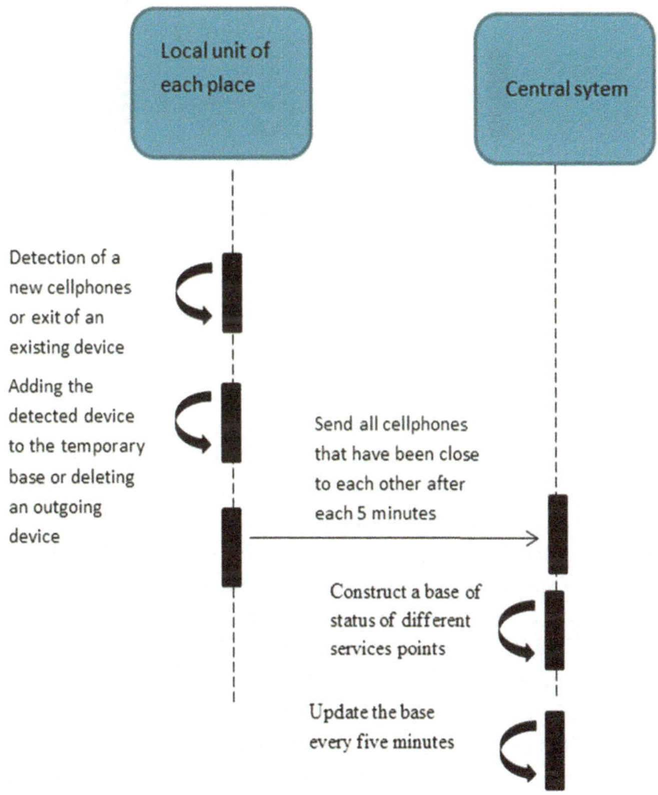


**Fig. 1** Different steps of building the database having status of all area of interest

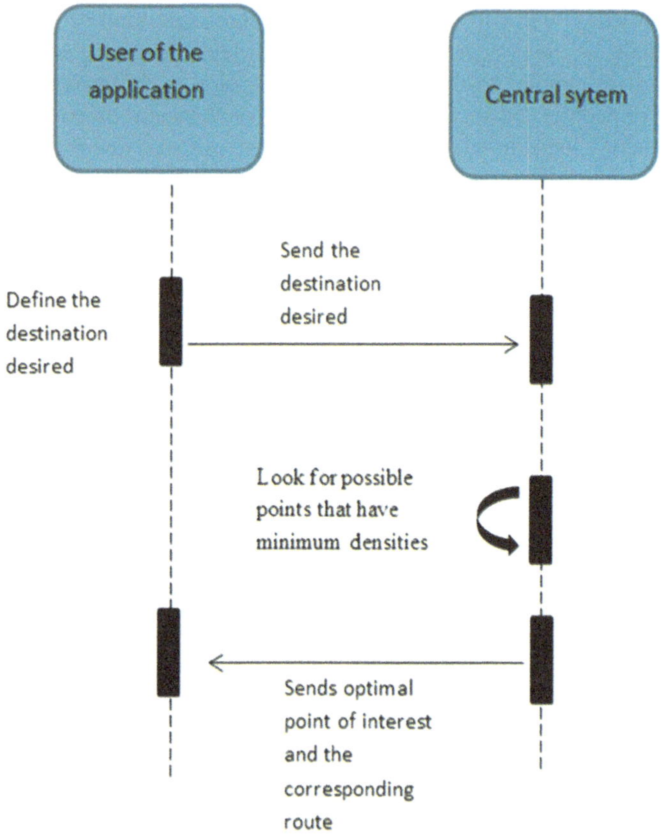


**Fig. 2** Different steps of choosing the best point of service for the user

itinerary to arrive safely its desired place. This approach is described in Fig. 2. The usefulness of establishing a database of all cellphones that have been near to each other is very useful, when someone catches a positive test for covid-19; after that, the application warns users who were near to the positive test diagnosis. The suggested system is summarized in Fig. 3. Our application allows picking the best place of service of a person and choosing an itinerary to arrive the best destination by picking the area of service that contains low number of people, then warning users who were near to positive covid-19 test relying on utilizing big data tools to have a fast processing in real time. For the experiment, we catch Casablanca City card, Morocco containing various areas; we took one thousand stores, one thousand banks branches, and one thousands of each different supermarket, wafacash, and tesshilat services. And we simulated the experience in our computer characterized by: Processor: Intel Core i7. RAM: 16 GB of DDR4 2400 Hz RAM, and Nvidia GeForce GTX 1070 8 GB in Graphics card.

**Fig. 3** Different steps of the methodology

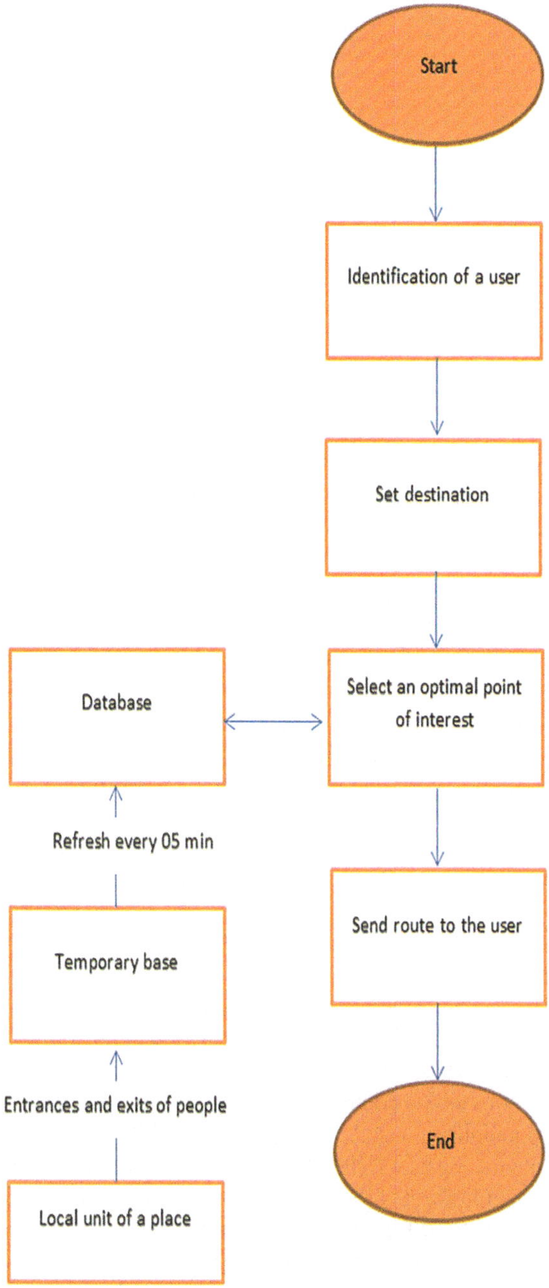

**Table 1** Affectation of people

| Timestamp | User identifier | Bank identifier |
|---|---|---|
| 12:00 | 11,12,13 | 12 |
| 12:10 | 14,15 | 13 |
| 13:00 | $n, n + 1$ | $k$ |

**Table 2** Density degree of different banks

| Timestamp | Bank identifier | Density (t) | Density ($t$ + 15 min) |
|---|---|---|---|
| 12:00 | 11 | High | Medium |
| 12:00 | 12 | Low | Medium |
| 12:10 | 13 | Low | Medium |
| 13:00 | $K$ | Low | Low |

For the experiment below, we selected some users who picked the same bank as destination; after that, the application will distribute them to banks having lower number of people; Table 1 exhibits the assignment of peoples to the bank assigned by the application relied on the database of places of number of people. Table 2 points the number of people degree of various banks and their degree after 15 min. At 12:00, bank of id eleven is characterized by a high density, so incoming people will be sent by the application to the bank of identifier twelve so the bank eleven is alleviated after some time, so the application can redirect next.

person to it. Table 3 indicates users who were diagnosed positive for covid-19 and users who were near to them. User of identifier one was near to person of id 7, 10, 14, 31, and then these users are warned to make their diagnostic of the infection; in case of these persons are diagnosed positive, we look for the entry on the database that have identifier $x$ of the infected user to warn users who were near to user $x$. Sending users to various banks with a manner to balance the number of people in different locations will permit to lighten people density who were near to infected ones, and warning users earlier will alleviate the spread of the infection and anticipate the treatment earlier to recover rapidly.

On account of the recurrent modification in the locations state within the city, there is lot of modifications on the database instantly. The suggested application offers a tool to control congestions and to limit the probability to have a high number of people on a such area of service, and it permits to spot positive tests betimes by diagnosing users who were near to positive person; it also permits to have as a gender of equilibrium of loads in the sense of our application fill up areas by users, and when it starts to be full of people, it redirects the coming peoples to a location of the same kind which is less filled than the others, and the application promotes this impact of equilibrium of load in an automatic way. Thanks to this application, we catch nearly all potential connection between users, and then we handle and limit the propagation of covid-19. These several modifications within the database allow our system more precise about the considering density of people on various locations and to distribute them to different place of services. The architecture is organized of central batch dataset storage and treatment techniques and a dispatching data storage technique for instantaneous processing to handle and treat this huge dataset.

The main tools utilized on our experiment, we employed apache storm to process data in real time and dispatching and rapid flow processing. And for handling the huge dataset, we employed Hadoop MapReduce which builds the database that is used by speed layer characterized by a fast treatment layer instantaneously. The use of storm is very important; it allowed us to process the data in real time, and it is the key to succeed and achieve the objectives of our method. We concluded according to the experimental result that the suggested application model is a good alternative that relies on big data tools, allowing disseminating data in real-time processing for smart affecting control of users to handle the number of people in various locations through city.

## 4 Conclusion

Places of interest engender huge dataset that is very difficult to monitor, annd it implies that employing big data tools is compulsory to mine interesting data from the huge data. The propagation of covid-19 reveals a peril to the safety of our life and to find an alternative to the problem of controlling the danger of high people density; this chapter develops a techniques of big data to ameliorate redirecting handling, and we managed to build an instantaneous number of people detection system with parallel data treatment, which permit to process rapidly in terms of time. The application allows logging precisely the people density being in each place of service, and it permits to overcome the density and conduct people to the safest place desired to attain and the establish the route to reach their destination, which lead to restrict congestion in several places and avoid the danger of covid-19 spread. The results of our simulation demonstrate that the application limit congestion accurately and so control covid-19 spread and not only this; but, the results show good latency and better precision. Utilizing the database of people number and merging this database with machine learning mechanism would be our next research focus to overcome and limiting people number of all locations and showing good results. The results of our method are good, but it need more better accuracy; we are working on the next paper on a method that will ameliorate this methodology by adding more factors in the purpose of improving latency and having a good insight over density in city to handle the distribution of people in different places.

## References


1. M. Nathan, J. Warren, *Big Data: Principles and Best Practices of Scalable Real-Time Data Systems* (Manning Publications Co., New York, 2015)
2. X.L. Wei, Z. Feng, History, current status and future of big data management systems. J. Softw. **30**(1), 127–141 (2019). https://doi.org/10.13328/j.cnki.jos.005644
3. S.S. Bukhari, J. Park, D.R. Shin, Hadoop based *Demography Big Data Management System.* in Presented at 19th IEEE/ACIS International Conference on Software Engineering, Artificial Intelligence, Networking and Parallel/Distributed Computing (SNPD) (2018). https://doi.org/10.1109/snpd.2018.8441032
4. S. Lin, B. Luo, J. Luo, Z. Wang, Y. Wei, *Guizhou Big Data Management System and Key Technology.* in 2018 26th International Conference on Geoinformatics (IEEE, 2018), pp. 1–7. https://doi.org/10.1109/GEOINFORMATICS.2018.8557109
5. H.Y. Chen, *Design and Realization of a Distributed Big Data Management System.* in Advanced Materials Research, vol. 1030 (Trans Tech Publications Ltd), pp. 1900–1904. https://doi.org/10.4028/www.scientific.net/amr.1030-1032.1900
6. H. Matallah, G. Belalem, K. Bouamrane, Towards a new model of storage and access to data in big data and cloud computing. Int. J. Ambient Comput. Intel. (IJACI) **8**(4), 31–44 (2017)
7. A. Raj, R. D'Souza, A Review on Hadoop Eco System for Big Data. (2019). https://doi.org/10.32628/CSEIT195172
8. S. Brunswicker, E. Bertino, S. Matei, Big data for open digital innovation–a research roadmap. Big Data Res. **2**(2), 53–58 (2015). https://doi.org/10.1016/j.bdr.2015.01.008
9. L. Liu, Emerging Study on the Transmission of the Novel Coronavirus (COVID-19) from Urban Perspective: Evidence from China. Cities, 102759
10. M.R.H. Mondal, S. Bharati, P. Podder, P. Podder, Data analytics for novel coronavirus disease. Inf. Med. Unlocked 100374 (2020)
11. S.J. Fong, G. Li, N. Dey, R.G. Crespo, E. Herrera-Viedma, Composite Monte Carlo decision making under high uncertainty of novel coronavirus epidemic using hybridized deep learning and fuzzy rule induction. Appl. Soft Comput. **1**, 106–282 (2020). https://doi.org/10.1016/j.asoc.2020.106282
12. S.J. Fong, N. Dey, J. Chaki, AI-Enabled *Technologies that Fight the Coronavirus Outbreak.* in Artificial Intelligence for Coronavirus Outbreak (Springer, Singapore, 2020), pp. 23–45

# Digital Twins Based LCA and ISO 20140 for Smart and Sustainable Manufacturing Systems

**Mezzour Ghita, Benhadou Siham, Medromi Hicham, and Hafid Griguer**

**Abstract** Currently, several facilities around the world, as part of government's sustainable development strategies, are turning to the development of smart value chains that can leverage efficiently all of countries available and local resources. Smart factories vision embracing the fourth industrial revolution of manufacturing plants introduced a completely renewed industrial organization based on collaboration between human intelligence and capabilities and machines intelligence and computing capacities, manufacturing plants horizontal and vertical integration, and with particular interest for our paper end-to-end engineering. This collaborative endeavour has been translated in the field by a set of technologies for instance advanced simulation tools through digital twins. The use of these new resources and productivity enhancement has not been without consequences on natural ecosystems, which are increasingly subject to industrial competitiveness pressure. To counter the adverse environmental effects of industrial and technological growth, some manufacturers are developing simulation-based life cycle assessment approaches. Over the last few years, several research communities have explored the potential of simulation-based LCA method for the optimization of the environmental impact of production systems through the application of advanced artificial intelligence algorithms. However, only a limited number of these attempts have seen their practical implementation. Currently, digital twins' technologies are rapidly expanding due to

M. Ghita (✉) · B. Siham · M. Hicham
National and High School of Electricity and Mechanic (ENSEM), HASSAN II University, Casablanca, Morocco
e-mail: mezzourghita1@gmail.com

Research Foundation for Development and Innovation in Science and Engineering, Casablanca, Morocco

B. Siham
e-mail: benhadou.siham@gmail.com

M. Hicham
e-mail: Hmedromi@yahoo.fr

M. Ghita · H. Griguer
Innovation Lab for Operations (ILO), Mohammed VI Polytechnic University (UM6P), Ben Guerir, Morocco

 
A. Joshi et al. (eds.), *Sustainable Intelligent Systems*, Advances in Sustainability Science and Technology, https://doi.org/10.1007/978-981-33-4901-8_8

the advantages they offer for real-time simulation, multidimensional replication of industrial systems and end-to-end engineering. Through this work, we propose a generic solution based on digital twins' technologies and ISO 20140 for real-time life cycle assessment and manufacturing systems sustainable optimization.

**Keywords** Digital twins · Life cycle analysis · Sustainable manufacturing · Smart manufacturing · Simulation-based life cycle analysis

## 1 Introduction

Businesses nowadays, due to the fierce competitiveness of the market, are being compelled to revolutionise their organisational and operational practices and strategies. Increased acceleration of industrial performance is forcing medium and large companies to respond to several functional and contextual requirements [1]. Functional requirements involve eradicating all sources of waste and bottlenecks across value chains, forcing companies to review, for example, their policies regarding the management of their material resources and labor force. Contextual requirements include companies' social and environmental responsibility towards their respective economic, regulatory and social ecosystems [2].

Current businesses operate within ever-changing ecosystems and are continually influenced by this change, which in some cases gives rise to unpredictable hazards, political and economic risks to which many industrial infrastructures are susceptible specifically in connection with the current context of increasing emphasis on ongoing integration of sustainable development concerns within industrial ecosystem strategies. The relationship between economic development and environmental and social changes has been explored through the literature by numerous theories consisting on the eco-economic theory of decoupling [3]. Decoupling refers to a variation over time in the coefficient of proportionality, consisting of some form of desynchronization of trends observed between given variables through time. In this case, these variables are defined in relation to economic development indicators and environmental and social indicators. There are two types of decoupling, absolute and relative. Relative decoupling is represented by a relative decrease in the proportionality between the variables under study, whereas absolute decoupling represents an opposite evolution of the two variables [4]. The latter type of decoupling is targeted by a set of countries that have implemented green economy strategies, and in this context, it consists of ensuring accelerated performances and economic development while maintaining social development and enhancing ecological fingerprint and optimizing resources use [5]. Eco-economic decoupling has been also defined as one of Sustainable Development Goals (SDGs) established by United Nations in their 2030 agenda. This year has known numerous changes due to the pandemic which had great impacts on both economic growth and environment. Lockdown period has known significant decrease in air pollution and GHG emissions within the country; however, it has created a relevant recession and increased unemployment rates. The results concluded from this period highlighted important dependencies between social, economic and environmental impacts that all contributes to sustainability manufacturing and sustainability achievement.

For large companies, risks and infringements related to environmental concerns, despite the availability of assessment and evaluation tools and technologies, can hinder the continuity of the company's industrial activity if companies cannot find a balance between accelerating its productivity and respecting the requirements established by countries environmental regulation bodies [6]. For small- and medium-sized enterprises that consider the environmental assessment process cost-intensive, the pressure is steadily increasing, especially in view of the absence of an infrastructure for assessing and identifying the environmental profile of their production systems and their value chain in general [7].

In many countries, in order to achieve this balance, a regulatory system consisting of well-defined measures has been put in place, forcing all structures to reinforce this decoupling. Among these measures are the control of greenhouse gas emissions, mandatory international group obligations to build wastewater treatment plants, waste management programmes and the intensification of renewable energy use. Some companies have opted for different long-term solutions based on environmental profiling of their value streams and supply chains. One of these approaches is life cycle analysis (LCA) method. LCA consists on defining, evaluating and analysing complex systems environmental profile according to a defined framework that takes into consideration diverse factors, steps and elements involved in product, process and services lifecycle. Since its initiation in early 1990, LCA has known numerous applications within various domains as instance food production industries, retail, mining and automotive industry [8]. The international organisation for standardization, ISO, in 2006 published the ISO 14040 standard that defined principles and framework for life cycle analysis and environmental management and as a result structured LCA method application within industrial ecosystems. Studies carried out in order to analyse and enhance LCA practical applications within manufacturing plants have highlighted a set of challenges and critics concerning the method efficiency, scalability and accuracy [9]. In order to overcome these challenges, some communities have experienced the fusion of LCA method and process simulation. This established combination has enabled to enhance method data collection and extraction methods, assessment models, indicators and results accuracy and the method knowledge management process, but it has also given birth to new complex research challenges on:

- How to develop a dynamic and real-time simulation of complex industrial processes that can help LCA inventory analysis and impact assessment while taking into consideration machine states, product variants and supporting systems by amalgamating process simulation models and LCA results calculations?
- How can simulation help to develop a holistic view of systems and their changing environment that reinforce data interoperability, multidimensional modelling and key performance environmental indicators generation?

Recently, the emergence of real-time manufacturing plants monitoring technologies and advances simulation tools such as digital twins has enabled to provide efficient and accurate frameworks for manufacturing plants processes performances optimization and simulation that is increasingly developing thanks to the combination between digital twins concept, industrial Internet of things technologies and advanced data analysis techniques based on artificial intelligence [10]. Digital twin has enabled manufacturing plants to integrate real-time decision making and forecasting into their supply chains for accurate, scalable and efficient industrial performances monitoring. Nevertheless, the implementation of digital twins does not go without challenges that vary between technical challenges related to the enhancement of the concept's potential and non-operational issues mainly cost effectiveness and interoperability [11].

In light of the aforementioned issues concerning LCA method implementation and for the exploration of digital twin challenges within phosphates valorisation and fertilizers production, this paper introduces a digital twin-based LCA solution for manufacturing plants sustainable optimization. The paper introduces the underlying contributions:

- The analysis of Moroccan country eco-economic decoupling and the definition of a three-dimensional perspective for sustainable manufacturing,
- The proposition of a new digital twins' architecture in order to reinforce manufacturing processes life cycle analysis and sustainability,
- The structuring of a social and environmental LCA based scheme for systems of systems environmental profiling and manufacturing systems eco-design,
- The definition of a new interoperable and secure framework through digital twins for life cycle inventory information and knowledge flow management and transparency reinforcement within manufactories.

The rest of the paper is organized as follow, and the Sect. 2 introduces the vision of sustainable manufacturing, the contribution that smart manufacturing has added to this vision efficient deployment across industrial plants and takes the example of Moroccan countries as a practical use case with many challenges to overcome. The Sect. 3 presents a state of the art on LCA method development and simulation-based LCA applications and highlights some of their limitations with regards to manufacturing systems complexity and processes sustainability development through the new industrial context. The Sect. 4 presents the concept of digital twins, its existing reference models and architectures and highlights its potential for process simulation, optimization and sustainability. Sect. 5 proposes a new framework for LCA simulation-based application through digital twins and across manufacturing systems and product life history in order to encounters the existing methods limitations. Through Sect. 6, the developed framework is discussed and compared to the proposed digital twins' architectures for sustainability and smart manufacturing discussed earlier. The Sect. 7 discusses the obtained results and concludes the paper.

## 2 Sustainable Manufacturing Perspective, Opportunities and Challenges

### 2.1 *Sustainable Manufacturing Perspective View and Main Challenges*

In 1987, the United Nations published a report entitled "Our Common Future". The report became the first pillar towards achieving sustainable development around the world. The vision promoted by the report connected for the first-time eco-friendly growth, social development and economic prosperity, by setting out a roadmap for the efficient use of resources to effectively meet the needs of today's generations without adversely impairing the self-sufficiency of future generations. This three-dimension vision was referred to by John Elkington in 1990 as triple bottom line. Throughout industrial development a set of regulations, scientific methods including LCA for our paper, and standards helped to tailor this triple bottom line within manufacturing environments and linked it to manufacturing systems processes optimization [12]. Industrial progress and the pressure it has exerted on the environment despite the measures put in place to minimize this impact has prompted the UN to launch further initiatives through various organizations and countries among them in 2015 the 2030 Agenda for Sustainable Development which was structured in January 2016 by the definition of 17 Sustainable Development Goals (SDGs) [13] that deal with all of sustainable development pillars and propose guides for their deployment and monitoring with a set of indicators including wider areas such as clean sanitation, clean energy, inclusive economic growth and inequalities reduction. Since the introduction of its indicators, a growing number of manufacturing plants are now trying to map their social and environmental strategy and policies to SDG. This involvement of both research and industrial communities has given rise to a new concept that is sustainable manufacturing. Sustainable manufacturing consists on developing efficient, cost-effective and inclusive manufacturing processes, product and services that are based on integrated and sustainable technologies for the establishment of connected, ergonomic and eco-friendly value creation networks, factories and societies. In [14], the authors tried on the background of a bibliographic review to establish a link between sustainability and manufacturing as a result it was defined within the paper by three elements circular economy, climate resiliency and low carbon, and finally resources efficiency.

Circular economy is a theory that emerged in response to the economic and industrial challenges that the world has experienced through its evolution and their impact on its ecosystem and its vital resources [15]. Figure 1 maps circular economy concepts to product, production lifecycles and value chain streams. The figure highlights the integration of circular feedback loops and introduces the integration of 6R into the product lifecycle. The principle of circular economy is to maintain the continuity of the value chain by creating a continuous feedback loop between the different

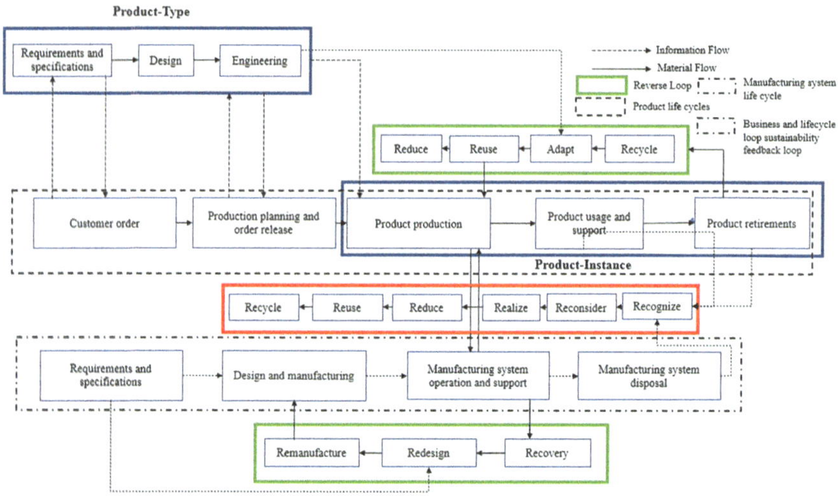


**Fig. 1** Product, manufacturing system and value stream under sustainability perspective

actors and stakeholders of the supply chain, and through the valorisation of individual elements of the chain and their synergies and flows by means of a 3R (Recycle Reuse Reduce) vision. It supports one of the main SDGs which is sustainable production and consumption. A growing number of industries have been inspired by this vision to develop green value chains and organizational practices that will allow the creation of green added value from their product's residuals or by optimizing their current production system through practices that encourage the identification and streamlining of the various waste sources occurring during the production process.

Embracing this vision enables companies both to maximize profit and to integrate SDGs into their industrial plant policies and strategies. Numerous standardization efforts have been made to encourage this integration, citing for example the Global Reporting Initiative (GRI), which consists of capitalizing on the efforts made worldwide to achieve sustainability goals by offering online guides to reporting on the achievement of sustainability objectives in line with the specific objectives and strategies of industrial companies in different fields [16]. The initiative's online portal offers about more than hundred reports from different industries around the world, among the African actors contributing to this initiative, the OCP Group. This capitalization is of major importance to accompany the industrial sectors towards the transformation of their value chain and for the evaluation of their contribution to achieve sustainable manufacturing that presents a number of technological and contextual challenges despite the research efforts made to date to develop ergonomic, cost-effective and adaptable assessment approaches to the different constraints of industrial ecosystems. The environmental issue has been the focus of multiple research communities, giving rise to several approaches for its mitigation. Among these approaches, in 1992, [17] proposed a holistic approach referred to as environmental footprints, which consists of evaluating for each organization its contribution to the degradation of ecological ecosystems through a number of parameters, mainly greenhouse gas emissions and $CO_2$. Currently, its derivative which is carbon footprint method is adopted by a large group of manufacturers and is subject to certification audits in compliance with other standards such as ISO 14 064—1, ISO 26000 and ISO 5000 for energy systems management.

The European Union can be considered as one of the continents that managed to create an integrated ecosystem for the management of climate change and environmental impacts. Thanks to a series of regulations, international charters, research projects and initiatives, the continent is in the process of reducing its impact and decoupling its economic, technological and industrial evolution from its environmental footprint and its impact on society and biodiversity [18]. In 2020, France has succeeded in closing down the majority of its coal-fired power plants, accounting for 2% of the country's $CO_2$ emissions and 30% of its electricity production emissions. The continent was among the first to initiate the implementation of a real-time energy performance and environmental performance tracking system, and the MORE project, which was completed in 2017, allows for distributed and connected monitoring of energy consumption across an industrial network of production facilities. In addition to energy efficiency, the continent is also at the path of achieving its social and economic sustainability by conscious production and consumption [5].

The continent is also taking great strides towards a technological revolution and the strengthening of its digitalization and scientific potential. Another continent whose technological and scientific revolution has permitted to tackle the emerging environmental constraints is the Asian continent, mainly China, which is classified as the second largest contributor to global $CO_2$ emissions [19]. In order to reduce this impact, China has been working for years to strengthen its sustainable development cycle through a number of energy efficiency, waste management and optimization of water and rare metals consumption. Its position as a world economic and technological leader has increased its environmental bill, as the country has become aware of this impact and, under the various regulations imposed by the international community, has been trying to re-orientate its value chains towards a more sustainable production, consumption and distribution. Water consumption and air pollution have also both known drastic growth due to industrial and social revolutions which prompted a lot of research and manufacturing communities to develop renewable resources for wastewater management mainly desalination stations and wastewater treatment plants.

In 2017, the world GHG emission increased by 1.2% with a major contribution of $CO_2$ and methane emissions, total $CO_2$ emissions to 1.5 trillion tonnes with a historical contribution of United states, the world largest contributor in $CO_2$ emissions estimated to 25% followed by China and European Union, and a total worldwide energy consumption estimated currently to approximately 249,000,000 MWh worldwide with 212,000,000 MWh from non-renewable resources and 37,600,000 MWh from renewable resources. China being the first consumer with a world share of 23.8% [20].

The average temperature rise of the world was estimated to 1.1 °C, global average sea level has risen over the past 100 year by 178 mm, and ice sheets has lost more than 426 gigatonnes per year. According to scientists, the current environmental situation could lead to various catastrophes if main contributors to its degradation among which are industrials do not take efficient and practical actions on the field. The last year has known multiple catastrophes, as instance the Amazonian forest fire that has significant impacts on world ecological ecosystem [21].

Waste management, land usage, social impacts and industries clean contributions to economic growths are all drivers that should be taking into consideration for these three elements development across industrial ecosystems.

## 2.2 *Smart Manufacturing as a Driver Force Towards Sustainability Achievement*

Since 2015, the world is witnessing its fourth industrial revolution. A revolution that has reinforced the shift towards smart factories and smart manufacturing by the means of advanced technologies driven by augmented human intelligence, vertical and horizontal interoperability, and smart lifecycle and value streams. Successor of automation, industry 4.0 has elevated industrial ecosystems to new automation levels that consist on merging IT growth and OT evolution. This new vision has enabled manufactories that adopted it to increase their production and enhance their internal and external networks and to integrate a new perspective reducing data silos and centralized management practices and has conducted several research communities to question about the potential of this revolution for sustainable manufacturing development [2].

OECD as we have seen in the previous section defined decoupling as the ratio between driver forces and environmental impacts. Eco-economic decoupling is amongst the economic theories that explore the strong correlations between economic development translated through production capacities, resources efficient use and optimized value-added creation, and environmental impact evolution across economies and throughout industrial ecosystems. Smart manufacturing has contributed mainly to optimized value creation through the introduction of artificial intelligence, simulation and advanced connectivity across industrial plants which has affected positively resource uses and circular economy vision integration and as a result contributed to sustainability triple dimensions.

Recently, skeleton models have been developed in order to detect the synergies that exist between different proposed reference architecture models of smart factories. The proposed analysis suggested two dimensions which are time and space. Time consists on lifecycle phases and value stream evolution within factories, whereas space explores the business functional layers and factories hierarchical level. By exploring each of these two dimensions through smart factories pillars, each reference model has proposed its vision of what could help to implement smart factories and fill gap that were identified considering these two dimensions between classical factories and this new vision. Spatial and temporal parameters are significant drivers for sustainability within industrial factories. One of the standards that has highlighted this double dimension is ISO 20140 for manufacturing systems energy efficiency and environmental impacts assessment. The standard addresses environmental assessment of manufacturing systems through their lifecycles, their hierarchical structures and their value stream with relationship to product lifecycle. The assessment framework proposed by the standard gives three categories of influences, direct, indirect and construction, reconfiguration and retirement (CRR) influences. Direct influences result from manufacturing systems direct mode operation, indirect influence is accounted from standby modes and with reference to support systems operation, and finally, CRR considers long-term influence resulting from manufacturing systems waste release or recycled and remanufactured component. In addition to taking into consideration these three main aspects, the standard discusses manufacturing system special hierarchical structure that takes effects at aggregation and global assessment steps [22]. Time and space variabilities and concerns hinder the distributed and smart vision of intelligent manufacturing, that is what explains the increasing interests put for their mitigation.

In this part, three of social, environment and economic sustainability challenges which are transparency efficiency and profit sharing are explored through the perspective of three smart manufacturing reference models. Assets visibility, information flow transparency throughout the system hierarchy and business functional layers are an important asset for sustainability management in manufacturing plants, taking into account the multitude and disparity of existing flows. To achieve the three R's of the 6R, recognize, reconsider and realize methodology, we need to have a clear view of the different waste streams that exist in the production process, and we need to be able to detect their sources in order to reduce their impact and propose a more optimal and sustainable alternative. Efficiency refers to the ratio between system inputs and outputs and reflects resource usage profiles throughout its life history, product life cycle and manufacturing value stream. Spatial and temporal components influence this efficiency in the short and long term. Profit sharing refers to prosperity within manufacturing which consists on both company value added and profit creation, and factories contribution to its social network economic prosperity. Figure 2 represents this three-dimensional vision.

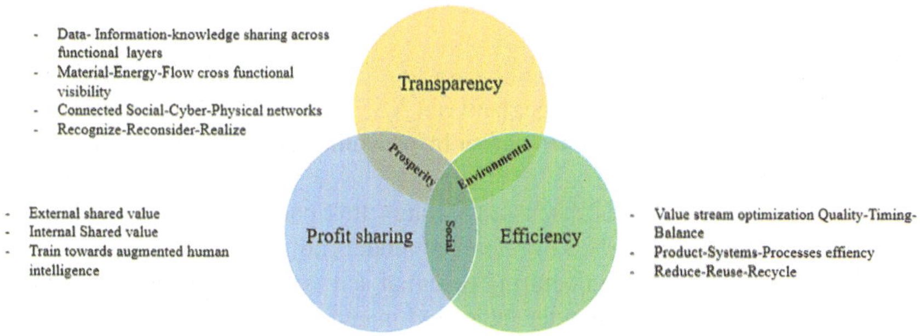


**Fig. 2** Three-dimensional perspective for sustainable manufacturing

This part aims to uncover the significant contribution of smart manufacturing for mitigating challenges resulting from these three dimensions.

Several models were proposed from different countries for smart manufacturing implementation which resulted on different perspective views of smart factories. One of the first and widely known model is reference architecture model for Industry 4.0 RAMI4.0 which was proposed by Germany works group on industry 4.0 implementation across the country. The group defined their vision of factories throughout three dimensions that are functional layers, hierarchical levels, life cycle and value stream. Through this perceptive, the group tried to propose a set of guidelines, technologies and standard that will serve four main objectives, horizontal and vertical integration, interoperability, digital lifecycle consistency and augmented human intelligence through advanced technologies and artificial intelligence [23]. To date, the framework has known several applications and the development of a set of technologies that can reinforce its adaptation notable ones concerned the proposition of common semantic models and standard modelling techniques and languages such as B2MML and automation ML, and some relevant domain ontologies for data models exchanges across functional levels, the proposition or the adaptation of exiting communication protocols for broader uses within industries and with the aim of reinforcing vertical integration as instance OPC UA and MTConnect [24]. RAMI 4.0 can be a relevant framework for sustainability as its offers numerous standard models that can link factories various stakeholders and enhance data-information-knowledge sharing across functional units and between shop floor networks and office floor networks. According to the research carried out on the analysis of the model and its flexibility to cover a number of industries and production systems, some research raises the ambiguity of the model with regard to the definition of the link between hierarchical levels and functional levels, which creates a gap at the level of its possible implementation. One concept that has emerged in relation to this model is that of asset administration shell (AAS) [25]. Asset administration shell is the digital representation of real physical or logical objects that gathers all the information and data on the object during its life cycle and through different functional levels by standard models, interoperable and exchangeable that can be communicated by the mean of complaint communication technologies throughout factories hierarchical architecture. Given its multidimensional aspects, AAS complements the vision of RAMI 4.0 regarding the link between the functional and hierarchical dimensions; however, the concept does not cover the extended vision of the life cycle which involves a number of actors and essential steps to reach the level of efficiency required by the two perspectives smart and sustainable factories. The merger between AAS and RAMI 4.0 allows to achieve transparency aspect that we defined earlier by providing asset with a data-driven and model-based identity.

The second model that is discussed through this part is National Institute of Standards and Technology (NIST) reference model [26]. NIST model is based on the exploration of smart manufacturing potential across life cycle perspectives and their intersection through computer-integrated manufacturing (CIM) pyramid. The proposed reference model explores business, product and production lifecycles with the eyes of the connected, distributed and intelligent view that smart manufacturing promotes and a set of relevant standards that can bring together this different life-cycles in order to create an interactive and dynamic network that will enable fast innovation, proactivity, resiliency of manufacturing systems and sustainability of manufactured product with regards to business constraints and plants changes. The model represents CIM pyramid as an intersection between the developed lifecycles. The connection between CIM layers was discussed in details within ISA standards that propose some significant guidelines in order to implement efficient and connected IT and OT architectures through a set of standards that each details one the level of CIM and its supposed compliant and efficient synergies and communication with other levels. ISA 95 that has been particularly used with relationship to smart manufacturing addresses manufacturing and operation management (MOM) level which is the third level involved in production planning, quality assurance, production resource management, recipe management, scheduling, operational performance analysis, data collection and dispatching [27]. ISA 88 and 95 addresses the first and second level with focus on sensing and controlling domains and ISA 101 for control HMI physical and cognitive ergonomics. Ergonomics has been recognized as a build block for smart automation and augmented human intelligence of operators' agents. NIST vision combined with ISA standards can help mitigate main efficiency challenges through real-time extraction and aggregation of operational data and the integration of advanced simulation tools along process and product life cycles and factories supply chains and through its combination with AAS models concretize digital thread vision that consists on developing a digital manufacturing test bed.

The last reference model that we will discuss is industrial Internet reference (IIRA) model. The pace in communication and information technologies and the development of industry 4.0 perspective through a set of concepts among which is the Internet of things has motivated research communities and standardization organization to explore the potential implementation of highly connected industrial networks. IIRA by adopting ISO 42020 viewpoints perspective has proposed an architectural and conceptual framework for tailoring industrial Internet of things. Some research efforts have tried to map RAMI 4.0 into IIRA in order to unveil its conceptual framework and reduce the gaps related to the definition and identification of the links between the different levels. IIRA gives a detailed description of the functional view through defining main functions that can be mapped to CIM pyramid for OT operation, cross functions that defines some essential elements for industrial Internet as instance connectivity and distributed data management and systems characteristics that is connected to systems critical nature such as trustworthiness. The interesting parts that IIRA architecture introduced are business view that will help to mitigate profit sharing. Business view within IIRA is defined by a set of interesting concepts for sustainability; its main function and concern is to define stakeholders intended values according to decision makers and stakeholders' vision. The resulting values are then translated to key objectives for systems engineers that map systems functionalities to fulfil these values. The definition of these concepts is based on vision and value-driven model.

The exploration of these different reference models, their values and limitations has enabled us to detect the main guide lines for the development of a new framework for sustainability within manufacturing driven by smart manufacturing.

### 2.3 The Fourth Industrial Revolution within Moroccan Country and its Potential Contribution to Sustainability Goals Fulfilment

Morocco in recent years is undergoing a twofold revolution. The first was initiated through the ambition of its governor and the government as well as public and private institutions to achieve energy independence and mitigate the environmental impacts due to climate change and the global regulative context. The second revolution was initiated mainly by its industrial ecosystems, and through the Ministry of Industry, Trade, Investment and the Digital Economy which decided to initiate the country to the fourth industrial revolution that the world is currently experiencing. The combination of Morocco in both of these areas has given rise to several laws, technologies, new practices and nascent ecosystems of innovation and research. To date, the country has recorded several achievements in this field with the implementation of different solar complexes NOOR, sea water desalination plants, wastewater treatment plants, internationally certified complexes and plants, and various laws dedicated to the mitigation of environmental impacts have been implemented not only in the manufacture plants but also in different sectors. The country has so far developed the first pillars of environmental governance. Following this energetic and environmental evolution, the country has initiated a digital transformation of its administrative sectors, encouraging dematerialization throughout its value chains and data sharing through dedicated platforms. The country's efforts in the transport sector are also considerable and are reinforced year after year by the efforts of a large community of researchers and specialists and key players in the sector such as the National Railways Office (ONCF), which is constantly developing its technological potential and its ecological footprint through its technologies. In 2016, the office has estimated its carbon emissions of 27.10 g $CO_2$/tonne-km for 32 million tonnes of goods and 27.50 g $CO_2$/ passenger-kilometres for 40 million passengers. The office has also put in place a plan to mitigate environmental impacts and its social responsibility based on a combination of measures including the decrease of external costs to adapt its services in accordance with the standard of living of a large segment of citizens, the reduction of energy consumption and the lowering of pollutant emissions. The Cherifian phosphate office is recognized as one of Morocco's leaders in the field [28]. OCP has succeeded in a few years to create a network of connected plants, equipped with the latest technologies and driven by sustainable and efficient processes that allow the group not only to keep its place as an important exporter of phosphate, but also at the national level as a valuable business and social growth driver force. In its latest GRI sustainability report, the group has presented the balance sheet of its sustainability activities [29]. A promising report that shows the group's contribution to the sustainable creation of added value notwithstanding the constraints of the sectors in which the group operates. The chemical and mining sectors are both sectors with high environmental impacts. The group, thanks to the development of a circular economy perspective and the implementation of comprehensive production management systems, as well as to the efforts of a large research community represented by a number of institutions, has achieved a remarkable level of energy self-sufficiency and is planning a series of new technologies to mitigate all the environmental impacts related to its activities [30]. OCP is contributing with 90% to the exports of the chemical and parachemical sector in Morocco with 5.62 billion equivalent in revenue. The group generates benefits representing 52% of the sector's revenues, 67% of the sector's investments and has counted in 2020 18,906 employees [31]. OCP in accordance with SDG goals has developed a series of commitments that were integrated horizontally and vertically across its complexes that we cite mainly responsible and inclusive management, sustainable production and shared value creation. The group has succeeded in covering 86% of its energy needs by clean energy and 30% of water needs by unconventional water. In 2007, the group has developed a certified system for carbon footprint evaluation, and he is intending to reinforce research and development in the filled of carbon capture and valorisation and renewable energy storage. Figures 3, 4 and 5 represent, respectively, GDP and Gross National Income growth GNI between 1990 and 2017, value added of industry, manufacturing, services and agriculture on GDP evolution for the period from 1990 to 2017, and contribution of renewable energies on total energy consumption from 1991 to 2016.

**Data Collection, Cleaning and Normalization**

In order to evaluate eco-economic decoupling and the impact of economic activities and growth on environmental impact evolution and social development, we conducted an analysis of decoupling between Gross Domestic Product (GDP), Human Development Index (HDI) and some environmental impacts indicators mainly energy

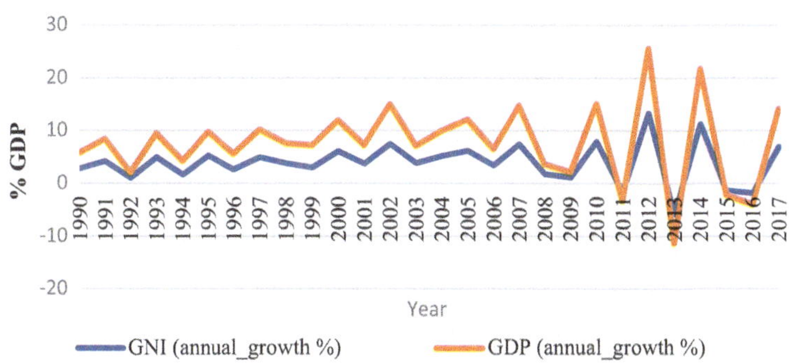


**Fig. 3** Morocco's GDP and GNI evolution from 1990 to 2017 according to World Bank countries economic development indicators databases

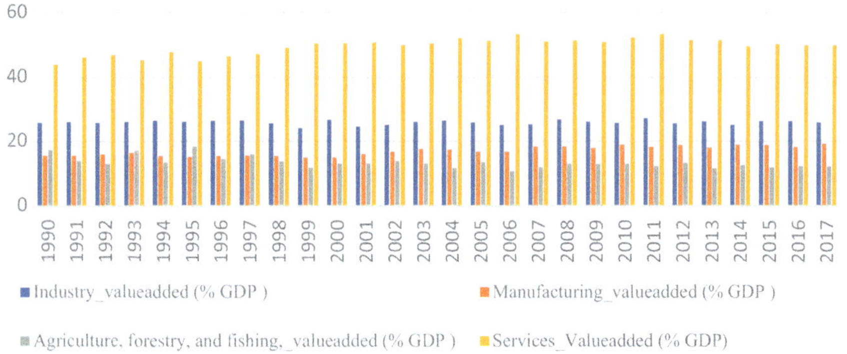


**Fig. 4** Morocco's industry, manufacturing, services and agriculture value-added evolution on countries GDP between 1990 and 2017 according to WHO statics for sustainable development and World Bank data

**Fig. 5** Morocco's total energy consumption and renewable energies consumption evolution between 1991 and 2016

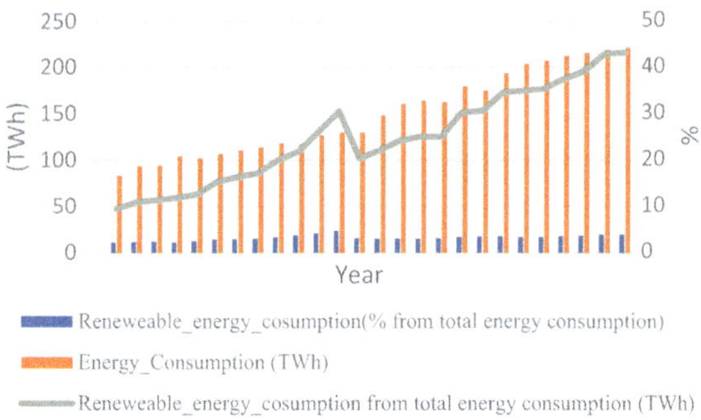


consumption, CO2 emissions, water efficiency, NH3 emissions, NOx emissions and total GHG emissions. Data were collected from international and national sources for tracking sustainable development indicators achievements across world countries and countries economic and social indicators. GDP and HDI evolutions for Moroccan country were extracted from World Bank database and WHO open databases; GHG emissions and water efficiency evolution and energy consumption were extracted from SDG online tracker [32], UN portal for sustainable goals progress [33] and World Bank databases. Collected data were normalized according to Eq. (1).

$$\text{Indicator intensity} = 1 + \frac{\text{Indicator value}_{\text{end of period}}}{\text{Indicator value}_{\text{end of period}} - \text{Indicator value}_{\text{start of period}}} \quad (1)$$

**Indicators Calculation and Results**

In order to evaluate the eco-economic decoupling, we based our analysis on proposed OECD method that defines decoupling ratio according to driver force and environmental impacts. Driver forces are represented by GDP intensity and HDI, whereas environmental impact through four indicators, water use intensity, energy consumption intensity, GHG intensity and $CO_2$ $NO_x$ $NH_3$ intensity. Equation (2) and Eq. (3) represent decoupling ratio and decoupling factor.

$$\text{Decoupling ratio} = \frac{\text{Environemental Impact intensity}}{\text{Driver Force intensity}} \tag{2}$$

$$\text{Decoupling Facteur} = 1 - \text{Decoupling ratio} \tag{3}$$

**Results Interpretation and Relevant Insights**

Through Fig. 6a, b, c and d, we can see that the economic evolution of the country until 2008 registered a low decoupling for all environmental indicators and a large increase in the country's GDP which was characterized by the increase in the contribution of the manufacturing sector which started to take more space with the introduction of new automotive and aeronautical operators. In 2009, as a result of the global economic crisis, the country's GDP declined and stabilized from +7.31% of annual growth to + 1.91%. The following years saw a decline in GDP and a positive increase in effective emissions and a relative coupling between the evolution of $CO_2$ emissions and GDP. During this period, the country has also seen the initiation of a number of projects aimed at energy efficiency and the preservation of natural resources including water. These private and public projects have enabled the development of new renewable energy sources, including solar energy, photovoltaic energy and cogeneration for industry.

The energy production sector is one of the country's largest energy consumers. Water efficiency has also seen a slight increase thanks to the implementation of new devices for the management of industrial and wastewater and to strict laws that regulate this use. In agriculture in recent years, a number of projects have been launched to reduce the risk of water stress which has long been a threat to the country. Agriculture is considered to be the largest consumer of water within Morocco. The period following 2008 has also seen a decrease in the contribution of agriculture to the GDP, which fell from 15.09% in 1990 to 11.37% in 2016. The evolution of the decoupling factor in particular for energy consumption and $CO_2$ emissions which have increased from 0.053350725 in 1991 to 0.215227358 in 2018, the last value that we have been able to extract, and from 0.01292902 1991 to 0.22442962 in 2016, reflects the country's ambition to develop an efficient and green economy.

The country during this period also experienced a number of initiatives for its social development, which concerned different aspects such as public healthcare and education. Figure 7 shows the evolution of decoupling factor between human development indicator and environmental impacts for the defined periods. The results

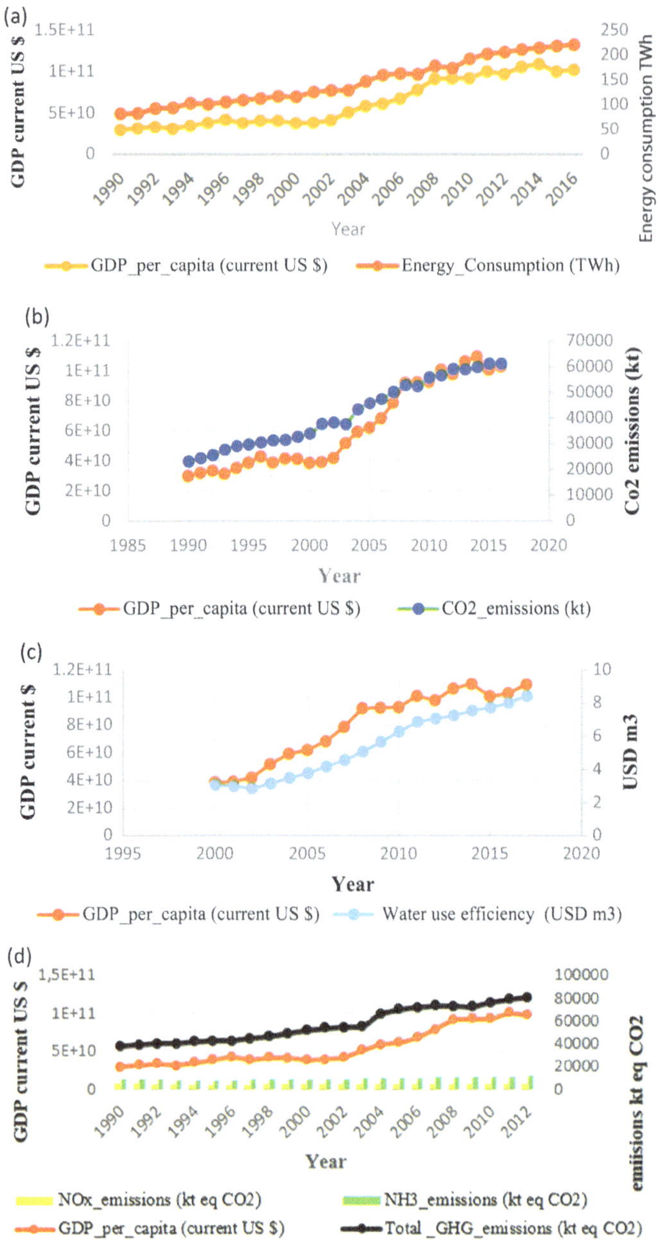


**Fig. 6** Morocco's environmental impacts and GDP evolution **a** country's GDP in current US $ and energy consumption in TWH **b** country's GDP in current US $ and $CO_2$ emission in kt **c** country's GDP in current US $ and water efficiency in USD $m^3$ **d** country's GDP in current US $ and $NH_3$ $NO_x$ and total GHG in kt eq $CO_2$

obtained globally show a positive decoupling for $NH_3$ and $NO_x$ emissions, and a negative decoupling for global emissions, $CO_2$ emissions and energy and water use efficiency. The evolution of the country's GDP shows an evolution of services and their contribution to the GDP, a contribution which is also reflected by developments in social indicators, among them trends of the country's population, which in turn generate growth of resource extraction and concomitant economic activities. Not forgetting transport, which is one of the elements for social development, is an active agent of climate change and emigration in Morocco, and its development is not without consequences on country's environmental profiles. The development of ICT and communication networks ensures connectivity and active communication within the society and access to electricity; all its measures contribute to environmental degradation if the country in question does not put in place plans that could create a balance between the social added value and the impacts on the country's ecological ecosystems, thus enhancing its decoupling or keeping it at acceptable levels.

## 3 State of the Art

### *3.1 LCA Method Roots, Industrial Applications and Limitations for Manufacturing Systems Sustainability*

Life cycle analysis (LCA) method has been initiated in 1990 as a new approach based on life cycle perspective for environmental impact assessment and contributes to product, process and service environmental profiles identification [34]. During its ongoing evolution, the method has been subject to several changes that have involved the integration of new dimensions to the evaluation process, including cost, social, economic and energy aspects [35]. In 2006, the international standardization organization published the ISO 14040 standard which defines the conceptual and organizational framework for the development and implementation of the approach [36]. This has allowed to broaden the scope of applicability for the method and identify it as one of the approved methods for environmental analysis of manufacturing systems. Currently, the method is derivated on several other broader approaches such as LCA sustainability assessment which includes social, financial and economic drivers in the evaluation process, while some are more specific such as exergy LCA for the valuation of systems and processes energetic sustainability [37].

ISO 14040 sets out the method in four phases which are goal and scope definition, inventory analysis, impact assessment and finally for gathering and auditing the different phases in the interpretation phase. Goal specification consists on defining analysis motivations, intended usage and contributors that can include analysis stakeholders, influencers and systems actors as main contributors to systems behaviour and evolution. Scope definition includes clear and structured description the system subject of the analysis, its life cycle phases, functional units and boundaries. Scope

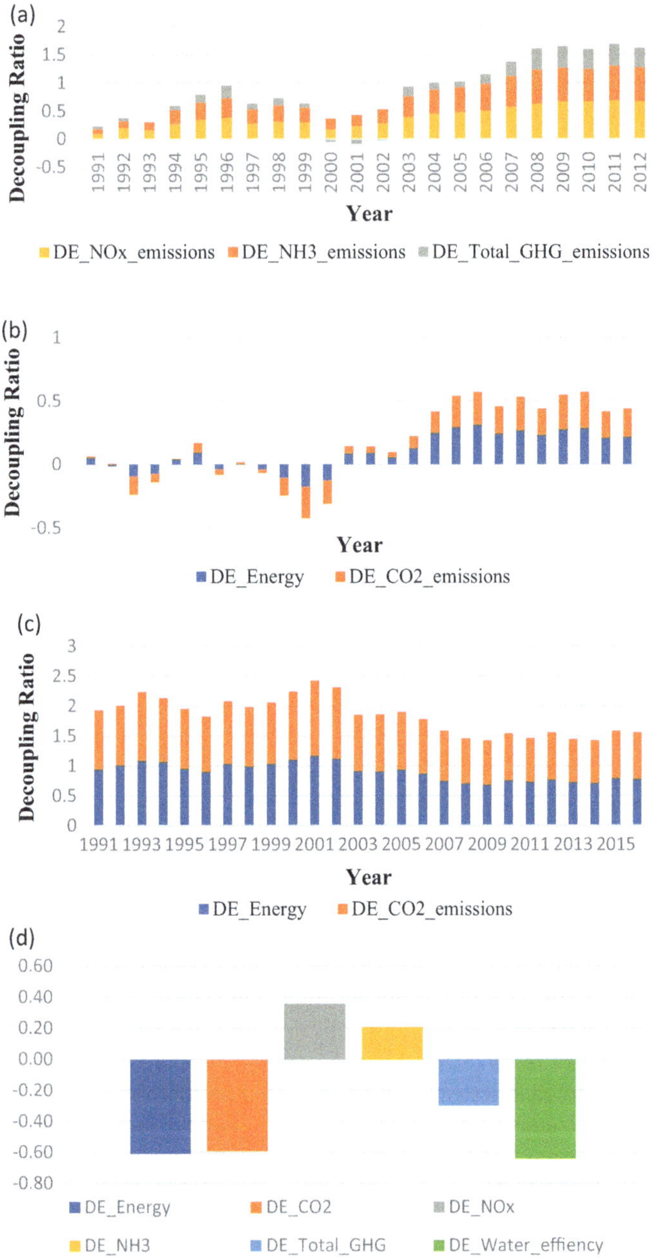


**Fig.7** Morocco's decoupling factors **a** Decoupling of GDP growth and greenhouses emissions. **b** Decoupling of GDP and energy consumption intensity and $CO_2$ emissions intensity. **c** Decoupling of HDI and different defined environmental impacts

definition is strongly related to some additional elements that are not explicitly included in the standard but contributes to accurate definition of the overall application of the method. These elements include systems modelling framework and granularity in accordance to study's goals, adopted hypothesis and defined borders for the study, method's data management techniques, retained impact assessment approach and system model verification and validation methods approved by study's stakeholders that have particular expertise with regards to systems functions and domain of interest [38]. Functional unit, systems modelling and boundaries definition and life cycle phases description play key roles within this phase. Functional unit is referred to by the standard as effect quantitative characterization and the main reference to which defined systems inputs and outputs are normalized for LCA evaluation. The second phases as defined by the standard is life cycle inventory analysis that mainly consists on documenting systems flows inputs and outputs in accordance with defined goals and through its lifecycle data collection and related environmental data acquisition. Data collection for LCA analysis represents a turning point as it contributes to assessment accuracy, scalability and impact assessment methods flexibility. Inventory process involves data extraction, gathering, calculation, aggregation and normalization according to defined functional unit [39]. The third phase is life cycle impact assessment (LCIA); it is intended to exploit inventory analysis results in order to provide qualitative or quantitative impacts analysis based on a set of defined and representative indicators and assessment categories. Research made on LCA method implementation has pointed out the importance of the integration within this phase of data quality evaluation mechanism that could enhance obtained results accuracy and significance and reduce their uncertainty and sensitivity [40]. LCIA is organized into three main steps selection, grouping through categories and characterization. The first step into assessment process consists on attributing LIC inputs to LCIA categories for assessment, and systems performances for each category are evaluated according to defined valuation factors that give assessment results [41]. The last phase is results interpretation, and this step summarizes study's outputs and gives recommendations, relevant observations, concluding remarks and identified limitations across the different phases and regarding the settled goals.

LCA method has been applied to various sectors and for different purposes, and this extension has made it possible to define its potential and detect some of its challenges. LCA contributions to sustainability and green processes development and implementation have been achieved across different industries and domains, including construction and infrastructure sector, food and kindred industry, mining and quarrying of non-metallic minerals sector, transportation equipment industry, electricity production and for the purposes of this paper agricultural production more particularly mineral fertilizers production [42]. Construction sector has evolved significantly in recent years with the introduction of new simulation, multidimensional modelling and project management tools. This evolution has led to the emergence of new requirements among decision makers, including constraints related to the environmental footprint of buildings and their energy efficiency throughout their lifecycle [43]. LCA method is widely used for building life cycle evaluation and environmental assessment, raw materials and supplier's selection and energy efficiency management [44]. Food supply chain is among the wide spread industrial applications of LCA. A lot of LCA software providers offer complete frameworks for the evaluation of food supply chains environmental impacts, by giving databases and material and energy flows inventories of real industrial applications [45]. A recent research published on 2019 applied LCA to beef industry on the USA. Based on a structured framework related to the industry's processes modelling approach, unlike previous research on the field, the authors of the paper presented an analysis that included social and environmental impacts and concluded their assessment process with a sensitivity analysis [40]. Mining and ore processing industries are among the sectors that are currently undergoing significant development of environmental assessment tools as a result of international regulations affecting the field. Across the world, several mining communities exploited the potential of LCA method to improve their process and product ecological foot print throughout their life cycle [46]. Recent research in the field are focalizing their interest on processes energy efficiency and waste management which can give an increasingly holistic view. A recent research discussed the method application for the development of batteries technologies which is also a promising field for LCA, especially with the advances in hybrid batteries development and eco-friendly vehicles.

These multiple applications especially in the context of manufacturing systems has revealed some of the limitations regarding method's efficiency and scalability for applications in the deployment of sustainable manufacturing systems under the auspices of the new industrial revolution. Some research communities have recently started to explore these challenges and risks through feedback from different application areas, which has identified a number of issues but also a significant potential for continuous improvement of the method. In this, authors based on failure mode and effect analysis (FMEA) dealt with the challenges and risks resulting from this application and provided a series of recommendations in order to improve LCA application that focused on data quality and goal and specification phase [9]. A similar analysis was conducted by [35], which explored the challenges of the method in the field of sustainable chemical industry. Feedback on the various applications carried out enabled them to identify a number of avenues for improvement suggesting the merging of LCA and risk analysis methods and the strengthening of the social and economic aspect at the level of impact analysis. As we were able to conclude from the analysis of this researches and cases of application for LCA within manufacturing sustainability framework, four categories of limitations can be defined according to each of LCA method's phase.

The first challenges concern the initial phase that consists on defining the system and its elements, in particular system boundaries, which has been the subject of a number of research projects attempting to propose approaches that would allow this definition to be structured. Industrial systems today are becoming increasingly complex, and their interactions with a number of external elements make their modelling and description more complicated for LCA studies which require a limitation of study scope. The exclusion of a number of boundaries could impact the final outcome of the study and therefore omit elements whose environmental impact is important especially for studies that focus on production and usage phases of the system. A second challenge detected concerns the definition of the functional unit that constitutes a key element of the analysis. The functional unit plays a primordial role in all phases of the process, and especially in the last phase of interpretation, a biased choice of this element could lead to uncertain and incomplete conclusions covering only a part of the life cycle of the system, thus making it difficult to compare different alternatives for improving the performance of the system under study.

As far as life cycle inventory is concerned, most of the limitations relate to data collection process. The limitations that arise from the data collection process are multiple and vary according to the scope of the method. The first problem concerns the quality of the data collected for the study based on identified inflows and outflows. Due to confidentiality and privacy constraints imposed by the industrial decision makers, several data relevant to the study are missing or have been estimated on the basis of the assumptions made during the first phase. The errors related to its estimates have a direct impact on the results of the evaluation and therefore may lead to the implementation of ineffective measures and the selection of inappropriate options given the industrial context of the study. To cope with this constraint, some studies have been interested in automatic data extraction using advanced data management tools, and information retrieval based on networks of intelligent sensors and cognitive sensing. These tools have in turn given rise to new concerns among industrialists and practitioners in the field. These concerns involve data aggregation methods, deployment costs, the energy bill of the solution and its technical and semantic interoperability. The existing tools for performing LCA analysis feature already developed scientific and industrial databases of real case studies that can be used to perform local analyses, yet a large number of these databases are incomplete or inaccessible for more global studies.

Impact analysis process is the driver of the method in the sense that it makes it possible to valorise the different previous phases by defining specific performances profiles corresponding to the objectives of the stakeholders involved in the study. The challenges of this process are particularly related to the definition of indicators and their relevance and reliability. Many existing methods used in LCA analyses are concerned with environmental impact and especially for various applicative use cases to energy efficiency concerns of manufacturing systems, but there are few structured and standardized approaches dealing with societal aspects. As discussed in the previous section, the current situation, characterized by the pandemic and the health and economic crisis, has demonstrated the need to consider all aspects of the life cycle within the method, including social and economic dimensions in order to achieve sustainability and sustainable manufacturing.

The current application of LCA, as we have seen, focuses only on the development of one aspect to the detriment of other relevant aspects. These neglected aspects can constitute key elements for sustainability enhacement and in the long term may influence the results of the environmental assessment of the systems through the LCA approach [47]. Table 1 represents the different references of development efforts to encounter LCA phases limitations.

**Table 1** Development efforts for encountering LCA process-related challenges

| LCA phases | Limitations | Deployed solutions, tools and methods | A* | Nascent challenges | Opportunities under smart manufacturing context |
|---|---|---|---|---|---|
| Goals and specifications definition | Dynamic and static representations of manufacturing processes for hotspots detection | Complex system simulation Tools | [48–51] | Simulation of machines time varying states, and detailed models, products variants databases information and supporting systems integration | Advanced PLM simulation through digital twins industrial Internet of things technologies with the integration of heterogeneous data analysis |
| | Framework agility and flexibility | Manufacturing systems simulation tools and related decision-making techniques | [52–55] | Networking, and interoperability for collaborative development | Collaborative project management tools through the cloud |
| | Spatial and temporal variability handling | Processes material energy and flows simulation tools and geospatial intelligence | [56–59] | Material, flow and energy simulation integration into a common and modular platform | Advanced PLM simulation through digital twins |
| | Systems boundaries definition and functional units' selection | Complex systems engineering approaches | [60, 61] | Systems increasing complexity, functional dependencies and difficult integration of non-operational concerns | Digital twins for complex systems engineering and in combination with model-based engineering |
| | Models verification and validation | Sensitivity analysis | [62, 63] | Data availability for multidimensional models testing and training | Artificial intelligence and data mining integrated tools |
| | Holistic view and life cycle phases handling | Complex systems engineering approaches | [64, 65] | Modelling granularity and systems increasing complexity | Digital thread development methods through digital twins |

(continued)

**Table 1** (continued)

| LCA phases | Limitations | Deployed solutions, tools and methods | A* | Nascent challenges | Opportunities under smart manufacturing context |
|---|---|---|---|---|---|
| | Models maintainability and interoperability | Collaborative and agile development approaches | [66, 67] | Models global exchange, semantic development and tools technical interoperability | Digital thread through digital twins and semantic modelling for interoperable KPI generation |
| Data collection and LCI | Measurement and calculation process variability | Sensitivity analysis, real-time data extraction and probabilistic approaches for uncertainty evaluation | [68] | Security, confidentiality, and privacy issues especially with regard to social inputs | Artificial intelligence-based tools for data confidentiality and privacy |
| | Data extraction and KPIs inputs generation | Open repositories and storage and analysis of production process real-time data | [69, 70] | Security, semantic and syntactic interoperability for data in motion and at rest exchange | Smart sensor networks for production phases, domain ontologies and semantic modelling for lifecycle covering |
| | Data quality and availability data confidentiality and privacy | Open repositories and databases production systems and discrete event simulation results<br>Open licences for data exchange and standardization effort for data governance | [58, 71, 72] | Knowledge capitalization constraints and context awareness<br>Security and confidentiality constraints for industrial data sharing and national LCA standards adoption for global and local applications | Advanced PLM simulation through digital twins and advanced data analysis<br>Artificial intelligence-based tools for data confidentiality and privacy |

(continued)

**Table 1** (continued)

| LCA phases | Limitations | Deployed solutions, tools and methods | A* | Nascent challenges | Opportunities under smart manufacturing context |
|---|---|---|---|---|---|
| | Impacts interdependencies evaluation | Weighting and normalization techniques based fuzzy logic and domain experts, systems users consulting behavioural mapping of stakeholders' product usage profiles | [73, 74] | Communication and consultation constraints, difficulties for qualitative data characterization | Artificial intelligence-based tools for data analysis and features extraction |
| LCIA | Social and economic impacts integration | Social-LCA and life cycle costing methods | [75–79] | Industrial costs data and information confidentiality, distributed systems networking and security, and social impacts temporal and spatial variability | Compare and comply technologies based on natural language processing and federated learning for costs models generation |
| | Impact categories definition and selection, and qualitative impacts characterization | Fuzzy logic and multicriteria analysis, big data analysis and machine learning for temporal and spatial variabilities reduction | [79, 80] | Experts remote support and domain-specific knowledge capitalization and sharing through connected networks | Artificial intelligence and expert systems application within industrial sectors |

(continued)

**Table 1** (continued)

| LCA phases | Limitations | Deployed solutions, tools and methods | $A^*$ | Nascent challenges | Opportunities under smart manufacturing context |
|---|---|---|---|---|---|
| Interpretation | Dynamic and interactive visualization of results and interfaces cognitive ergonomics<br>continuous reporting and monitoring of results through usage phase | Interactive dashboard and real-time LCA for dynamic KPI's visualization<br>Industrial data management architecture and data at rest analysis through artificial intelligence and results archiving and reporting | [81, 82] | Smart KPI's generation and sharing, data in motion real-time analysis systems complexity for life cycle data inventory | Collaborative notebooks for real-time and smart access to advanced analysis cognitive modelling for ergonomic interfaces digital twins' and digital thread for lifecycle data sharing and geospatial technologies for regional and global LCA analysis |
| Cross cutting concerns | Stakeholders trustworthiness | Secure and anonymized data sharing through scientific communities and LCA practitioners | [83, 84] | Shared data quality and cross domain availability of social and economic LCI results and LCIA models learning outputs | Agile development approaches for static LCA application in product early phases federated learning for distributed machine learning and models exchanges |
| | Tools cost effectiveness | Open source software's and open accessible databases of industrial data and LCA + LCC analysis real and simulated applications | [85–87] | Data availability and data quality | Costs hotspots detection through digital twins and interoperable and sustainable tools for life cycle data archiving |

(continued)

**Table 1** (continued)

| LCA phases | Limitations | Deployed solutions, tools and methods | A* | Nascent challenges | Opportunities under smart manufacturing context |
|---|---|---|---|---|---|
| | Deployment tools scalability | Connected platform through Web-based collaborative platforms and portable LCA software's architectures | [88] | Security and costs concern for tools development, storage and sharing through cloud platforms Cloud platforms additional environmental impacts on manufacturing systems life history | Knowledge bases and standard models' repositories blockchain and digital twins for PLM data archiving and sharing |
| | Standardization and mainstreaming all along product supply chain | Large-scale LCA analysis for bio-economic and integration of ecosystems analysis | [89–92] | Data interoperability, industrial data governance | Good practices capitalization and industry-specific models' integration within industries large structures |

### 3.2 LCA Based Simulation—A Potential Solution with Nascent challenges-

In recent years, in order to address the limitations that have been identified following the application of the method in various fields and in the manufacturing field particularly, researchers have resorted to merging the methodological framework for the deployment of the method with manufacturing processes and systems simulation dynamic simulation. This fusion relied on the development of systems simulation platforms that would replicate in a virtual environment systems behaviour and dynamic states and consequently reproduce the different results perceived in the field for a more efficient and reliable data extraction during data collection phase and also for a predictive analysis of environmental impacts in a broader context that includes systems' productivity and operating constraints within their environment and in contact with their users.

This mix proven its efficiency for dealing with some of LCA limitations.

The first one is model's uncertainty that can result from initial assumptions and parameters estimations based on linearity hypothesis, and process simulation can help to verify the proposed models under realistic conditions that take into consideration systems states and dynamic changes as well as expert domain feedback. Modelling with simulation can as well enhance LCI sensitivity analysis results and as a consequence guarantee developed models maintainability throughout systems life cycle [93].

The second limitation that concerns data quality and availability can be dealt with through the use of simulation results according to multiple operational scenarios and situations established by systems engineering teams in accordance to productivity requirements forecasting and decision maker's needs. Available simulation tools that deals with production various production aspect can gives LCA unbiased inputs across systems life cycle especially during usage phases and more recently due to the advanced simulation technologies through virtual prototyping for conception phases [94].

Last but not least are spatial and temporal variability resulting from systems environments, machine states, product variants and some relevant supporting systems. The majority of LCA studies that does not rely on process simulation may neglect supporting systems that for some manufacturing can represent a high potential for environmental impacts monitoring as instance waste management systems for phosphorus processing and land usage for mining sector.

Despite being a promising alternative, this combination gives rise to new challenges especially with manufacturing systems nascent complexity and industrial environments continuous changes due to the introduction of new technologies and organizational methods as for instance the particular use case of energy grids and renewable energy systems that are extremely intermittent. One of these challenges remains on developed tools capabilities to mirrors both systems dynamic and static behaviour and to merge efficiently and securely between both process simulation and LCA calculation for cross functional sharing of results across factories. Sustainability as we have seen through previous parts is a wide concept that includes numerous and heterogenous elements as a results LCA based simulation solution in order to contribute to it should integrate a holistic vision of manufacturing systems that can be faithful response to decision makers increasing needs in the field. Developed model's interoperability is considered as a relevant challenge also as for LCA to reach its main goals, it should rely on multiple scenarios that takes into consideration various constraints. For LCA to cover a manufacturing systems life cycle and their interconnection as well as impacts interdependencies, simulation should provide multidimensional models that take into consideration temporal and spatial variability which needs advanced computing and networking capacities. As we have seen, sustainability is a broad and multidimensional construct, and LCA based simulation development framework should take this dimensionality deeply into consideration. Table 1 summarizes challenges that encounters current LCA based simulation methods and applications and opens up the discussion on smart manufacturing opportunities for LCA based simulation.

## 4 Digital Twins towards Manufacturing Systems Sustainability

### *4.1 Digital Twins Background, Applications and Reference Architectural Frameworks*

Digital twin origins dates back to 2003, when Dr. Grieves presented the concept in one of his talks about product life cycle management as an efficient alternative that can combine simulation capacities, multidimensional data and information models of complex systems in order to merge virtual and physical worlds for virtual prototyping, processes optimization, product health management and various other applications [95]. The idea was after developed by several communities working on different advanced process simulation, connectivity and optimization tools. The potential of the concept has enabled it in few years to cross the door of manufacturing companies by a set of technologies that extended its application field to some critical domains such as healthcare [96].

Currently, digital twins are undergoing extensive adaptation in several areas. Defined by Gartner as one of the technologies in full expansion since 2018 and taking into account the numerous technologies offered by the market and the need to have a reference framework that could bring together the experience and knowledge acquired to date on the subject, the scientific community of the International Organization for Standardization decided to develop a standard that would have as its main objective to define the principles, structure and reference model of digital twins of manufacturing systems [97]. The standard series proposition that was approved on January 2018 is currently under development ISO/DIS 23247 which is entitled digital twin framework for manufacturing and is part of the framework of standards aimed at the integration of IT and OT for a smarter integrated automation. The proposed reference architecture is based on references layers proposed through ISO 30141 standards that define IOT reference architecture [98]. ISO 30141 defines IOT systems through six domains that are physical domain, sensing and controlling domain, operation and management domain, application and services domain, resource and interchange domain, and user's domain. Digital twin entity is represented in ISO/DIS 23247 by the operation and management domain, application and services domain, resource and interchange domain and the observable physical thing which is defined through the first domain, whereas communication between them is established through sensing and controlling domain that is referred to as data collection and device control entity. An additional cross system entity is defined for data assurance, information exchange and security support across the architecture. The system communicates with its users through user's entity that is represented by users' interfaces tailored to user's interests. The framework addressed main development and deployment phases focusing on discrete manufacturing systems and manufacturing systems development, without given details about the representation of product life cycle and supply chain value stream, which are both relevant elements within manufacturing system life cycle particularly for manufacturing system operation and support. Human agents and social networks contribution in the framework deployment and development is only discussed through a functional view.

One of the structured conceptual frameworks proposed for digital twin development is the administration shell concept that introduces as we have seen in previous section the concept of assets digital representation [99]. The concept different applications have proved its contribution to digital twin development and implementation by offering to digital twin the appropriate infrastructure and context to operate within factories control and monitoring architecture and for product life cycle management. The structure proposed that consist of developing standard models of assets under a unified and interoperable format in early phases of assets life cycle. The developed sub-models will be then gathered within distributed and accessible repositories constituting AAS body and as a results DT-type models library. Each model according to its domain of interests and perspective is assigned a property constituted by function for their calculation and data necessary for their evaluation. The different models of one AAS are controlled through AAS manager that is connected to AAS head. AAS head is the digital footprint of the asset and its AAS that defines it according to a unique identifier proposed according to universal identification standards for assets and serves for connection assets to their administration shell but also to preserve model AAS security. Once the asset planned is developed, it is linked to its administration shell by compliant communication interfaces. Through its functioning, the asset communicates its operational data and historical data to the version of AAS adapted to stakeholders needs which creates a dynamic version of the AAS that represents DT-instance. This representation of digital twins through asset life cycle generates what is referred to as the digital thread. Digital thread is a digital test bed for covering digitally all of asset life cycle phases and their development [100].

In addition to AAS, one of the research efforts that has linked industry 4.0 and industrial IOT to digital twin's development framework is Eclipse proposed framework for digital twin Ditto [101]. Ditto is an open source platform for digital twin networks development and deployment based on a microservices architecture and links between physical connected things and a distributed platform of their digital replica. Eclipse has developed this framework with the aim of providing an accessible and open source interface for digital twin experts and developers, and it is referred to as device as a service. The framework offers a set of services that enables fluent communication between the thing which is the real asset and the digital twin that gathers all its data models. The framework offers several advantages and helps to mitigate some of digital twin adopters concerns that are security and cyber physical connectivity, and communication of large networks of digital twins, leaving detailed description and development of the models that will constitute the replica to digital twin experts.

More recently research within digital twin field has focusing on developing intelligent networks of digital twins that will help mitigate the challenges of both smart and sustainable manufacturing visions.

In [102], the authors based on the main structure of smart manufacturing reference models and dimensions proposed a generic and intelligent digital twin architecture for sustainable manufacturing. The proposed conceptual framework of the architecture that was addressed to discrete manufacturing systems is based on three main axes, sustainable and intelligent manufacturing service, sustainable and intelligent manufacturing equipment and sustainable and intelligent manufacturing systems. Sustainable and intelligent manufacturing service are based on production, process and after sales services that refer to supply chain value stream. Sustainable and intelligent manufacturing equipment is constituted through intelligent manufacturing unit and intelligent manufacturing production line. The last module sustainable and intelligent manufacturing system helps to connect to the discrete manufacturing system life cycle through sustainable design, production, logistics, sales and services with reference to ISA 95 levels for control architecture of discrete manufacturing. These three components are connected with a data-driven repository for social-economic-environmental impact assessment and based on different advanced technologies as instance cloud technologies, IIOT and 5G for communication. The framework has covered a set of concepts for traceability and some elements that constitutes efficiency basis in brief without given details on the methods that will be deployed for environmental assessment and the contribution of the proposed advanced technologies for achieving static and dynamic life cycles assessment. The proposed foundation pyramid of the system makes human at the top of the pyramid, but it do not give clear links between the platform main units and human contribution in the platform; the authors refer to it briefly in architecture description through collaborative manufacturing without giving clear key functionalities for achieving this connection with other elements of the foundation. Human and machine collaboration is represented as a main element; however, the authors had not addressed ergonomic aspect particularly cognitive ergonomic as well as occupational health and safety constraints that were taken in consideration in researches aiming the integration of smart manufacturing [103]. The proposed framework focused on asset with its broad meaning efficiency aspects mainly information efficiency, operation efficiency, maintenance efficiency and services efficiency and highlighted the role that artificial applied to these domains can bring to reinforce digital twin intelligence and sustainability aspect. Profit sharing is not discussed in the paper despite being a relevant element for manufacturing system and maintenance efficiency. Some main challenges for digital twin integration were little discussed as instance trustworthiness and security, models and digital twin platform internal services synergies management in order to achieve DT autonomy and resiliency as main characteristics highlighted by the three previously discussed models.

The analysis of the different reference models has enabled us to explore digital twin potential for implementing smart manufacturing, and their analysis through the three dimensions that we have proposed in the previous section has unveiled the opportunities that they can offer for sustainability integration within smart manufacturing.

## 5 Digital Twins Based LCA

### *5.1 LCA Renewed Framework Under the Auspices of Sustainable and Smart Manufacturing*

Analysis of digital twins' scope of application and in particular in Morocco and analysis of the impact and potential of LCA and in particular dynamic LCA through simulation allowed us to identify several intersection points between the digital twins and the LCA approach for the implementation of the sustainability vision that we have described throughout the previous sections and the strengthening of the eco-economic decoupling within emerging economies aspiring to integrate both smart manufacturing and sustainable development such as that of Morocco [104].

The new framework that we propose through Fig. 8 exposes two essential blocks, the first block is that of dynamic LCA that offers us a continuous and proactive life cycle analysis of environmental, social and economic impacts across three levels of product life cycle, manufacturing systems and value chain life cycles. The LCA method proposed by the standard is reinforced by a number of new building blocks made possible through advanced smart manufacturing technologies. The second block of the framework is the life cycles of the real and virtual physical environment combined with reverse loops reinforcing the circular vision of sustainable development, green production and conscious consumption.

The first lifecycle is the manufacturing system lifecycle which is extended by the integration of digital twins whose missions are digital prototyping, digital commissioning and digital design. Developed in parallel with the traditional life cycle of the product type, it will allow to create a digital certificate to the real system well

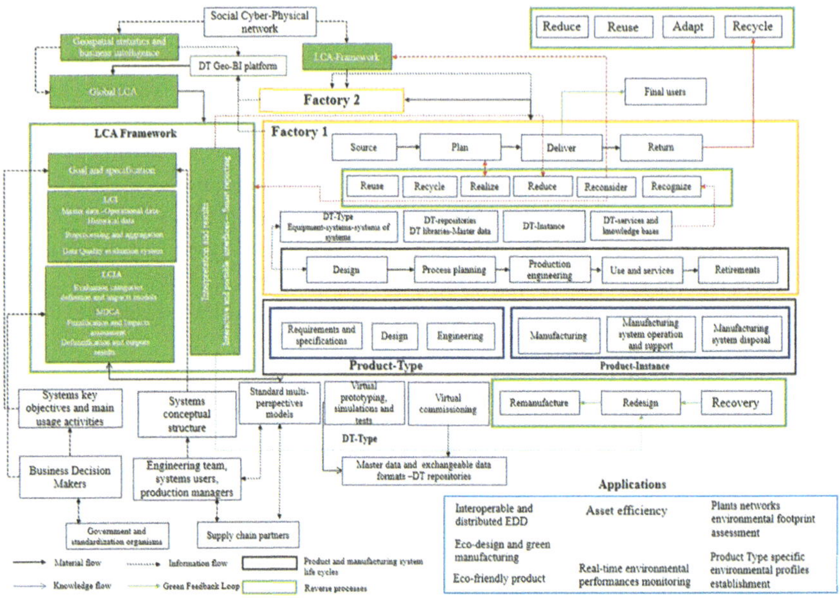


**Fig. 8** Digital twin-based dynamic LCA framework for sustainability

before their actual realization to develop what is called digital twin type. Digital twin type which is continuously connected to the network of stakeholders allowing a collaborative development gives rise to standard and common models of real assets. These models are communicated to the LCIA and to goals and specifications steps of dynamic LCA for the definition of complex system structures and physical system life cycle data meta structures from a multiperspective viewpoint. After the production phase, the system moves from the product type to the product instance and goes out of the manufacturer's field to integrate the integrator's field and thus bring its added value to the product life cycle. At this stage through LCA, system analysis already has in its possession an impact profile, containing reference information of its energy consumption, its emissions and a number of parameters defined, evaluated and characterized by a network of experts in collaboration with the stakeholders involved in the production of the final finished product and taking into account a number of criteria defined by the multicriteria analysis.

In the course of its operation, the instance records in real time the data related to its operation and communicates them to the DT repositories and to the dynamic LCA learning models that serve to evaluate the risks and to predict the deviations from the profiles designed for each equipment and contained in the digital twin type. This link between the type and the instance allows the detection of the hotspots and the recognition of the operational points subject to deviation both from the performances required by the client and from the environmental and social performances required by the system. The purpose of the global LCA analysis is to connect a network of sites with different conditions and scenarios and to create a number of learning and evaluation models and to compare between these models without violating the security and confidentiality limits defined by the stakeholders and the system manufacturer, thus reinforcing the application of federated learning within the framework of digital twinning to ensure intelligent and secure processing. Each of these two blocks is applied to each manufacturing site in the same region and communicating between or within the same country through a geospatial business intelligence interface hosted on the cloud that allows communicating the performance of several manufacturing sites and combining them with the geo-statistics of the region and its multidimensional spatial view integrating cost, regulation and biosphere of the region in question [105]. This dual framework has been developed with the aim of guaranteeing the three elements of transparency, efficiency and profit sharing. The main objective of the digital twin is to enhance transparency and efficiency of value and life cycles, while LCA's mission is to ensure the connection between its life cycles, to guarantee profit sharing and to enhance long-term efficiency for the eco-design of complex industrial systems.

## 5.2 *Digital Twins Proposed Architecture for Manufacturing Sustainability*

Our analysis of the different opportunities available to developing countries to overcome the obstacles that they face in integrating industry 4.0 has enabled us to develop a specific perception of digital twins. The state of the art that we developed on industry 4.0 revealed that one of the elements that hinders advanced technologies development and adaptation by emerging industries despite the numerous opportunities that actually these technologies have to flourish within these countries industrial environments is new concepts understanding, cost effectiveness and integration constraints. This new framework would allow an emerging industry to overcome all the technological, cultural and economic limitations of a digital twin development project and comes to introduce for emerging industries an optimal purely adapted and holistic digital twin's deployment architecture. The architecture is represented through five main different layers that are context layer, perception and interrogation layer, mirroring and cognitive layer, intelligence layer and finally services layer. Figure 9 represents the overall structure of the architecture. Those layers interact with three main elements for digital twins' deployment that are physical twin and living ecosystem, and physical system social network and digital twin social network. Physical system social network is constituted by physical twin stakeholders that can be users of the system, users directly influencing the systems within its environment, users influenced indirectly by the systems and its value stream, and domain experts. Digital

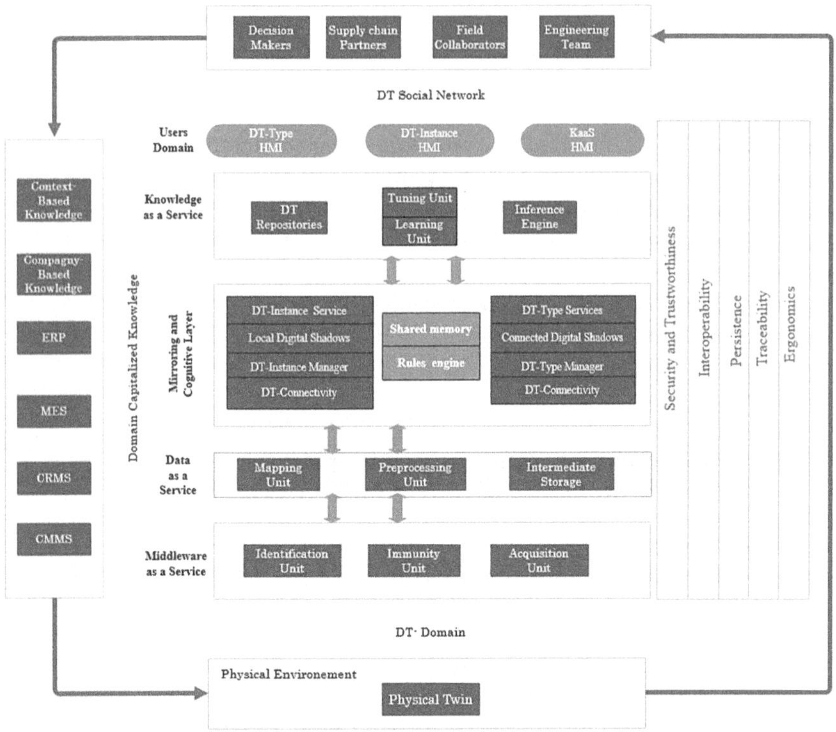


**Fig. 9** Digital twin proposed architecture for sustainability

twin social network is composed by digital twin digital users and digital twin physical users. Digital twin physical users are digital twin architects, engineering team, field collaborators, technologies providers and decision makers.

## Perception and Interrogation Layer

This layer consists of two elements which are middleware as a service and data as a service. The development goal of this layer is the establishment of an interactive and secure system of exchange between field agents and the dynamic components of the digital twin by allocating the necessary mechanisms for data acquisition, interpretation and intelligent data transfer to the appropriate digital twin agents whilst meeting the various concerns raised by context and functional analysis.

Middleware as a service's mission is to ensure the oriented and secure acquisition of physical twin data, the identification of industrial layer elements and the reliability of transactions with the physical twin. Three units constitute this element, immunity unit, identification unit and acquisition unit. Immunity unit plays the role of a security mechanism based on a contract system established with real-world assets interacting with digital twin agents. The virtual and physical components according to their exchanges establish a security contract to authorize this connection. This unit is responsible for the identification of physical twins and requests coming from the physical environment through a dedicated authentication system connected to the authorization system. The third unit is an acquisition unit. The third unit is an acquisition unit. This unit receives the data output from the two previous units and assigns specific indices to them. According to those indices, acquisition unit receives the output and transfers it to the second elements which is data as a service.

Data as a service element is constituted by three units mapping unit, pre-processing unit and intermediate storage unit. Mapping unit plays the role of a data interpreter unit. As it arranges data according to a predefined data meta structure and interoperable format, it refers to exchangeable data formats defined according to users' responsibilities and concerns towards the developed digital twin architecture. This unit also plays the role of an interactive interface with the physical environment order to enhance interpretation capacities; this unit is provided by environment communicated data. Pre-processing unit works on data curation and conversion in order to meet data consumers technical and non-technical requirements. Intermediate storage stores pre-processed data based on their type and defined three classes of digital twin data types that are operational data, master data and historical data. These two layers enhance the twin's interoperability with its real-world compatriot and provide two dynamic interfaces for communication with the real world. These interfaces are of great added value for the twin since they allow the accommodation of different inputs from the real world to the virtual world.

**Mirroring and Cognitive Layer**

The mission of this layer is to reliably mimic real complex systems, their dynamics and connections and to offer the user an interactive interface for monitoring while capitalizing on the experiences, knowledge and dimensions related to the system within its environment and during its life cycle. It is made up of two agents, a cognitive agent and a hybrid agent, and these two agents provide the digital twin architecture with the vision of intelligence and autonomy required by smart factory model. The figure shows the two agents in this layer.

The first agent within this layer is digital twin type. Digital twin type is a hybrid agent composed by connected digital shadow, DT connectivity, DT manager and DT-type services. Connected digital shadows are standard models of physical twins that represents physical systems commonly known representations through semantic models and knowledge graphs. These models are defined according to a set of usage views and functions based on agent developed skills set and beliefs. DT-type services represented the set of expertise's and skills that the agent acquired through its design and interactions with physical twins. The last two elements are building blocks of DT-type proper operation and resources interchange mechanisms, its constituted by communication and control unit. Communication unit represents DT connectivity module build upon a set of communication interfaces and protocols with digital and physical users. DT manager is the internal control mechanism of the digital twin type. Based on our analysis of digital twin iceberg model, this unit is composed by four modules that are access management for digital twin exchange external and internal agents identification, self-defence for digital twin security and virtual world cross-cutting security, self-configuration for digital twin persistence and self-operations optimization for self-organizing and the optimization of resources management. The mission of those units is to ensure digital twin autonomy.

The second agent is digital twin instance; it is a hybrid agent. Digital twin instance is constituted by local digital shadows, DT-instance services, DT connectivity with digital and physical users and finally DT-instance manager. Local digital shadows are simulated models of physical twin as they evolve in their hosting environments. They are based on a detailed exploration of physical systems through their various feedback loops and their perception of the world that encompasses temporal and spatial considerations that field interpreters ensure at a first level. Their main mission is to model and simulate real systems functioning through different levels of abstraction required for performances smart tracking and systems smart control and command. DT-instance services are responsible for shadows behaviour generating and sensory processing through a set of designed services and established connections with data as service layer. DT connectivity is ensured by global communication skills of DT-instance agent that enables it to interact with digital twins' network, peer systems and physical users. DT-instance manager is similar to DT-type manager; however, it represents one added functionality that is referred to as value judgement. The mission of this added feature is to reduce services and shadows cognitive biases and guarantee its compliance with DT-instance stakeholders' values.

**DT-Gateway**

This module's role is to establish a link between the two agents of the previous layer by creating a continuous feedback loop allowing the two aggregates to share information on their functioning and on the functioning of the assets in the field. The set of rules that dominate the feedback loop are defined in this module, which plays a similar role as a gateway. The common information shared between the type and the instance is defined at the level of the DT-gateway and contained in a shared memory between the two twins. Some complex prognostication functions for their execution require the combination of several information about the system and essentially the results of the real-time simulation oriented according to one of the perspectives defined; thus, the advantage of this layer is to allow its results to be transferred from the field twin to the connected twin.

**Intelligence Layer**

This last layer is based on a knowledge as a service system that allows the architecture to develop, in addition to the interoperability and autonomy layers, an intelligence layer contained in an intelligent decision system structure. This system consists of three units. This last layer is based on a knowledge as a service system that allows the architecture to develop, in addition to the interoperability and autonomy layers, an intelligence layer contained in an intelligent decision system structure. This system consists of three units.

The first unit is DT repositories; this unit gathers together the entire knowledge produced at the level of its two architecture agents as well as a capitalization of the knowledge of domains of interest for the architecture. Two types of knowledges interact in this layer, heuristic knowledges and knowledges resulting from the capitalization of external know-how related to the different business domains of the company's strategic activities. The second unit that interacts at the level of this layer is the inference engine; it plays the role of a new rules' engine capitalizing a number of expertise for the generation of a number of recommendations and new rules for intelligent decision making at a higher level of abstraction. This unit is linked to a third unit which is a tuning unit. Tuning unit acts in collaboration with inference engine to reduce cognitive biases and uncertainties due to unexpected system changes. This tuning unit is in constant connection with DT-type's learning mechanism.

**Cross-cutting Concerns Within Architecture Functional View**
As we have seen through the analysis of digital twins through ISO 42010, a set of concerns constitute the framework for the development and realization of an adaptive and generic digital twin architecture. Some concerns are common to all layers of the architecture and in particular those pertaining to its functional view. These common concerns are interoperability, security and trustworthiness, ergonomics, persistence and traceability for digital twin knowledge management. Interoperability is one of the essential features for the running of the different digital twin aggregates and therefore represents an indispensable skill across architecture different layers. Interoperability through layers can be regarded across global communication interfaces, governing policies and procedures, smart and continuous feedback loops and finally layers unit's data meta structure. The second function within this category is security and trustworthiness that constitute key elements within digital twin and asset administration shell scope. Security as we have detailed through previous section can be regarded through confidentiality, integrity and availability, trustworthiness which represents user's perception towards the architecture and architecture defined transparency mechanism in order to ensure it. Ergonomics that was detailed across the paper is considered essentially through cognitive ergonomics that related to models, knowledge representations, and interfaces usability. Cognitive ergonomics comes to reinforce digital twins and architecture semantic interoperability. Persistence concerns layers compatibility and continuous functional suitability with the various heterogeneous data sources of the physical environment and domain capitalized knowledge. Persistence can be defined through architecture unit's smart configuration and resources management mechanisms. Traceability consists on ensuring architectures layers functions, inputs and outputs knowledge management through smart archiving mechanisms.

## 6 Discussion and Assessment of the Proposed Architecture Through Smart Manufacturing Filters and Defined Sustainability Dimensions

The solution that we propose aims to guarantee the three missions which we have defined for sustainability that are transparency, efficiency and profit sharing.

Transparency is achieved through the digital twin structure that presents an end-to-end view of complex industrial systems throughout their lifecycles and connects them to each other and to a wider network of stakeholders in a secure manner. The model that we propose draws on the security aspects developed by Ditto framework and adds a layer of expertise and domain-specific description of assets. This vision also allows us to integrate the principles conveyed by ISA 95 and ISA 101 for ergonomics by developing cognitive, role-based models and integrating them with local control in a flexible and intelligent way, thus filling the safety and OHS gaps of the other proposed models. The solution has been developed with reference to the requirements of the administration shell structure, which also allows to bridge the safety and interoperability gaps that often hinder the effective integration of the twins in the field and through safety and cyber safety critical industrial control architectures.

Efficiency is perceived through the different services and functions of the digital twin type and digital twin instance functions that respond to the main aspects of efficiency by integrating the vision of sustainability and preservation of resources, optimization of production systems and proactivity. Digital twin instance allows to increase energy and environmental awareness layer systems as close as possible to the physical systems and through the first control layers.

Profit sharing through geospatial intelligence offers an intelligent and continuous evaluation of the functioning and operation of the systems from a cost effectiveness point of view while continuously indexing the production responsible hotspots of costs target deviations set both externally and internally.

Our proposed solution takes as its main purpose the fusion between smart automation techniques that were promoted through smart factories alongside sustainability requirements that results in the proposition of renewed manufacturing and monitoring tools within different fields. Digital twins reinforced by LCA analysis at early stages but also during usage phase can revolutionize existing methods for sustainability management and transparency integration through systems and products life cycles. The identification of current architectures loopholes that may result from their large-scale implementation outside research laboratories and limited contexts has benefited maturity assessment of digital twins applications, thus the need to integrate digital twin instance perspective that takes into consideration its hosting context and digital twin type perspective that integrates extended considerations related to domain-specific concepts. This combination of two different and communicating intelligent and autonomous agents aiming to enhance overall system performances enable to develop a smart and distributed systems that reacts proactively with its evolving environment and creates a dynamic impact sphere.

## 7 Conclusion and Future Works

Digital twins are currently a concept widely adopted by several manufacturers from different sectors, our analysis of the concept and its application in concrete case studies in Morocco has allowed us to identify different potential applications for the concept that could help exploit its generic vision that combines a number of principles and technologies to face the different challenges that the world is currently facing. Among the projects carried out in different countries around the world is the implementation of smart manufacturing. As we have seen throughout the paper, the smart manufacturing vision does not stop at the integration of artificial intelligence and advanced communication and information tools and cobotics, but its scope is expanding towards another vision whose implementation in today's world has become indispensable, that of sustainable manufacturing. To merge these two visions and come out with an efficient solution, we proposed a framework that combines digital twins and dynamic LCA. Through this merger, we aimed to overcome the limitations of LCA and its requirements in terms of life cycle data and spatial–temporal variability through digital twins and to augment the digital twins with a dynamic environmental assessment layer through LCA to address three challenges of sustainability traceability, efficiency and profit sharing. Architecture implementation through the industrial tissue may result in some limitations that revolve mainly around the establishment of an equilibrium between operational constraints and computational capacities of the different layers and services of the architecture. The second concern relates to data governance, although paper proposes a generic architecture and different cross-functioning layers for managing layers inflows and outflows resiliency, security and integrity; the increasing pressure on the different layers due to data in motion acquisition and treatment can give birth to some biases and uncertainties especially for critical industrial applications. Our future work will focus on the application of the solution on concrete case studies in the chemical and parachemical industry for asset efficiency and in the power generation sector for sector eco-economic decoupling.

**Acknowledgements** Work carried out within the framework of the cooperation agreement for technological and scientific development concluded between the UM6P and the FRDISI.

## References

1. W.S. Alaloul, M.S. Liew, N.A.W.A. Zawawi,I.B. Kennedy, Industrial revolution 4.0 in the construction industry: challenges and opportunities for stakeholders. Ain Shams Eng. J. S2090447919301157 (2019). https://doi.org/10.1016/j.asej.2019.08.010
2. W. El Hilali, A. El Manouar, *Smart Companies: How to Reach Sustainability During a Digital Transformation.* in Proceedings of the 3rd International Conference on Smart City Applications–SCA '18. (ACM Press, Tetouan, Morocco, 2018) pp. 1–6
3. E. Conrad, L.F. Cassar, Decoupling economic growth and environmental degradation: reviewing progress to date in the small island state of malta. Sustainability. **6**, 6729–6750 (2014). https://doi.org/10.3390/su6106729
4. P. Tapio, Towards a theory of decoupling: degrees of decoupling in the EU and the case of road traffic in Finland between 1970 and 2001. Transp. Policy **12**, 137–151 (2005). https://doi.org/10.1016/j.tranpol.2005.01.001

5. E. Sanyé-Mengual, M. Secchi, S. Corrado, A. Beylot, S. Sala, Assessing the decoupling of economic growth from environmental impacts in the European Union: A consumption-based approach. J. Cleaner Prod. **236**, 117535 (2019). https://doi.org/10.1016/j.jclepro.2019.07.010
6. S. Stavropoulos, R. Wall, Y. Xu, Environmental regulations and industrial competitiveness: evidence from China. Appl. Econ. **50**, 1378–1394 (2018). https://doi.org/10.1080/00036846.2017.1363858
7. D.C.A. Pigosso, A. Schmiegelow, M.M. Andersen, Measuring the readiness of SMEs for eco-innovation and industrial symbiosis: development of a screening tool. Sustainability. **10**, 2861 (2018). https://doi.org/10.3390/su10082861
8. F.D. Pero, M. Delogu, M. Pierini, Life Cycle Assessment in the automotive sector: a comparative case study of Internal Combustion Engine (ICE) and electric car. Procedia Struct. Integrity. **12**, 521–537 (2018). https://doi.org/10.1016/j.prostr.2018.11.066
9. I. Djekic, M. Pojić, A. Tonda, P. Putnik, D. Bursać Kovačević, A. Režek-Jambrak, I. Tomasevic, scientific challenges in performing life-cycle assessment in the food supply chain. Foods. **8** (2019). https://doi.org/10.3390/foods8080301
10. C. Cimino, E. Negri, L. Fumagalli, Review of digital twin applications in manufacturing. Comput. Ind. **113**, 103130 (2019). https://doi.org/10.1016/j.compind.2019.103130
11. M. Ghita, B. Siham, M. Hicham, Digital twins development architectures and deployment technologies: moroccan use case. Int. J. Adv. Comput. Sci. Appl. (IJACSA). **11** (2020). https://doi.org/10.14569/IJACSA.2020.0110260
12. B. Mota, A. Carvalho, M.I. Gomes, A. Barbosa-Póvoa, *Design and Planning Supply Chains with Beneficial Societal Goals*. in Computer Aided Chemical Engineering (Elsevier 2019), pp. 439–444
13. *Sustainable Development Goals*, https://www.undp.org/content/undp/en/home/sustainable-development-goals.html
14. A. Moldavska, T. Welo, The concept of sustainable manufacturing and its definitions: A content-analysis based literature review. J. Cleaner Prod. **166**, 744–755 (2017). https://doi.org/10.1016/j.jclepro.2017.08.006
15. P. Rosa, S. Sassanelli, A. Urbinati, D. Chiaroni, S. Terzi, Assessing relations between Circular Economy and Industry 4.0: a systematic literature review. Int. J. Prod. Res. **58**, 1662–1687 (2020). https://doi.org/10.1080/00207543.2019.1680896
16. *Global Reporting Initiative*, https://www.globalreporting.org/Pages/default.aspx
17. W.E. Rees, *Ecological Footprints and Appropriated Carrying Capacity: What Urban Economics Leaves Out: Environment and Urbanization* (2016). https://doi.org/10.1177/095624789200400212
18. J. Vogler, H.R. Stephan, The European Union in global environmental governance: Leadership in the making? Int. Environ. Agreements. **7**, 389–413 (2007). https://doi.org/10.1007/s10784-007-9051-5
19. H. Ritchie, M. Roser, *$CO_2$and Greenhouse Gas Emissions. Our World in Data* (2017)
20. *Worldometer—real time world statistics*, https://www.worldometers.info/
21. *Climate Change: Vital Signs of the Planet*, https://climate.nasa.gov/
22. 14:00–17:00: ISO 20140–3:2019, https://www.iso.org/cms/render/live/en/sites/isoorg/contents/data/standard/06/46/64674.html
23. M.A. Pisching, M.A.O. Pessoa, F. Junqueira, D.J. dos Santos Filho, P.E. Miyagi, An architecture based on RAMI 4.0 to discover equipment to process operations required by products. Comput. Ind. Eng. **125**, 574–591 (2018). https://doi.org/10.1016/j.cie.2017.12.029
24. M. Ghazivakili, C. Demartini, C. Zunino, *Industrial Data-Collector by Enabling OPC-UA standard for Industry 4.0*. in 2018 14th IEEE International Workshop on Factory Communication Systems (WFCS) (IEEE, Imperia, Italy, 2018), pp. 1–8
25. C. Toro, A. Seif, H. Akhtar, Modeling and connecting asset administrative shells for mini factories. Cybernet. Syst. **51**, 232–245 (2020). https://doi.org/10.1080/01969722.2019.1705554
26. robert.lipman@nist.gov: *Enabling the Digital Thread for Smart Manufacturing*, https://www.nist.gov/el/systems-integration-division-73400/enabling-digital-thread-smart-manufacturing

27. A. Semmar, N. Machkour, R. Boutaleb, H. Bnouachir, H. Medromi, M. Chergui, L. Deshayes, M. Elouazguiti, F. Moutaouakkil, M. Zegrari, Modeling Input Data of Control System of a Mining Production Unit Based on ISA-95 Approach, in *Smart Applications and Data Analysis*. ed. by M. Hamlich, L. Bellatreche, A. Mondal, C. Ordonez (Springer International Publishing, Cham, 2020), pp. 47–55
28. Harbal, A.: *La question environnementale au Maroc* (2017)
29. SDD—GRI Database, https://database.globalreporting.org/organizations/7288/
30. *Livre Blanc : La Transformation Digitale Au Maroc*. AUSIM MAROC, https://www.ausimaroc.com/livre-blanc-la-transformation-digitale-au-maroc/
31. *CHIMIE-PARACHIMIE | Ministère de l'Industrie, du Commerce et de l'Économie Verte et Numérique*, https://www.mcinet.gov.ma/fr/content/chimie-parachimie
32. *Goal 6: Clean Water and Sanitation–SDG Tracker*, https://sdg-tracker.org/water-and-sanitation
33. *SDG Indicators*, https://unstats.un.org/sdgs/indicators/database/
34. E. Nieuwlaar, Life Cycle Assessment and Energy Systems, in *Encyclopedia of Energy*. ed. by C.J. Cleveland (Elsevier, New York, 2004), pp. 647–654
35. G. Mondello, R. Salomone, *Chapter 10—Assessing Green Processes Through Life Cycle Assessment and Other LCA-Related Methods*. in Studies in Surface Science and Catalysis, ed by A. Basile, G. Centi, M.D. Falco, G. Laquaniello A. Basile, G. Centi, M.D. Falco, G. Laquaniello (Elsevier 2019), pp. 159–185
36. 14:00–17:00: ISO 14040:2006, https://www.iso.org/cms/render/live/fr/sites/isoorg/contents/data/standard/03/74/37456.html
37. R. Basosi, M. Cellura, S. Longo, M.L. Parisi, (eds.) *Life Cycle Assessment of Energy Systems and Sustainable Energy Technologies: The Italian Experience* (Springer International Publishing, 2019)
38. Y. Liu, A. Syberfeldt, M. Strand, Review of simulation-based life cycle assessment in manufacturing industry. Prod. Manuf. Res. **7**, 490–502 (2019). https://doi.org/10.1080/21693277.2019.1669505
39. S. Suh, G. Huppes, Methods in the Life Cycle Inventory of a Product, in *Handbook of Input-Output Economics in Industrial Ecology*. ed. by S. Suh (Springer, Netherlands, Dordrecht, 2009), pp. 263–282
40. S. Asem-Hiablie, T. Battagliese, K.R. Stackhouse-Lawson, C. Alan Rotz, A life cycle assessment of the environmental impacts of a beef system in the USA. Int. J. Life Cycle Assess. **24**, 441–455 (2019). https://doi.org/10.1007/s11367-018-1464-6
41. F. Torabi, P. Ahmadi, *Battery Technologies*. in Simulation of Battery Systems (Elsevier, 2020), pp. 1–54
42. M.L. Brusseau, *Chapter 32 - Sustainable Development and Other Solutions to Pollution and Global Change*. in Environmental and Pollution Science, ed. by M.L. Brusseau, I.L., Pepper, C.P. Gerba, Third Edition. (Academic Press, 2019), pp. 585–603
43. C.C. Wang, S.M.E. Sepasgozar, M. Wang, J. Sun, X. Ning, Green performance evaluation system for energy-efficiency-based planning for construction site layout. Energies. **12**, 4620 (2019). https://doi.org/10.3390/en12244620
44. D. Husain, R. Prakash, Ecological footprint reduction of built envelope in India. J. Building Eng. **21**, 278–286 (2019). https://doi.org/10.1016/j.jobe.2018.10.018
45. openLCA Nexus: The source for LCA data sets, https://nexus.openlca.org/
46. L. Hermann, F. Kraus, R. Hermann, Phosphorus Processing—potentials for higher efficiency. Sustainability. **10**, 1482 (2018). https://doi.org/10.3390/su10051482
47. C.G. Machado, M.P. Winroth, E.H.D.R. Silva da, Sustainable manufacturing in Industry 4.0: an emerging research agenda. Int. J. Prod. Res. **58**, 1462–1484 (2020). https://doi.org/10.1080/00207543.2019.1652777
48. F. Guarino, M. Cellura, M. Traverso, Costructal law, exergy analysis and life cycle energy sustainability assessment: an expanded framework applied to a boiler. Int. J. Life Cycle Assess. (2020). https://doi.org/10.1007/s11367-020-01779-9

49. Life Cycle Assessment in the minerals industry, Current practice, harmonization efforts, and potential improvement through the integration with process simulation. J. Cleaner Prod. **232**, 174–192 (2019). https://doi.org/10.1016/j.jclepro.2019.05.318
50. P. Stasinopoulos, N. Shiwakoti, M. Beining, Use-stage life cycle greenhouse gas emissions of the transition to an autonomous vehicle fleet: a system dynamics approach. J. Cleaner Prod. 123447 (2020). https://doi.org/10.1016/j.jclepro.2020.123447
51. L.F. Morales-Mendoza, C. Azzaro-Pantel, Bridging LCA data gaps by use of process simulation for energy generation. Clean Techn. Environ. Policy. **19**, 1535–1546 (2017). https://doi.org/10.1007/s10098-017-1349-6
52. C. Brondi, E. Carpanzano, A modular framework for the LCA-based simulation of production systems. CIRP J. Manuf. Sci. Technol. **4**, 305–312 (2011). https://doi.org/10.1016/j.cirpj.2011.06.006
53. G.M. Zanghelini, E. Cherubini, S.R. Soares, How Multi-criteria decision analysis (MCDA) is aiding life cycle assessment (LCA) in results interpretation. J. Cleaner Prod. **172**, 609–622 (2018). https://doi.org/10.1016/j.jclepro.2017.10.230
54. M.J. Hermoso-Orzáez, J.A. Lozano-Miralles, R. Lopez-Garcia, P. Brito, Environmental criteria for assessing the competitiveness of public tenders with the replacement of large-scale LEDs in the outdoor lighting of cities as a key element for sustainable development: case study applied with Promethee methodology. Sustainability. **11**, 5982 (2019). https://doi.org/10.3390/su11215982
55. M. Budzinski, M. Sisca, D. Thrän, Consequential LCA and LCC using linear programming: an illustrative example of biorefineries. Int. J. Life Cycle Assess. **24**, 2191–2205 (2019). https://doi.org/10.1007/s11367-019-01650-6
56. K. Allacker, V. Castellani, G. Baldinelli, F. Bianchi, C. Baldassarri, S. Sala, Energy simulation and LCA for macro-scale analysis of eco-innovations in the housing stock. Int. J. Life Cycle Assess. **24**, 989–1008 (2019). https://doi.org/10.1007/s11367-018-1548-3
57. S. Kim, G.-H. Kim, Y.-D. Lee, Sustainability life cycle cost analysis of roof waterproofing methods considering $LCCO_2$. Sustainability. **6**, 158–174 (2014). https://doi.org/10.3390/su6010158
58. B. Löfgren, A.-M. Tillman, Relating manufacturing system configuration to life-cycle environmental performance: discrete-event simulation supplemented with LCA. J. Cleaner Prod. **19**, 2015–2024 (2011). https://doi.org/10.1016/j.jclepro.2011.07.014
59. R. Geyer, D.M. Stoms, J.P. Lindner, F.W. Davis, B. Wittstock, Coupling GIS and LCA for biodiversity assessments of land use. Int. J. Life Cycle Assess. **15**, 454–467 (2010). https://doi.org/10.1007/s11367-010-0170-9
60. N. Perry, J. Garcia, Sustainable Design of Complex Systems, Products and Services with User Integration into Design, in *Designing Sustainable Technologies, Products and Policies: From Science to Innovation.* ed. by E. Benetto, K. Gericke, M. Guiton (Springer International Publishing, Cham, 2018), pp. 365–369
61. S. Payen, C. Basset-Mens, F. Colin, P. Roignant, Inventory of field water flows for agri-food LCA: critical review and recommendations of modelling options. Int. J. Life Cycle Assess. **23**, 1331–1350 (2018). https://doi.org/10.1007/s11367-017-1353-4
62. C.-Y. Baek, K. Tahara, K.-H. Park, Parameter uncertainty analysis of the life cycle inventory database: application to greenhouse gas emissions from brown rice production in IDEA. Sustainability. **10**, 922 (2018). https://doi.org/10.3390/su10040922
63. M. Ziyadi, I.L. Al-Qadi, Model uncertainty analysis using data analytics for life-cycle assessment (LCA) applications. Int. J. Life Cycle Assess. **24**, 945–959 (2019). https://doi.org/10.1007/s11367-018-1528-7
64. K. Tokimatsu, L. Tang, R. Yasuoka, R. Ii, N. Itsubo, M. Nishio, Toward more comprehensive environmental impact assessments: interlinked global models of LCIA and IAM applicable to this century. Int. J. Life Cycle Assess. (2020). https://doi.org/10.1007/s11367-020-01750-8
65. M.R. Giraldi-Díaz, L. De Medina-Salas, E. Castillo-González, R. León-Lira, Environmental impact associated with the supply chain and production of grounding and roasting coffee through life cycle analysis. Sustainability. **10**, 4598 (2018). https://doi.org/10.3390/su10124598

66. P. Kerdlap, J.S.C. Low, S. Ramakrishna, Life cycle environmental and economic assessment of industrial symbiosis networks: a review of the past decade of models and computational methods through a multi-level analysis lens. Int. J. Life Cycle Assess. (2020). https://doi.org/10.1007/s11367-020-01792-y
67. T. Schaubroeck, Both completing system boundaries and realistic modeling of the economy are of interest for life cycle assessment—a reply to "Moving from completing system boundaries to more realistic modeling of the economy in life cycle assessment" by Yang and Heijungs (2018). Int. J. Life Cycle Assess. **24**, 219–222 (2019). https://doi.org/10.1007/s11367-018-1546-5
68. Y. Leroy, D. Froelich, Qualitative and quantitative approaches dealing with uncertainty in life cycle assessment (LCA) of complex systems: towards a selective integration of uncertainty according to LCA objectives. Int. J. Design Eng. **3**, 151–171 (2010). https://doi.org/10.1504/IJDE.2010.034862
69. V. Bellon-Maurel, M.D. Short, P. Roux, M. Schulz, G.M. Peters, Streamlining life cycle inventory data generation in agriculture using traceability data and information and communication technologies—part I: concepts and technical basis. J. Cleaner Prod. **69**, 60–66 (2014). https://doi.org/10.1016/j.jclepro.2014.01.079
70. I.T. Herrmann, A. Jørgensen, S. Bruun, M.Z. Hauschild, Potential for optimized production and use of rapeseed biodiesel. Based on a comprehensive real-time LCA case study in Denmark with multiple pathways. Int. J. Life Cycle Assess. **18**, 418–430 (2013). https://doi.org/10.1007/s11367-012-0486-8
71. V.E. de Oliveira Gomes, D.J. De Barba, J. de Oliveira Gomes, K.-H. Grote, C. Beyer, *Sustainable Layout Planning Requirements by Integration of Discrete Event Simulation Analysis (DES) with Life Cycle Assessment (LCA).* in Advances in Production Management Systems. Competitive Manufacturing for Innovative Products and Services, ed. by C. Emmanouilidis, M. Taisch, D. Kiritsis (Springer, Berlin, Heidelberg, 2013) pp. 232–239
72. T. Henriksen, J.W. Levis, M.A. Barlaz, A. Damgaard, Approaches to fill data gaps and evaluate process completeness in LCA—perspectives from solid waste management systems. Int. J. Life Cycle Assess. **24**, 1587–1601 (2019). https://doi.org/10.1007/s11367-019-01592-z
73. S. Schwarzinger, D.N. Bird, T.M. Skjølsvold, Identifying consumer lifestyles through their energy impacts: transforming social science data into policy-relevant group-level knowledge. Sustainability. **11**, 6162 (2019). https://doi.org/10.3390/su11216162
74. J. Pohl, L.M. Hilty, M. Finkbeiner, How LCA contributes to the environmental assessment of higher order effects of ICT application: A review of different approaches. J. Cleaner Prod. **219**, 698–712 (2019). https://doi.org/10.1016/j.jclepro.2019.02.018
75. A.N. Azimi, S.M.R. Dente, S. Hashimoto, Social Life-cycle assessment of household waste management system in Kabul City. Sustainability. **12**, 3217 (2020). https://doi.org/10.3390/su12083217
76. Z. Jin, J. Kim, C. Hyun, S. Han, Development of a model for predicting probabilistic life-cycle cost for the early stage of public-office construction. Sustainability. **11**, 3828 (2019). https://doi.org/10.3390/su11143828
77. M. Zimek, A. Schober, C. Mair, R.J. Baumgartner, T. Stern, M. Füllsack, The third wave of LCA as the "decade of consolidation." Sustainability. **11**, 3283 (2019). https://doi.org/10.3390/su11123283
78. A.M. Herrera Almanza, B. Corona, Using social life cycle assessment to analyze the contribution of products to the sustainable development goals: a case study in the textile sector. Int. J. Life Cycle Assess. (2020). https://doi.org/10.1007/s11367-020-01789-7
79. S. O'Keeffe, D. Thrän, Energy crops in regional biogas systems: an integrative spatial LCA to assess the influence of crop mix and location on cultivation GHG emissions. Sustainability. **12**, 237 (2020). https://doi.org/10.3390/su12010237
80. R. Kc, M. Aalto, O.-J. Korpinen, T. Ranta, S. Proskurina, Lifecycle assessment of biomass supply chain with the assistance of agent-based modelling. Sustainability. **12**, 1964 (2020). https://doi.org/10.3390/su12051964

81. H.E. Otto, K.G. Mueller, F. Kimura, Efficient information visualization in LCA. Int. J. LCA. **8**, 183 (2003). https://doi.org/10.1007/BF02978468
82. H.E. Otto, K.G. Mueller, F. Kimura, Efficient information visualization in LCA: Application and practice. Int. J. LCA. **9**, 2 (2004). https://doi.org/10.1007/BF02978531
83. openLCA Nexus: The source for LCA data sets, https://nexus.openlca.org/databases
84. G. Sonnemann, B. Vigon, C. Broadbent, M.A. Curran, M. Finkbeiner, R. Frischknecht, A. Inaba, A. Schanssema, M. Stevenson, C.M.L. Ugaya, H. Wang, M.-A. Wolf, S. Valdivia, Process on "global guidance for LCA databases." Int. J. Life Cycle Assess. **16**, 95–97 (2011). https://doi.org/10.1007/s11367-010-0243-9
85. J. Liu, Z. Huang, X. Wang, Economic and environmental assessment of carbon emissions from demolition waste based on LCA and LCC. Sustainability. **12**, 6683 (2020). https://doi.org/10.3390/su12166683
86. M.L. Kambanou, Life cycle costing: understanding how it is practised and its relationship to life cycle management—a case study. Sustainability. **12**, 3252 (2020). https://doi.org/10.3390/su12083252
87. R. Baum, J. Bieńkowski, Eco-efficiency in measuring the sustainable production of agricultural crops. Sustainability. **12**, 1418 (2020). https://doi.org/10.3390/su12041418
88. J. Sanfélix, F. Mathieux, C. de la Rúa, M.-A. Wolf, K. Chomkhamsri, The enhanced lca resources directory: a tool aimed at improving life cycle thinking practices. Int. J. Life Cycle Assess. **18**, 273–277 (2013). https://doi.org/10.1007/s11367-012-0468-x
89. A. Siebert, A. Bezama, S. O'Keeffe, D. Thrän, Social life cycle assessment indices and indicators to monitor the social implications of wood-based products. J. Cleaner Prod. **172**, 4074–4084 (2018). https://doi.org/10.1016/j.jclepro.2017.02.146
90. L. Jarosch, W. Zeug, A. Bezama, M. Finkbeiner, D. Thrän, A regional socio-economic life cycle assessment of a bioeconomy value chain. Sustainability. **12**, 1259 (2020). https://doi.org/10.3390/su12031259
91. J. Veselka, M. Nehasilová, K. Dvořáková, P. Ryklová, M. Volf, J. Růžička, A. Lupíšek, Recommendations for developing a BIM for the purpose of LCA in green building certifications. Sustainability. **12**, 6151 (2020). https://doi.org/10.3390/su12156151
92. J. Sherry, J. Koester, Life cycle assessment of aquaculture stewardship council certified atlantic Salmon (Salmo salar). Sustainability. **12**, 6079 (2020). https://doi.org/10.3390/su12156079
93. L.F. Morales-Mendoza, C. Azzaro-Pantel, J.-P. Belaud, A. Ouattara, Coupling life cycle assessment with process simulation for ecodesign of chemical processes. Environ. Progress Sustainable Energy. **37**, 777–796 (2018). https://doi.org/10.1002/ep.12723
94. R. Gaha, A. Benamara, B. Yannou, *Proposition of Eco-Feature: A New CAD/PLM Data Model for an LCA Tool*. in CMSM 2017: The Seventh International Congress Design and Modelling of Mechanical Systems (Hammamet, Tunisia, 2017)
95. M. Grieves, J. Vickers, Digital Twin: Mitigating Unpredictable, Undesirable Emergent Behavior in Complex Systems, in *Transdisciplinary Perspectives on Complex Systems*. ed. by F.-J. Kahlen, S. Flumerfelt, A. Alves (Springer International Publishing, Cham, 2017), pp. 85–113
96. Y. Liu, L. Zhang, Y. Yang, L. Zhou, L. Ren, F. Wang, R. Liu, Z. Pang, M.J. Deen, A novel cloud-based framework for the elderly healthcare services using digital twin. IEEE Access. **7**, 49088–49101 (2019). https://doi.org/10.1109/ACCESS.2019.2909828
97. 14:00–17:00: ISO/CD 23247-1, https://www.iso.org/cms/render/live/en/sites/isoorg/contents/data/standard/07/50/75066.html
98. F. Coallier, ISO/IEC JTC 1/SC41 IoT and Related Technologies. 46
99. The Structure of the Administration Shell: Trilateral Perspectives from France, Italy and Germany. 64
100. E.J. Tuegel, P. Kobryn, J.V. Zweber, R.M. Kolonay, *Digital Thread and Twin for Systems Engineering: Design to Retirement*. in 55th AIAA Aerospace Sciences Meeting (American Institute of Aeronautics and Astronautics, Grapevine, Texas, 2017)
101. Digital twins • Eclipse Ditto • a digital twin framework, https://www.eclipse.org/ditto/intro-digitaltwins.html

102. B. He, K.-J. Bai, Digital twin-based sustainable intelligent manufacturing: a review. Adv. Manuf. (2020). https://doi.org/10.1007/s40436-020-00302-5
103. J. Jiao, (Roger), F. Zhou, N.Z. Gebraeel, V. Duffy, Towards augmenting cyber-physical-human collaborative cognition for human-automation interaction in complex manufacturing and operational environments. Int. J. Prod. Res. **0**, 1–23 (2020). https://doi.org/10.1080/00207543.2020.1722324
104. M. Ghita, B. Siham, M. Hicham, A. Abdelhafid, D. Laurent, Digital twins: development and implementation challenges within Moroccan context. SN Appl. Sci. **2**, 885 (2020). https://doi.org/10.1007/s42452-020-2691-6
105. M. Ghita, B. Siham, M. Hicham, A.E.M. Abdelhafid, D. Laurent, *Geospatial Business Intelligence and Cloud Services for Context Aware Digital Twins Development*. in 2020 IEEE International conference of Moroccan Geomatics (Morgeo) (2020), pp. 1–6

# Integrated Smart System for Robotic Assisted Living

**Marius Pandelea, Isabela Todiriţe, Corina Radu Frenţ, Luige Vlădăreanu, and Mihaiela Iliescu**

**Abstract** The life of people with disabilities, health problems, and of the elderly ones could be improved by robotic systems integrated within virtual, intelligent, and portable platforms. The concept of the integrated smart system presented by this paper is focused on three major research directions that are: safe navigation (trajectory and stability) of anthropomorphic walking robots for customized service; voice-assisted implementation of the operating system by the visually impaired people (customized for the Romanian language); project and prototype design of biomechanical upper limb prosthesis. The proposed system architecture is fit for people with social and medical problems, standing as significant support for their active and assisted life. Advantages of this new integrated system, over others similar available on the market, consist in its performance and characteristics, complexity and connection of the component subsystems, accessibility and, not the least, relatively affordable price.

**Keywords** Robotic assisted living · Anthropomorphic walking robots · Control · Integrated smart system · Intelligent assistance · Hand prosthesis

M. Pandelea · I. Todiriţe · C. Radu Frenţ · L. Vlădăreanu · M. Iliescu (✉)
Department of Mechatronics and Robotics, Institute of Solid Mechanics of the Romanian Academy, Bucharest, Romania
e-mail: mihaiela.iliescu@imsar.ro

M. Pandelea
e-mail: marius.pandelea@imsar.ro

I. Todiriţe
e-mail: todirite.isabela@imsar.ro

C. Radu Frenţ
e-mail: corinagabrielaradu@gmail.com

L. Vlădăreanu
e-mail: luige.vladareanu@vipro.edu.ro

 
A. Joshi et al. (eds.), *Sustainable Intelligent Systems*, Advances in Sustainability Science and Technology, https://doi.org/10.1007/978-981-33-4901-8_9

# 1 Introduction

The excessive, but natural development of applications and equipment supported by artificial intelligence and robotics is based on the potential of the concepts, which represent opportunities and offer consistent help in many fields of activity.

In order to have a normal social life, safety, comfort and medical care, people with health problems need dedicated platforms and smart devices. The personalized interactions of the robots with the patients represent a work that ensures the monitoring of the arterial pressure [1], medication control and reminders [1], the avoidance of obstacles, and planning of daily activities [1, 15, 16].

The integrated smart system (ISS) concept, presented in this paper, fulfills needs of people with social and/or medical problems and focuses on three main research directions:

- locating obstacles and objects in crowded dynamic spaces, navigation, and performance stability control of anthropomorphic walking robots (AWRs) for customized assistance services;
- development and implementation on PC of a personalized operating system in Romanian language, with voice command and communication function, for people with visual impairments;
- design and prototype of a biomechanical prosthesis for upper limb, which will be used by a young teenager from the social assistance system of Romania.

The links and interaction of the main components of the proposed intelligent robotic assisted living system, within the smart mechatronic platforms (AWRs or prosthetic devices) are presented in Fig. 1.

The novelty of research presented by this paper consists in the concept of interaction between people (people with special needs), smart mechatronic systems (AWRs, upper limb prosthesis), and dedicated operating system, all in a versatile intelligent platform, aided by specific sensors signals and/or voice assistance (in Romanian language).

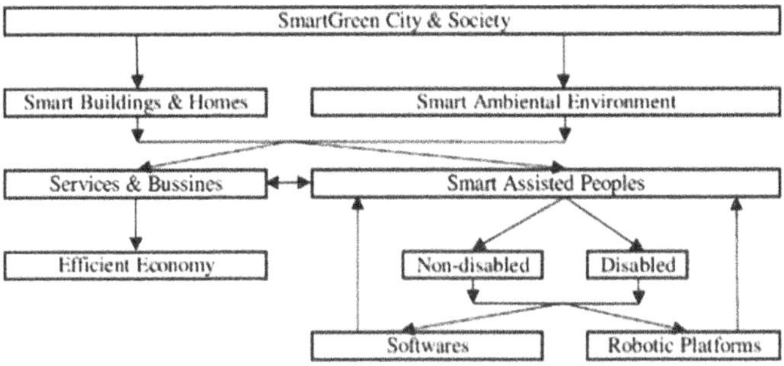


**Fig. 1** Robotic assisted living block scheme

Buildings, houses, and infrastructure also need intelligent processes, technologies, and equipment. As shown in Fig. 1, robotic assisted living is a concept achievable in the twenty-first century, by interconnecting smart components, software, devices, and robotic platforms in a smart environment, together with humans. More and more researches published recently focus on the concepts of active and assisted living (AAL) [3, 6, 14], socially assistive robotics (SAR) [1], or ambient-assisted living (AAL) [4]. These concepts are sustainable only through research and development of smart models, algorithms, and methods integrated into the designed systems, as demonstrated in the articles mentioned [7] propose a four-layer architecture for the assisted life of the elderly or of those with health problems, the mapping of the interior space being performed by a wheeled robot.

Visually impaired people, and not only them, have an acute need for social assistance easily achievable through dedicated equipment and applications through integrated robotic systems within virtual or portable platforms [10]. The integration of AWRs in smart systems used for human assistance in the cities is a topic often addressed [11].

Regarding the decisions control within the component of smart devices and robotic platforms, the most common are: PID control, fuzzy control [10, 14], expert systems control, neural networks control [8, 13], hybrid control, adaptive control, predictive control, but also cognitive control [1], neutrosophic control, and extenics control. It is also noted that many of the ISS are based on robots. The constructive types of robots encountered in tests and experiments in the real environment are either bipedal [5, 10], either wheeled [7] or multi-agents [3].

The learning process attached to smart systems is a technologically controlled overlap of previous knowledge, information, and experiences regarding objects and phenomena combined with soft and hard elements, so that the real surrounding systems and their environment can be understood [1, 12].

The diversity, performances, and types of sensors provide the perception system with precision and safety, so that current researches always study the sensors, as found in [2, 5, 9].

Researches [2, 5] use the simultaneous localization and mapping (SLAM) method for navigation, offering numerous variants and solutions, so that the location of the robot can be found precisely on the map of the unknown area that must be permanently generated and updated.

Current researches also show that Telehealth or Telemedicine [3, 6] are viewed with great confidence in order to solve the medical problems of long-distance patients.

People who suffer from the fact that they cannot use certain parts of their body (limbs) are in the attention of researchers [9, 14], who offer the solution of physical recovery or, as the case may be, wearing a prosthesis.

The structure of this article is as follows: Sect. 2 includes research on stability control and AWR navigation, with additional laser sensor, Sect. 3 shows the Romanian language customization of a PC operating system for visually impaired people, and Sect. 4 focuses on the design and the prototype of a human upper limb biomechanical prosthesis. Section 5 provides an overview of the research directions mentioned above and their integration within versatile platform, and Sects. 6 and 7 refer to discussions and, respectively, conclusions.

## 2 Safe Navigation with AWRs

The control represents the way in which the set of parameters that define a certain state of a mechanism or of the environment can be adjusted sequentially and admissibly. The control subsystem is a basic mean that acts continuously dynamically through corrective or proactive intervention measures imposed by intelligent control algorithms, coordinating processes, and actions in the context in which the system has not yet reached the proposed objective.

Determining a parameter of the subsystem to change in a certain way to ensure the purpose according to a well-defined and chosen strategy, or according to the knowledge acquired, or existing in the database, is part of the control. Modifying all parameters during the actions facilitates obtaining the desired results; the waiver of the action being similar to the achievement of the objective only after the correct evaluation of the final state. Control and decision operate through the analysis and synthesis of environmental information obtained by sensors.

The control scheme of the interaction between the disadvantaged person and the AWR from Fig. 2 shows that the processes and actions of the two parties are different, being done in different ways and with different resources; their interconnection being made in the real environment and by means of databases.

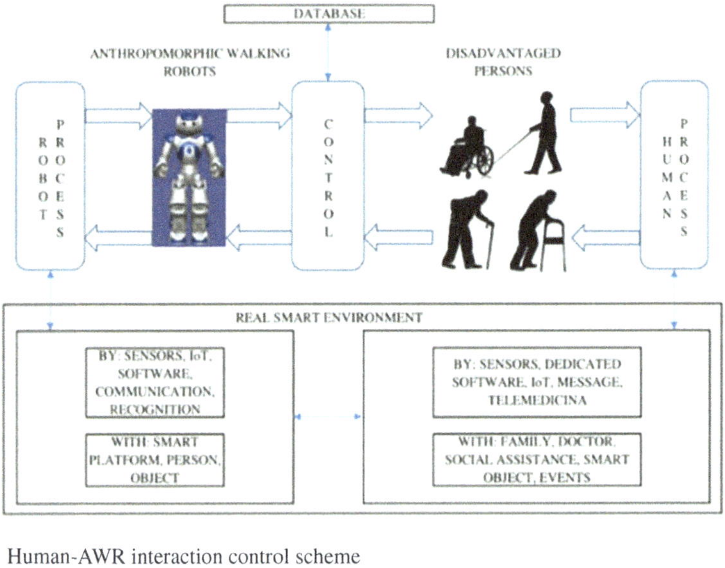


**Fig. 2** Human-AWR interaction control scheme

The flow of controlled processes between smart platforms (AWRs or prosthetic devices) and socially disadvantaged people or people with health problems, as shown in Fig. 3, considers navigation, communication, planning, prediction, monitoring, decision, observation, detection, surveillance, needs, movement, location, bypassing obstacles, recognition, and learning.

In Fig. 4, it is presented that the proposed scheme for the perception subsystem of AWR in which an additional Sharp GP2Y sensor has been integrated which can ensure much better qualitative images, with more precise determination of shapes and the distances between the robot and the object.

The operating mode of the laser sensor is as follows: it is attached to the walking robot arm and by swinging the arm horizontally and vertically, information on shapes and distances is obtained. In Fig. 5, the results of the obtained images are presented compared to the real ones for a school furniture body and for the school staircase. The scanned images have defining lines for the shape, and their resolution being poor due to improper calibration.

The research will study the influence of an additional sensor on locating obstacles and identifying the shapes of objects in crowded dynamic spaces, such as school

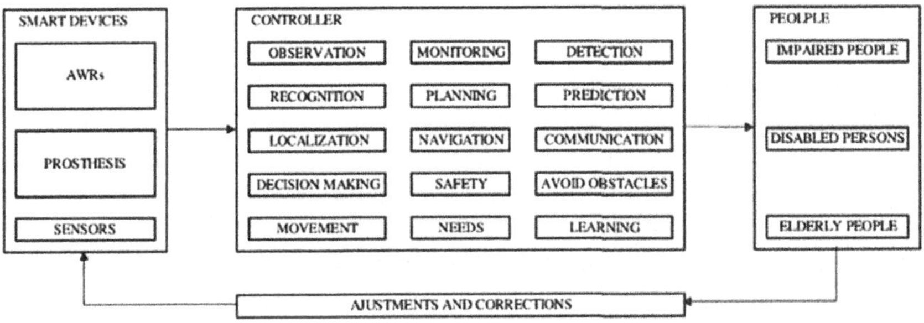


**Fig. 3** Human–smart device interaction control scheme

**Fig. 4** AWRs perception subsystem block scheme

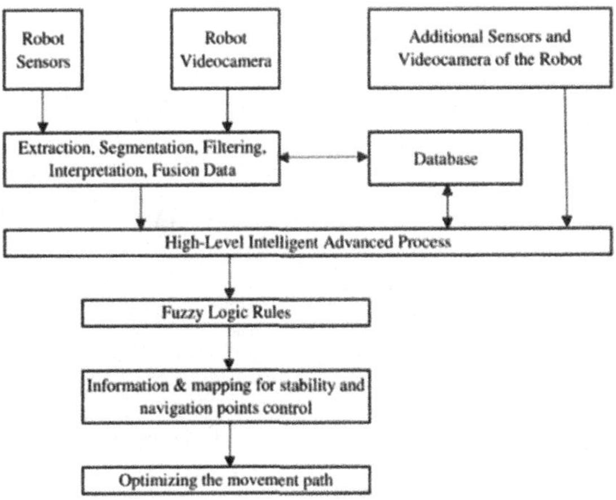

**Fig. 5** Real image and the scanned image of class furniture and staircase

classrooms, and safe navigation of AWRs in order to provide personalized assistance services.

## 3 Dedicated Integrated System

The integrated operating system, presented by this paper, has a modular design with customized modules depending on the people needs, as they have various needs in order to perform activities of a normal life. For example, the impaired people need assistance for orientation in a closed space and specific tools for training and learning all through the education process. These needs are specific to the educational cycle, meaning primary school, gymnasium, high school, and university.

The integrated system developed is based on GNU/Linux operating system, VINUX distribution (Fig. 6) and is aimed to people with visual and/or writing and reading deficiencies. This system can be used by the visually impaired native Romanians or, by the people who want to improve their Romanian knowledge, for the purpose of introduction to computer use. If these people are already at a medium, or

**Fig. 6** Screenshot desktop system

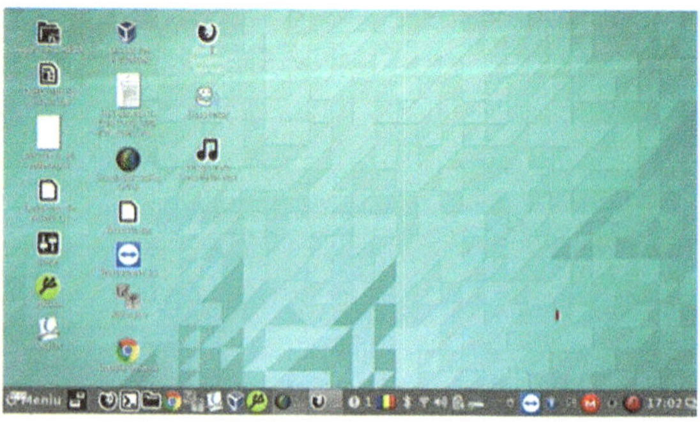

advanced level, the system could help them with professional development, internet navigation, find a good job and/or training in robotics, etc.

By its integrated components, TTS (text to speech), the visually impaired people can use the text editor with an implemented tool for spelling and grammar, with audible warning of misspelling by voice: "ortografiere greșită/wrong spelling" (Fig. 7).

The tools TTS (text to speech) and STT (speech to text) are integrated in the system, so that each of its application are assisted by them, with no need of activation or deactivation—as in the case of other similar but paid applications.

The Gespeaker application (Fig. 8) enables registration of the text sequences while they are written and, further after saving, the user can resume recording and listen to them, repeating after the voice that spells the text.

The integrated system developed is open source (Fig. 9), so that the users could get support from the Linux and Ubuntu communities.

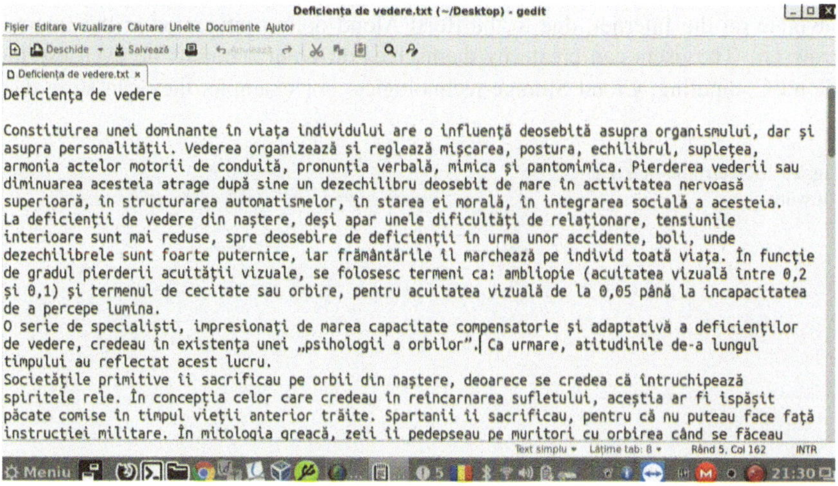


**Fig. 7** Screenshot text editor

**Fig. 8** Screenshot Gespeaker software

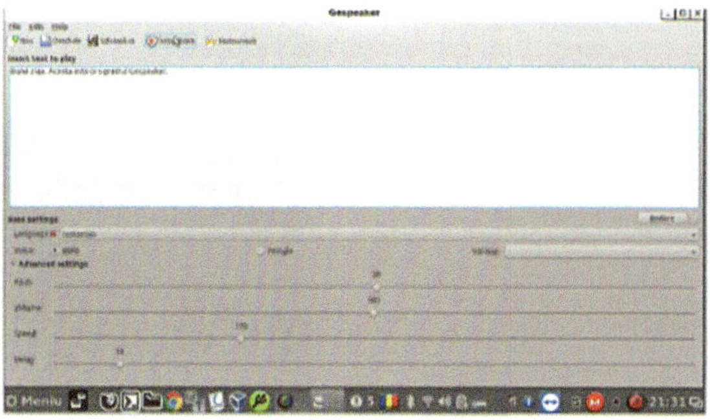

**Fig. 9** Screenshot—Ubuntu software center

The system offers the conversion from one format to the other, useful formats for visually impaired people, such as: daisy, epub, and pdf. (Fig. 10).

The system's interface is very similar to Windows, in order to help the beginners who are most familiar with the Windows operating system (Fig. 6).

This integrated system has tools for real-time remote communication, so that any user can be voice assisted and helped when it is required. The person can use the TeamViewer application with voice assistance—by indicating the ID and password (Fig. 11).

Another feature of the developed integrated system is that it enables the user to navigate on the Internet, due to the Read Aloud option integrated in the Internet browsers. The users can create by themselves different accounts or, even use the Cloud Computing, Cloud Storage technologies. For example, they can store data

**Fig. 10** Screenshot—Okular software

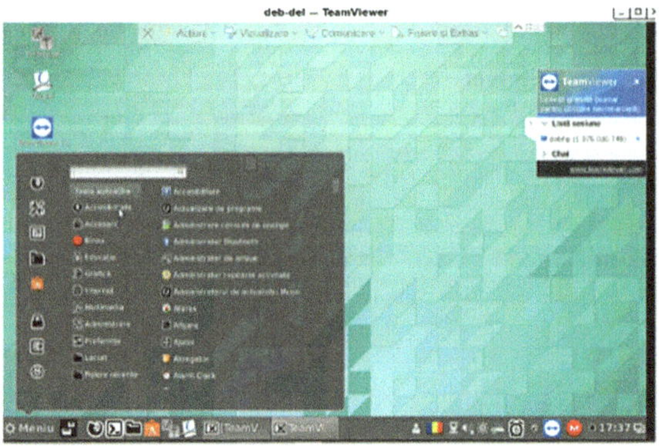


**Fig. 11** Remote assistance TeamViewer

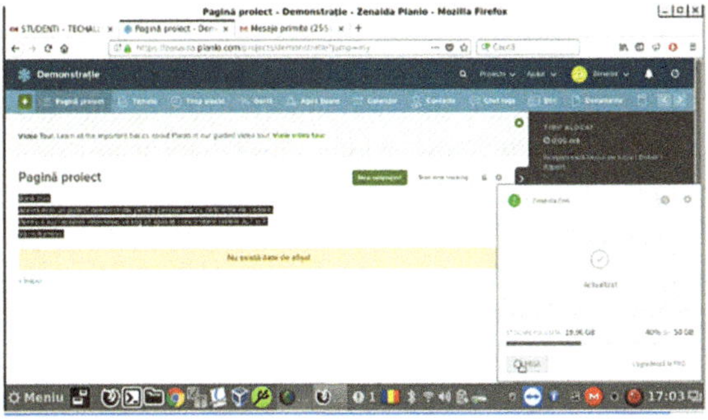


**Fig. 12** Screenshot—cloud computing and cloud storage

and access them later from any other devices like the mobile phone—when they do not have access to the computer (Fig. 12).

The concept of this integrated dedicated system is that it can be configured depending on the user needs. For a beginner user, who wants just to get minimum knowledge and abilities for operating the computer, then he/she could choose the management for desktop (Unity interface or Cinnamon—which looks similar to the Windows desktop), with/without the Write Type application for learning the keyboard, with applications that recognize specific formats for the visually impaired or, a standard configuration offered to users who want to get a medium level in operating the computer.

For users with a good level of operating, they will be offered an extended version of this system, for example, a version for programming the ARM microcontroller (Fig. 13). This configuration would help the visually impaired people to find a job in a team that develops application for microcontrollers and, thus, they would take the chance for social integration and to a relatively normal life.

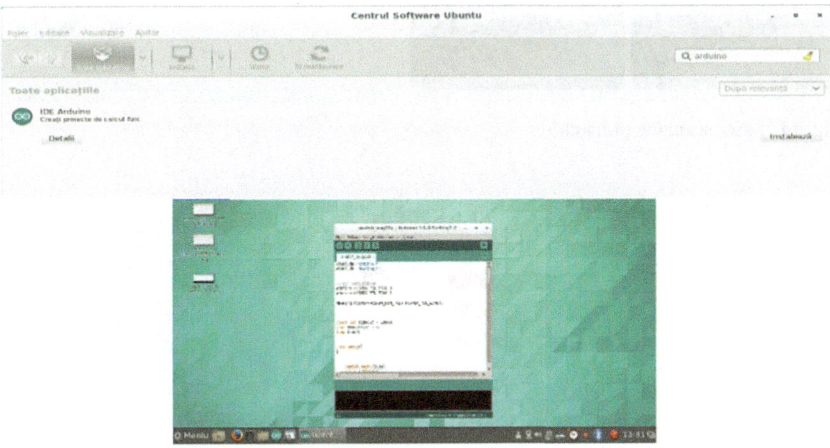


**Fig. 13** Screenshots IDE Arduino—installation and usage

## 4 Hand Prosthesis

The prostheses are aimed to improve people life, when from different reasons, they cannot use their limbs. Along time, there were so many attempts to improve these people's life, meaning design and make hand and foot prostheses.

The concept and design of a customized hand prosthesis aimed to improve the life quality of a teenager in the social protection system (of Romania) have been developed. The research was aimed to have an affordable price prosthesis, with good functionality, relatively low weight and, not the least, user friendly.

In order to achieve all the above mentioned, there has been used reverse engineering technique (Fig. 14) so that to get the dimensions of that teenager real hand. It is intended that the dimensions and look of prosthesis to be as much as possible alike to the real hand ones.

There were several variants of the prosthesis 3D model, depending on the mechanical system's components and on the micromotors and sensors intended to be used. One of the latest ones is presented in Fig. 15, where one can notice the pressure sensors on the fingertips.

In order to perform its intended tasks, there are used many sensors types, as follows: EMG sensors—for detection of muscles electric signals; pressure sensors—for gripping; acceleration and gyration sensors—for motion and positioning.

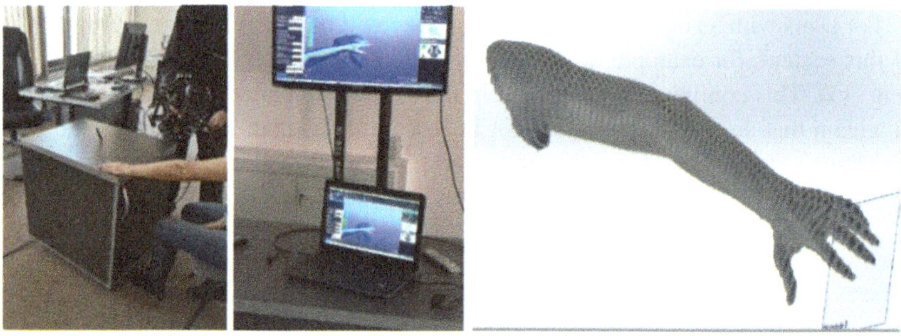

**Fig. 14** Laser scanning of hand

**Fig. 15** Hand prostheses design

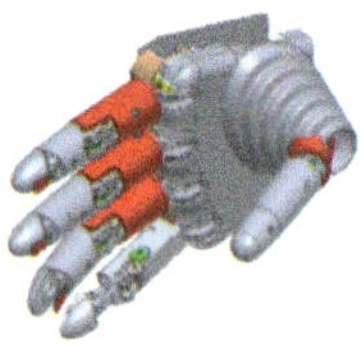

3D model

Index finger prototype - with pressure sensor

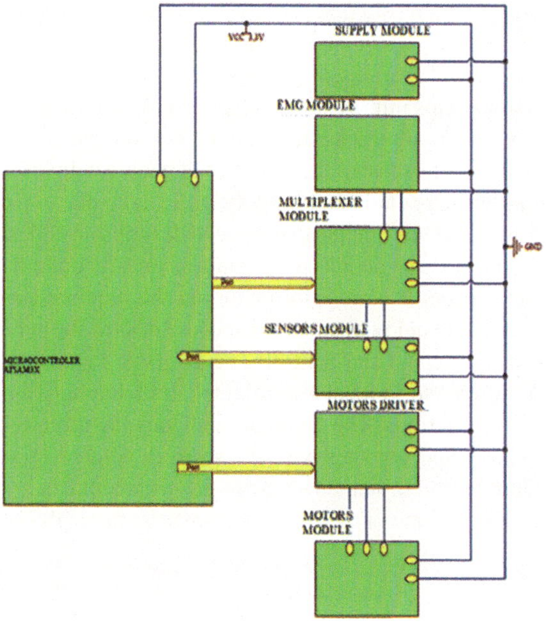


**Fig. 16** Command and control scheme

The command and control scheme for the customized hand prosthesis is shown in Fig. 16.

It can be noticed the ATSAMIX controller (with ARM microprocessor) that can be programmed, also, by a visually impaired person if, previously has attended courses (by the operating system developed and described in Sect. Sect.). Mainly, the basic principles learned while assisted training by this system are useful and stand as the background for new challenges in programming microcontrollers.

Further development of the integrated system will consist in attaching IoT modules, so that the person who has the hand prosthesis to be able get real-time information (audible, too) on the objects that are going to be touched and manipulated so that to be aware of their reality.

## 5 Integrated Smart System Concept

The proposed concept for robotic assisted life is a sustainable one and integrates three architectural directions for intelligent assistance. Even it seems that these directions are bounded, when fused in a complex integrated system, it results in useful and important part of the everyday life for visually impaired, elderly, and disabled people. Even it is the AWR, or the hand prosthesis, or the operating system, all are intended to support the activities of the people in need.

As far as we studied, there is not yet presented this kind of integrated smart system, that enable voice assistance in guidance, motion, therapy, teaching, and training of people with various needs (disability, elderly, visual impairment). The developed system helps improving, and they live by active assisted motion and travel, training, team work, not the least, social integration, etc.

Basically, most of the activities of people with disabilities, elderly, and with health problems are static and little dynamic, take place in close, or open space, in free or crowded environments that do need description and, many times, warning. The interaction of these people with smart devices is done through bidirectional interfaces. The sensors acquire information on the required parameters (obstacles, land type, heat, cold, and wet) and send specific data to the central unit for management, decision, command and control. The available resources, ongoing events, and risks are analyzed and in a balanced mode, all including people safety and environmentally friendly actions. The data base includes the three robotic laws mentioned by Isaac Asimov and together with the services offered by cloud computing represent the central point of human—robot—environment triad.

The scheme of the integrated smart system (ISS) developed is shown in Fig. 17.

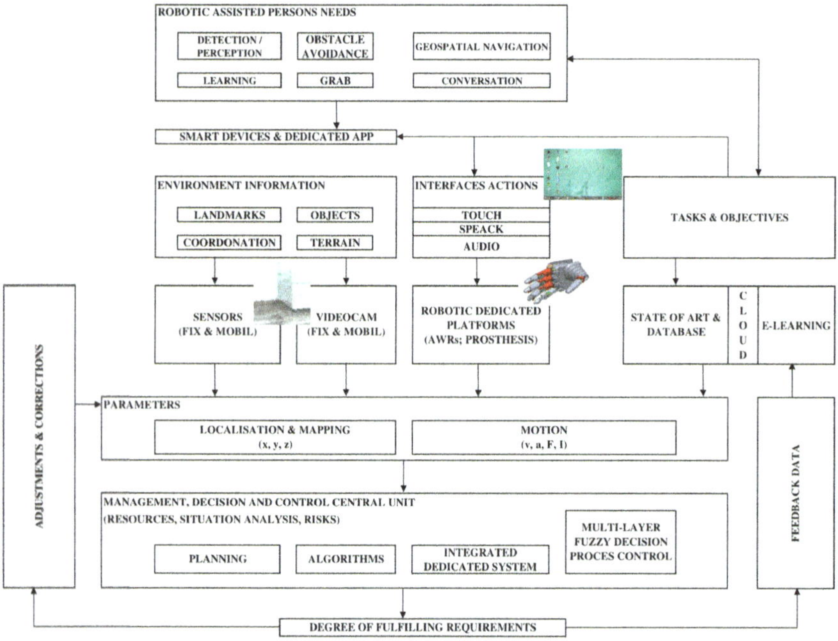


**Fig. 17** Integrated smart system

## 6 Discussion

Basically, the ISS proposed system is new, as it integrates by virtual platform, components (AWRs, mechatronic system—hand prosthesis and dedicated IT system) designed for the customized needs of people with different disabilities. Experiments with the high-performance laser sensor attached to the AWR arm and preliminary tests with the management for desktop (Unity interface or Cinnamon), with the Write Type application for learning the keyboard, proved good results.

The AWR and hand prosthesis are "intelligent" devices, so that their systems have to access data implemented by cloud storage platforms. The algorithms for fast data selection have to be optimized so that to enable fast response and execution time of the mechatronic structures, as well as accurate data set selection.

The dedicated integrated system has some limitations caused by the lack of standardization for webpage display. This is caused by the fact that when the user does access the webpage with advertising banner, with no standard dimensions or fixed position, the screen reader software, or the read aloud browser option, cannot distinguish the advertising from the page text. This causes discomfort and tiredness, not to mention the repetition of advertising at different pages.

## 7 Conclusion

The research presented by this study is aimed to integrate different types of smart equipment into the life of elderly, visually impaired, and people with disabilities. The people need and the customized device's signals are analyzed by control algorithm so that are assisted in the activities that help for, relatively, normal life. Improving the life quality by robotic assisted living is based on the interaction of people reactions, living environment signals, and equipment performances. One important issue is, also, to train the people so that to be willing and able to use the high-performance technologies of assisted living.

The architecture of the integrated smart system developed is based on three major directions that are: customized operating system for the visually impaired people to use the computer; safe navigation of AWRs for guidance or service; customized hand prosthesis with reliable, accurate motions, and affordable price (for a teenager in the social protection system). The developed system is complex, reliable, user friendly, and of high benefit for the life assistance of people with various health problems.

Further development of this research would involve testing and validation of the smart system prototype, software updates, and new modules integration, a smart customized mechatronic system for helping the visually impaired pupils to move within narrow spaces in the classroom.

## References


1. A. Umbrico, A. Cesta, G. Cortellessa et al., A holistic approach to behavior adaptation for socially assistive robots. Int. J. Soc. Robotics **12**, 617–637 (2020). https://doi.org/10.1007/s12369-019-00617-9
2. H. Chen, Z. Yang, X. Zhao, G. Weng, H. Wan, J. Luo, X. Ye, Z. Zhao, Z, He, Y. Shen, S. Schwertfeger, Advanced mapping robot and high-resolution dataset. Robot. Autono. Syst. **131** (2020) 103559. Elsevier B.V. https://doi.org/10.1016/j.robot.2020.103559
3. F. Lanza, V. Seidita, A. Chella, Agents and robots for collaborating and supporting physicians in healthcare scenarios. J. Biomed. Inform. (2020). https://doi.org/10.1016/j.jbi.2020.103483
4. T.E. Foko, N. Dlodlo, L. Montsi, *An Integrated Smart System for Ambient-Assisted Living* (2013). https://doi.org/10.1007/978-3-642-40316-3_12
5. F. Gomez-Donoso, F. Escalona, F.M. Rivas, J.M., Cañas, M. Cazorla, Enhancing the ambient assisted living capabilities with a mobile robot. Comput. Intell. Neurosci (2019)
6. A. Sapci, H. Sapci, Innovative assisted living tools, remote monitoring technologies, artificial intelligence-driven solutions, and robotic systems for aging societies: systematic review. JMIR Aging. **2**, e15429 (2019). https://doi.org/10.2196/15429
7. D. De Silva, J. Roche, X. Shi, A. Kondoz, *IoT Driven Ambient Intelligence Architecture for Indoor Intelligent Mobility* (2018), pp. 451–456. https://doi.org/10.1109/DASC/PiCom/DataCom/CyberSciTec.2018
8. K. Lin, Y. Li, J. Sun, D. Zhou, Q. Zhang, Multi-sensor fusion for body sensor network in medical human-robot interaction scenario. Inf. Fusion. **57**, 1 (2019). https://doi.org/10.1016/j.inffus.2019.11.001
9. S.-W. Pu, J.-Y. Chang, Robotic hand system design for mirror therapy rehabilitation after stroke. Microsyst. Technol. **26**(10), 1007 (2019)
10. M. Pandelea, I. Todirite, M. Iliescu, C*ustomized Assistive System Design for Visually Impaired People*. in Fourth International World Conference on Smart Trends in Systems, Security and Sustainability (WS4 2020). 27th - 28th July 2020 (London, UK, 2020)
11. M. Pandelea, L. Vladareanu, C.F. Radu, M. Iliescu, *Anthropomorphic Walking Robots Integration in Smart Green Systems*. in 8th International Conference on Smart Cities and Green ICT Systems, 3–5 May 2019 (Crete, 2019)
12. I. Todirite, C.F. Radu, M. Iliescu, *IT&C System Solution for Visually Impaired Romanian Teenagers*. in Fifth International Congress on Information and Communication Technology, 20–21 February 2020 (London, UK, 2020)
13. M. Iliescu, V. Vladareanu, M. Şerbănescu, M. Lazăr, Sensor input learning for time-of- flight scan laser. CEAI **19**(2), 51–60 (2017). ISSN 1454–8658
14. C. Radu (Frent), M.M. Rosu, M. Iliescu. *Design and Model of a Prosthesis for Hand*. in ModTech2020 International Conference, Modern Technologies in Industrial Engineering, June 24–27, 2020 (Romania, 2020)
15. N. Dey, A. Mukherjee, *Embedded Systems and Robotics With Open Source Tools* (CRC Press, 2018)
16. A. Mukherjee, N. Dey, Smart Computing With Open Source Platforms (CRC press, 2019)

# Latin American Smart University: Key Factors for a User-Centered Smart Technology Adoption Model

**Dewar Rico-Bautista, César A. Collazos, César D. Guerrero, Gina Maestre-Gongora, and Yurley Medina-Cárdenas**

**Abstract** The quality of administrative, academic, and extension processes is influenced by the management of technology in universities. *Smart technologies* such as *artificial intelligence (AI), cloud computing, the Internet of things (IoT), and big data* continue to emerge with great prominence. Universities must use tools that connect learning with the use of new technologies, bridging gaps with the outside world. Smart education is provided by an educational environment supported by smart technologies and devices. A *smart university* is primarily based on the integration of smart technologies into the educational process. The emergence of this concept allows the application of a large number of components that involve the adaptation of the traditional educational model using these technologies. Many of the ideas of adoption in Latin American universities generate demotivation toward change and do not imply the expected impact in their processes. One way to prevent this situation is to monitor the adoption process itself. The generation of tools that can measure the level of adoption by universities is a necessity as smart technologies are incorporated and become more powerful. Focusing on the users of the processes is an objective of adoption measurement. A User-Centered Smart University Model is

D. Rico-Bautista (✉) · Y. Medina-Cárdenas
Universidad Francisco de Paula Santander Ocaña, Sede el Algodonal Vía Acolsure, 546552 Ocaña, Colombia
e-mail: dwricob@ufpso.edu.co

Y. Medina-Cárdenas
e-mail: ycmedinac@ufpso.edu.co

C. A. Collazos
Universidad del Cauca, Calle 5 No. 4–70, 190003 Popayán, Colombia
e-mail: ccollazo@unicauca.edu.co

C. D. Guerrero
Universidad Autónoma de Bucaramanga, Avenida 42 No. 48–11, 680003 Bucaramanga, Colombia
e-mail: cguerrer@unab.edu.co

G. Maestre-Gongora
Universidad Cooperativa de Colombia, Calle 50 No. 40–74 Bloque A piso 6, 050016 Medellín, Colombia
e-mail: gina.maestre@campusucc.edu.co



A. Joshi et al. (eds.), *Sustainable Intelligent Systems*, Advances in Sustainability Science and Technology, https://doi.org/10.1007/978-981-33-4901-8_10

the goal of this transformation process, established on factors such as an individual's perception, safety and risks, and organizational support.

**Keywords** Adopting smart technology factors · Smart university · Technologies · User-centered

## 1 Introduction

The advent of technology has driven changes in society; today, it is common to see cities transformed into a network of connections that carry, analyze, and bring information from any process that takes place, establishing a relationship with the community, promoting the implementation of activities in an efficient and faster. Technology has become a fundamental tool in the development of daily activities, and over time, it has been necessary to make the application of technology in environments such as cities and organizations [22, 23, 70, 71]. In the current context, "smart technologies" refer to the combination of information and communication technologies (ICT) that include hardware, software, and communication systems capable of acquiring data, analyzing, predicting behavioral trends, and automatically adapting [31, 72]. They play a substantial role in the generation, exchange, dissemination, management, and access to knowledge. They are increasingly important and relevant in all areas of social life, but especially in education [51, 68].

The role of universities has changed as a consequence of the irruption of the technological revolution and globalization, meaning the need to learn how to manage technologies to strengthen their contribution to society. Technologies like big data, cloud computing, Internet of things (IoT), and artificial intelligence (AI), continue to emerge and make great advances [65]. In the results generated at the international level [59], a review of the literature from 1990 to 2017, 97% of the literature focuses on IoT and AI (more specifically, 55% of the literature focuses on AI and 42% of the studies focus on IoT). Additional smart technology, cloud computing, had 3% of the studies.

The transformation from a traditional university to a smart university involves more than using technology; it is about establishing the articulation between smart education, smart classrooms, smart learning, and smart campus when the previous concepts are unified [3, 24, 25]. A smart university is mainly based on the integration of information technologies into the educational process [53]. The appearance of this concept allows the application of a great number of components that imply the adaptation of the traditional educational model using smart information technologies, guaranteeing a collaborative learning environment by layers, fed by data that have a special treatment to generate solutions to the needs raised.

The adoption of technology can occur in some areas, but its incorporation in areas of educational research and innovation aims to create scientific knowledge that improves pedagogical practice through participation or influence in decision-making on the development of teaching–learning [27]. In this particular case, the adoption of technology in the field of learning will be taken into account, a growing issue because more and more institutions are opting for the development and implementation of smart technologies to strengthen and focus their educational processes. The incorporation of technological trends in universities and academic institutions has its reason for being in the needs faced by the community every day, born as a tool that aims to contribute to solving the problems raised.

The article is presented in three (3) parts that are: (i) method, (ii) proposed factors around the user-centered model, and (iii) conclusions about the object of study.

## 2 Method

The university is the main component for the development of the country, technological change and development cause certain modifications in the system, which are worth overcoming to find a smart environment, which contributes to solve problems and provide high-quality services, research as a fundamental pillar must be taught to students to complement the use of new technologies. To obtain a higher level of adoption of smart technologies in universities, the objective of measuring them must be user-centered, knowing for example their perceptions about perceived utility and ease of use [41]. A summary of the process with the results so far is presented in Table 1.

## 3 Characterization of Models by Smart Technology

The documents selected by smart technology associated with the most used adoption models are found in Table 2. For the four smart technologies, the use of the TAM model is highlighted over others such as UTAUT or UTAUT2.

## 4 User-Centered Smart Technology Adoption Model: First Approach

The change from a traditional university to a smart university must be carried out step by step, considering all areas. We could talk about a reengineering process to establish a smart environment [15, 46, 58]. The adoption of technology can occur in some areas, but its incorporation in areas of educational research and innovation aims to create scientific knowledge that improves teaching practice through participation or influence in decision-making on the development of teaching–learning [27]. The inclusion of technological trends in universities and academic institutions has its reason for being in the needs faced by the community every day, born as a tool that

**Table 1** Summary of the methodology

| Phase | Summary |
|---|---|
| 1.Conceptual framework for smart university [1, 50, 52, 58] | The concept of a smart university has evolved as follows:<br>• *Smart city versus Smart university* [21], analyzes the characteristics, similarities, and differences from the concept of the smart city associated with the smart university. It covers aspects related to sustainable urban development [29, 30, 36, 42, 43]. The second, in an equivalent but more limited way takes its characteristics to implement solutions in the processes of these organizations [2, 9, 25, 31, 60]<br>• *Smart campus versus smart university* [62] which compares the concepts of smart campus vs. smart university divided into four thematic areas: infrastructure; governance and management; services and education [13]<br>• *IoT is an emerging technology* [53], which visualizes the concept of the university from a perspective of the smart technology Internet of things [21, 37, 44]<br>• *Conceptual framework* [58], which shows the pillars of the smart university concept associated with the generation of a smart maturity model, as an evolutionary approach for a traditional university to progress toward various levels of smart university maturity<br>• *Strategic map* [55], which presents a proposal for a strategic map. As the technology matures, smart campuses and universities [17, 69] should be extended to management areas |
| 2. Characterization of adoption models by smart technology [55] | The methodology was adapted from Petersen [44], which can be further developed in Revelo et al. [49]. The selected documents were classified according to smart technologies: Artificial intelligence, big data, cloud computing, and the IoT [18, 51], see Table 2 |
| 3. Proposal of models for the adoption of smart technology [54, 56, 57] | Each model classified by smart technology has its advantages and disadvantages when it comes to fulfilling its function. The variables that best measure the degree of adoption are generated. The factors associated with each of the defined variables are formulated [14, 27, 33] |
| 5. Integration of smart technology adoption models | The integration of the four proposed models of smart technology adoption is generated |
| 6. User-centered smart technology adoption model: First approach | The product is generated in this document |

**Table 2** Models of smart technology adoption

| Smart technology | The model of adoption is more used by smart technology | Selected documents |
|---|---|---|
| IoT | TAM, UTAUT, UTAUT2, TPB, TRA | [1, 3, 4, 10–12, 16, 19, 24, 34, 38, 64, 66] |
| Cloud computing | TAM, UTAUT, DOI, UTAUT2 | [5, 8, 37, 39–41, 45, 47, 48, 61, 67] |
| Big data and artificial intelligence | TAM, UTAUT | [6, 63] |

aims to contribute to solving the problems raised. The adaptation of institutions with digital and technological tools is an extensive process, which is focused on improving the academic plans that the institution has, transforming traditional learning into flexible, collaborative, and oriented to the needs of each person, i.e., the relationship between the elements of education and technology is aligned to improve the skills and knowledge of students; this involves the development of new curricula that include technology as an important part in the implementation of academic activities.

Elements of smart technology adoption models, such as the intention to use, ease of use, and attitude toward use were considered by the application of the proposal in higher education environments. These elements were related to the process model of usability and accessibility engineering (see Fig. 1). In the implementations of technologies or smart tools, everything that has to do with usability and user experience represents a determining factor that influences the good adoption, and that directly

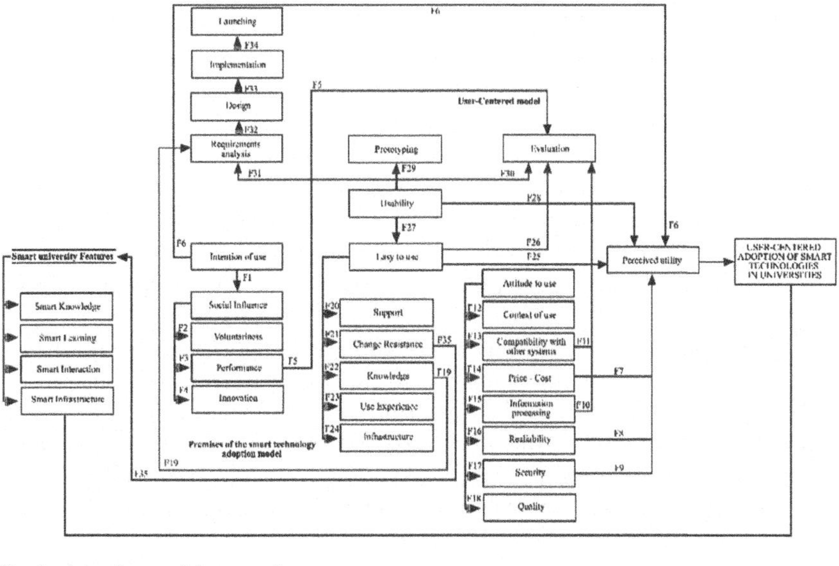


**Fig. 1** Adoption model proposal

affects the positive or negative results that are generated; in other words, it could be said that the user is in the center of the implementation and of the different facets of the development; in many occasions, this adoption is given to optimize processes or to make activities more agile, investing in tools and others, but very little is invested in the people who are going to make use of this [20].

Within the phases of the process model of the engineering of the usability and the accessibility are the analysis of requirements, one of the main elements is where the necessities or the fundamental characteristics to consider in the implementation of the technology become visible [32]. The prototyping and evaluation phase, where the user's perception of the execution of tasks performed together or through the system, is known, for example, the first time using the system without training, learning ease to perform activities, number of tasks performed, among others, are tests that allow us to know the user's perception [28]; the success of the product depends on the user feeling comfortable with the system, not on errors and in the case of errors, that it is not complicated to use it, that it recognizes the options and that the expected results are obtained. The degree of usability of an interactive system is directly related to the user interface, and the time it takes to recognize all the system's functionalities.

This characteristic, therefore, refers to the speed and ease with which people carry out their tasks through the use of the product they are working with [26]. The Factors raised are:

***F1. Social influence***. It directly affects the intentionality of the use of technology. It has to do with the influence received from friends, relatives, co-workers, or other educational institutions that use the technology and can be both positive and negative, but is vitally important when trying to adopt a new technology.

***F2. Voluntary***. There is a significant relationship between what is voluntarily adopted and what is imposed.

***F3. Performance in educational environments***. It is a two-way factor. When starting up, it can be seen as a benefit factor; the institution expects that by using new resources, the performance will be higher; however, it requires the help of other elements to ensure that this factor only acts positively.

***F4. Innovation***. Being at the forefront of technology has its advantages, such as increased performance in processes. Innovation has a close relationship with the perception of the individual; it is a positive factor that has benefits. It is related to the context of competitiveness, encourages universities to be in constant innovation, implementing strategies that keep the institution at the forefront of technology.

***F5. Relationship between performance and evaluation (Usability)***. During the evaluation, performance is an important factor for both technology and users.

***F6. Perceived utility***. A significant element in the adaptation of any technology is scalability and efficiency, which are expected results when using new technology.

***F7. Perceived utility-cost ratio***. According to the perceived utility of the implementation of the technology, the price/cost is justified.

***F8. Perceived utility-reliability***. The primary value of educational institutions has to do with the information. The use and adoption of technology ensure not only the protection of information but also its availability.

***F9. Perceived utility-security***. Information protection is a fundamental value and, therefore, a perceived utility when using technology.

***F10. Information processing—Evaluation***. During the evaluation of the system, one of the points to conclude the good adoption of the technology has to do with the treatment that this one gives to the information and how the user is capable of managing it across the new technology.

***F11. Compatibility with other systems—Evaluation***. When an institution has implemented other technologies, different from those that will be adopted, one of the main points to take into account is the compatibility and good management that can be established between both, reducing the rate of conflict that can generate poor management of information.

***F12. Context of use***. Has an impact on human resources. It is a factor that contributes to growth and the adoption of new technologies. In this point, it is important to make the technology known before it is implemented.

***F13. Compatibility with other systems***. Many times, higher education institutions have already invested in tools that are not compatible with the systems currently in place at the institution, and for this reason, they incur costs and generate a negative attitude when entering into the use of new technologies.

***F14. Price–cost***. Universities generally look for efficient tools at a moderate value, which will provide them with profits. There is a significant relationship between perceived utility and the price of that utility when one tool represents several utilities. It should be noted that the relationship between the utility and the price–cost of technology has to be measured from the implementation, operation, and maintenance.

***F15. Processing of information***. This has to do with the handling of information by the technology. At this point, the user/administrator has a great deal of influence, since it is, he who manages the information.

***F16. Reliability***. Smart technologies will always require a degree of reliability to be implemented in institutions.

***F17. Security***. The element of vital importance when in contact with information.

***H18. Quality***. The attitude toward use is favorable, as long as the system has quality in its results.

***F19. Requirements Analysis-Knowledge***. Requirements are of vital importance during any process, and knowledge about the technology provides clear and concise information that allows the generation of good quality requirements that will be useful during the development.

***F20. Support***. When there is assistance from the creators or implementers of the technology, the educational institution will feel more supported when adopting technology and therefore it will be easier to use.

***F21. Resistance to change***. It is an individual element; the institution must know the benefits that can be achieved when you are at the forefront of technology and not only that, the resistance to change is a negative value, which can reach the point of hindering the implementation and use of smart technology.

***F22. Knowledge***. It is mostly focused on IT knowledge. It is a value also related to the attitude toward the use if one knows the handling of the technologies the attitude at the moment of implementing and using is going to tend toward the positive thing.

***F23. Use experience***. The degree of adoption of technology can be based on the good or bad experiences the user has had in handling the technology. It has a direct impact on the ease of use because the experience simplifies the use.

***F24. Infrastructure***. It is a favorable point in the context of use. This technology does not represent a large investment in physical elements, and so institutions can access it more easily.

***F25. Ease of use—Perceived utility***. When there is the ease of use for users in making use of new technology, all that can be generated is profit; in other words, great adoption is achieved.

***F26. Ease of use—Evaluation***. One of the main values when evaluating a system, development, or implementation of technology has to do with usability, how easy it is for the user to perform tasks, activities, and how complex is the solution of errors.

***F27. Usability—Easy to use***. Taking up what was said before, usability is a fundamental value within the system; it has to do with the capacity that users see to interact and be able to carry out tasks. The model of usability within its phases throws improvements for the implementation or development is carried out in the best way.

***F28. Usability—Perceived utility***. When the improvement phase is complete, adoption is ready to generate maximum utility.

***F29. Prototyping***. They represent the mechanism that enables testing to be carried out before it is implemented within the institution. They are systems that simulate or have implemented parts of the final system [20].

***F30. Evaluation***. It constitutes a key point for obtaining usable and accessible interactive systems. The necessary techniques are applied to receive the necessary feedback from users and/or expert evaluators that will be reflected in the design of the interfaces. [20].

***F31. Requirements analysis***. Communication with users is a priority; they establish the services the system must provide and the restrictions under which it must operate.

***F32. Design***. This phase is related to the previous one, and to the following ones, they represent the complementary phases in the model of usability, and the requirements give a point of departure to the design.

***F33. Implementation***. It has to do with the adjustment of all the preliminary versions to the final system and the implementation of the same.

***F34. Launch***. It corresponds to the most critical phase in the process. And that is where success can be known. In the case of the user, it is the factor where it feels comfortable with the system.

Today's society is in constant evolution; innovation, technology, and learning are key factors that allow the generation of more effective and automated processes in all areas of an organization. New advances in technology offer great opportunities to develop new and better learning environments, to break the educational barriers that exist today [35]. The term smart university is a innovative concept that promotes the creation of good educational practices, having as a fundamental premise the use of technologies for process improvement. The adoption of technology in educational institutions is promising, but at the same time, it assumes a challenge since it challenges the traditional educational practices that are consolidated and executed by teachers today; despite the constant challenge, more and more institutions decide to start using this type of tools that facilitate not only learning inside the classroom but also outside it. In general, when talking about a smart university, it is not only about the education provided, but also emphasizes an institution capable of collecting information, through technologies, thus facilitating assertive decision-making in all areas, from knowledge to administrative.

## 5 Conclusion

The relationships between the main premises of the adoption models:

**Intention to use**. Based mainly on the social influence and secondarily on the voluntary nature of its use, performance, and innovation, all the above elements influence whether a person incorporates the new technology into his or her organizational environment.

**Easy to use**. It includes elements such as support, resistance to change, knowledge, user experience, infrastructure.

**Attitude toward use**. Includes, the context of use, compatibility with other systems, price–cost, security, information processing, reliability, quality. Of the above, the price–cost element represents a barrier value that hinders the adoption of new technologies,

In the adoption of new technology, it is extremely important to perceive the degree of usability that is perceived by the user. For this reason, the Process Model of Usability and Accessibility Engineering (MPIU + a) was included. All of the above elements lead to the use of technology, where one can verify the usefulness and perceived and finally conclude the adoption of the technology. The adoption of technology in educational institutions is promising. However, it also assumes a challenge since it challenges the traditional educational practices that are consolidated and executed by teachers today; despite the constant challenge, more and more institutions decide to start using this type of tool that facilitates not only learning inside the classroom but also outside it. Usually, when talking about smart universities, not only the education provided but also emphasizes an institution capable of collecting information through technologies, thus facilitating assertive decision-making in all areas, from knowledge to administrative. Organizations, including universities, need to incorporate "smart" technologies to take advantage of the capabilities they provide to transform their processes and promote new organizational models that allow them to adequately incorporate this new concept of a smart university.

Among the limitations of the project that have been presented so far are the lack of previous research studies on the subject, the lack of available data and dependence on having access to people, organizations, or documents. The proposed research will use both primary and secondary data; data will be collected using both qualitative and quantitative methods. Survey and user interview data will be collected from technology directors and decision-makers in higher education to investigate what technology adoption in education looks like, the barriers technology users face, and how it can be optimized to obtain full job benefits. A quantitative method will be used to measure knowledge about the level of maturity of technology adoption, while qualitative data will be collected to give a broader picture of how technology has been used in education. Both qualitative and quantitative data can be obtained from literature reviews, interviews, and surveys [7, 71].

## References

1. A. Abushakra, D. Nikbin, *Extending the UTAUT2 Model to Understand the Entrepreneur Acceptance and Adopting Internet of Things (IoT)* (Springer International Publishing, 2019). https://doi.org/10.1007/978-3-030-21451-7
2. A. Adamkó et al., *Intelligent and Adaptive Services for a Smart Campus*. in 5th IEEE International Conference on Cognitive Infocommunications, CogInfoCom 2014–Proceedings (2014), pp. 505–509. https://doi.org/10.1109/CogInfoCom.2014.7020509
3. A.M. Al-momani et al., A review of factors influencing customer acceptance of internet of things services. **11**(1), 54–67 (2019). https://doi.org/10.4018/IJISSS.2019010104
4. A.M. Al-momani et al., Factors that influence the acceptance of internet of things services by customers of telecommunication companies in Jordan. **30**(4), 51–63 (2018). https://doi.org/10.4018/JOEUC.2018100104
5. M. Al-ruithe et al, Sciencedirect procedia science direct current state of cloud computing adoption—an empirical study in major public sector organizations current state of cloud computing adoption—of Saudi Arabia (KSA) An empirical study in major public S Procedia Comput. Sci. **110**, 378–385 (2017). https://doi.org/10.1016/j.procs.2017.06.080
6. A. Alamri, Harnessing the power of big data analytics in the cloud to support learning analytics in mobile learning environment Comput. Human Behav (2018) https://doi.org/10.1016/j.chb.2018.07.002
7. S. Alqassemi. et al., *Maturity Level of Cloud Computing at HCT*. in ITT 2017—Information Technology Trends Exploring Current Trends Information Technology Conference Proceedings 2018-January Itt, 5–8 (2018). https://doi.org/10.1109/CTIT.2017.8259558
8. I. Arpaci, Antecedents and consequences of cloud computing adoption in education to achieve knowledge management. Comput. Human Behav. (2017). https://doi.org/10.1016/j.chb.2017.01.024
9. Y. Atif et al., Building a smart campus to support ubiquitous learning. J. Ambient Intell. Humaniz. Comput. **6**(2), 223–238 (2015). https://doi.org/10.1007/s12652-014-0226-y
10. F. Authors, Adoption of internet of things ( IOT ) based wearables for elderly healthcare—a behavioural reasoning theory (BRT) approach (2018). https://doi.org/10.1108/JET-12-2017-0048
11. F. Authors, An exploratory study of Internet of Things (IoT) adoption intention in logistics and supply chain management—a mixed research approach (2016)

12. P Brous et al., The dual effects of the Internet of Things (IoT): A systematic review of the benefits and risks of IoT adoption by organizations. Int. J. Inf. Manage (2019). https://doi.org/10.1016/j.ijinfomgt.2019.05.008
13. F.H. Cerdeira Ferreira, R. Mendes de Araujo, *Campus Inteligentes: Conceitos, aplicações, tecnologias e desafios*. Relatórios Técnicos do DIA/UNIRIO. **11**(1), 4–19 (2018)
14. R.A. Choix et al, *Factores determinantes en la adopción de tecnologías de información (TI) en las pymes*. VinculaTégica EFAN (2015)
15. M. Coccoli et al., Smarter universities: A vision for the fast changing digital era. J. Vis. Lang. Comput. **25**(6), 1003–1011 (2014). https://doi.org/10.1016/j.jvlc.2014.09.007
16. S. Das, The early bird catches the worm—first mover advantage through IoT adoption for Indian Public sector retail oil outlets. The early bird catches the worm—first mover advantage. J. Glob. Inf. Technol. Manag. **00**(00), 1–29 (2019). https://doi.org/10.1080/1097198X.2019.1679588
17. Z.Y. Dong et al., Smart campus: definition, framework, technologies, and services. IET Smart Cities. **2**(1), 43–54 (2020). https://doi.org/10.1049/iet-smc.2019.0072
18. T. Dybå et al., Applying Systematic Reviews to Diverse Study Types: An Experience Report. in Proceedings—1st International Symposium on Empirical Software Engineering and Measurement, ESEM 2007 (2007). https://doi.org/10.1109/ESEM.2007.21
19. E.E. Grandón et al., *Internet de las Cosas : Factores que influyen su adopción en Pymes chilenas Internet of Things : Factors that influence its adoption among Chilean SMEs* (2020)
20. T. Granollers i Saltiveri, *MPIu+a. Una metodología que integra la Ingeniería del Software, la Interacción Persona-Ordenador y la Accesibilidad en el contexto de equipos de desarrollo multidisciplinares* (2004).
21. C.D. Guerrero et al., IoT: Una aproximación desde ciudad inteligente a universidad inteligente. Rev. Ingenio UFPSO. **13**(1), 1–12 (2017)
22. S. Jose et al., Disruptive architectural technology in engineering education. Procedia Comput. Sci. **172**, 641–648 (2020). https://doi.org/10.1016/j.procs.2020.05.083
23. S. Jose et al., Nurturing engineering skills and talents, a disruptive methodology in engineering education. Procedia Comput. Sci. **172**, 568–572 (2020). https://doi.org/10.1016/j.procs.2020.05.069
24. Y. Kao et al., An exploration and confirmation of the factors influencing adoption of IoT-based wearable fitness trackers (2019)
25. Y. Khamayseh et al., *Integration of Wireless Technologies in Smart University Campus Environment: Framework Architecture* (2015) https://ovidsp.ovid.com/ovidweb.cgi?T=JS&PAGE=reference&D=psyc11&NEWS=N&AN=2015-00124-004, https://doi.org/10.4018/ijicte.2015010104
26. U. Lleida, de: Departament de Llenguatges i Sistemes Informàtics Universitat de Lleida Lleida, julio 2004. Screen (2004)
27. M.V. López Cabrera et al., Factors that enable the adoption of educational technology in medical schools. Educ. Medica. 20(xx), 3–9 (2019). https://doi.org/10.1016/j.edumed.2017.07.006
28. J. Lorés, T. Granollers, *Ingeniería de la Usabilidad y de la Accesibilidad aplicada al diseño y desarrollo de sitios web* (2004)
29. G. Maestre-Góngora, Revisión de literatura sobre ciudades inteligentes: una perspectiva centrada en las TIC. Ingeniare. **19**(19), 137–149 (2016)
30. G. Maestre-Gongora, R.F. Colmenares-Quintero, Systematic mapping study to identify trends in the application of smart technologies. Iber. Conf. Inf. Syst. Technol. Cist. 1–6 (2018). https://doi.org/10.23919/CISTI.2018.8398638
31. E.M. Malatji, *The Development of a Smart Campus—African Universities Point of View*. In: 2017 8th International Renewable Energy Congress, IREC 2017 (2017). https://doi.org/10.1109/IREC.2017.7926010
32. J. Mariano, G. Romano, *Introducción a la IPO* (Metro, 2008)
33. A.V. Martín García et al., *Factores determinantes de adopción de blended learning en educación superior*. Adapta ción del modelo UTAUT. Educ. XX1 (2014). https://doi.org/10.5944/educxx1.17.2.11489

34. M. Mital et al., Technological forecasting & social change adoption of internet of things in India : A test of competing models using a structured equation modeling approach. Technol. Forecast. Soc. Chang. 1–8 (2017). https://doi.org/10.1016/j.techfore.2017.03.001
35. A. Mukherjee, N. Dey, Smart Computing with Open Source Platforms. (2019). https://doi.org/10.1201/9781351120340
36. L Muñoz López et al., El Estudio y Guía Metodológica sobre Ciudades Inteligentes ha sido dirigido y coordinado por el equipo del ONTSI Deloitte (2012). https://doi.org/10.1017/CBO9781107415324.004
37. U. Nasir, Cloud computing adoption assessment model (CAAM). **44**(0), 34–37 (2011)
38. D. Nikbin, A. Abushakra, *Internet of Things Adoption: Empirical Evidence From An Emerging Country*. in Communications in Computer and Information Science. (2019). https://doi.org/10.1007/978-3-030-21451-7_30
39. F. Nikolopoulos, *Using UTAUT2 for Cloud Computing Technology Acceptance Modeling* (2017)
40. K. Njenga et al., Telematics and Informatics The cloud computing adoption in higher learning institutions in Kenya : Hindering factors and recommendations for the way forward. Telemat. Inf. (2018). https://doi.org/10.1016/j.tele.2018.10.007
41. P. Palos-Sanchez et al., Models of adoption of information technology and cloud computing in organizations. Inf. Tecnol. **30**(3), 3–12 (2019). https://doi.org/10.4067/S0718-07642019000300003
42. G. Perboli et al., A new taxonomy of smart city projects. Transp. Res. Procedia. **3**, 470–478 (2014). https://doi.org/10.1016/j.trpro.2014.10.028
43. F.M. Pérez et al., *Smart university: hacia una universidad más abierta*, https://dialnet.unirioja.es/servlet/libro?codigo=676751, (2016)
44. K. Petersen et al., *Systematic Mapping Studies in Software Engineering*. in 12th International Conference on Evaluation and Assessment in Software Engineering, EASE 2008 (2008). https://doi.org/10.14236/ewic/ease2008.8
45. P. Pinheiro, C. Costa, Adoption of Cloud Computing Systems. 127–131 (2014)
46. P. Pornphol, T. Tongkeo, Transformation From a Traditional University into A Smart University (2008). https://dl.acm.org/citation.cfm?id=3178167, https://doi.org/10.1145/3178158.3178167
47. P. Priyadarshinee et al., Understanding and predicting the determinants of cloud computing adoption: A two staged hybrid SEM—neural networks approach. Comput. Human Behav. (2017). https://doi.org/10.1016/j.chb.2017.07.027
48. R.D. Raut et al., Analyzing the factors influencing cloud computing adoption using three stage hybrid SEM-ANN-ISM (SEANIS) approach. Technol. Forecast. Soc. Change. (2018). https://doi.org/10.1016/j.techfore.2018.05.020
49. O. Revelo Sanchez et al., Gamification as a didactic strategy for teaching/learning programming: a systematic mapping of the literature. Rev. Digit. LAMPSAKOS. (2018). https://doi.org/10.21501/21454086.2347
50. D Rico-Bautista et al., Analysis of the potential value of technology: Case of universidad Francisco de paula santander Ocaña. RISTI—Rev. Iber. Sist. e Tecnol. Inf. **E17**, 756–774 (2019)
51. D. Rico-Bautista et al., Caracterización de la situación actual de las tecnologías inteligentes para una Universidad inteligente en Colombia/Latinoamérica. RISTI—Rev. Iber. Sist. e Tecnol. Inf. **E27**, 484–501 (2020)
52. D. Rico-Bautista, Conceptual framework for smart university J. Phys. Conf. Ser. (2019)
53. D. Rico-Bautista et al., *Smart University: A Review from the Educational and Technological View of Internet of Things*. In Advances in Intelligent Systems and Computing (2019), pp. 427–440
54. D. Rico-Bautista et al., *Smart University: Key Factors for An Artificial Intelligence Adoption Model*. in Advances in Intelligent Systems and Computing (2020)
55. D. Rico-Bautista et al., Smart University: Strategic map since the adoption of technology. RISTI—Rev. Iber. Sist. e Tecnol. Inf. 2020, **E28**, 711–724 (2020)

56. D. Rico-Bautista et al., *Smart University: Big Data Adoption Model.* in *2020 9th International Conference on Software Process Improvement, CIMPS 2020 - Applications in Software Engineering* (2020)
57. D. Rico-Bautista et al., *Smart University: IoT Adoption Model.* in Proceedings of the Fourth World Conference on Smart Trends in Systems, Security and Sustainability, WorldS4 2020 (2020)
58. D.W. Rico-Bautista, Conceptual framework for smart university. J. Phys. Conf. Ser. **1409**, 012009 (2019). https://doi.org/10.1088/1742-6596/1409/1/012009
59. Rjab, A. Ben, S. Mellouli, Smart cities in the era of artificial intelligence and internet of things. **1**, 1–10 (2018). https://doi.org/10.1145/3209281.3209380
60. M. Rohs, J. Bohn, Entry points into a smart campus environment-overview of the ETHOC system. Distrib. Comput. Syst. Work. 1–7 (2003)
61. H.M. Sabi et al., Conceptualizing a model for adoption of cloud computing in education. Int. J. Inf. Manage. **36**(2), 183–191 (2016). https://doi.org/10.1016/j.ijinfomgt.2015.11.010
62. B. Sánchez-Torres et al., Smart Campus: Trends in cybersecurity and future development. Rev. Fac. Ing. **27**, 47, (2018). https://doi.org/10.19053/01211129.v27.n47.2018.7807
63. F.P. Sejahtera et al., Information & Management Factors influencing effective use of big data : A research framework. Inf. Manag. 103146 (2019). https://doi.org/10.1016/j.im.2019.02.001
64. H. Shaikh et al., *A Conceptual Framework for Determining Acceptance of Internet of Things (IoT) in Higher Education Institutions of Pakistan.* in 2019 International Conference on Information Science Communication Technology (2019), 1–5
65. C. Shaoyong et al., *UNITA : A Reference Model of University IT Architecture.* in ICCIS '16 Proc. 2016 International Conference on Information System (2016), 73–77. https://doi.org/10.1145/3023924.3023949
66. B. Sivathanu, Adoption of internet of things (IOT) based wearables for healthcare of older adults—a behavioural reasoning theory (BRT) approach. J. Enabling Technol. (2018). https://doi.org/10.1108/JET-12-2017-0048
67. H. Vasudavan, *User Perceptions in Adopting Cloud Computing in Autonomous Vehicle* (2018), 151–156
68. M.C. Vega-Hernández et al., Multivariate characterization of university students using the ICT for learning. Comput. Educ. **121**, 124–130 (2018). https://doi.org/10.1016/j.compedu.2018.03.004
69. M.S. Viñán-Ludeña et al., *Smart University: An Architecture Proposal for Information Management Using Open Data for Research Projects. Advances in Intelligent Systems and* Computing, 1137 AISC, March, 172–178 (2020). https://doi.org/10.1007/978-3-030-40690-5_17
70. J. Vuorio et al., Enhancing User Value of Educational Technology by Three Layer Assessment (2017),220–226. https://doi.org/10.1145/3131085.3131105
71. M. Zapata-ros, La universidad inteligente La transición de los LMS a los Sistemas Inteligentes de Aprendizaje en Educación Superior The smart university. **57**, 10, 1–43 (2018)\
72. Applied Machine Learning for Smart Data Analysis. (2019). https://doi.org/10.1201/9780429440953

# Study of Technological Solutions in the Analysis of Behavioral Factors for Sustainability Strategies

María Cazares, Roberto O. Andrade, Julio Proaño, and Iván Ortiz

**Abstract** The management and consumption of resources in cities in sustainable way have been considered a relevant issue worldwide. For that reason, the New Urban Agenda defines a set of sustainable development objectives-SDG for the year 2030. On the other hand, consumerism behaviors by people have had a considerable increase since of the Second World War period. Achieving sustainability objectives requires the establishment of policies and standards as enabling elements, but they must go hand in hand with a cultural change on the people aligned with the strategies to achieve sustainability. The purpose of this study is the development of a comprehensive analysis of the cultural beliefs that influence in the behavioral aspects of people and how they could affect the achievement of the development of sustainability in cities and how some technologies like big data, IoT, and IA, can be used to analyze these behaviors and improve the decision-making processes in city managers.

**Keywords** Smart city · Sustainability · Cognitive maps · Social factors · Behaviors

## 1 Introduction

Cities have chosen the development of smart cities as a management and operation model to achieve their sustainability and resilience goals. The smart city definition is related to a type of urban development but based on sustainability that can satisfy its citizens' basic needs [14]. New improvements in artificial intelligence (AI), computing power, and the inclusion of emerging technologies such as IoT and

M. Cazares (✉) · J. Proaño
IDEIAGEOCA Research Group, Universidad Politécnica Salesiana, Quito, Ecuador
e-mail: mcazares@ups.edu.ec

R. O. Andrade
Facultad de Ingeniería de Sistemas. Escuela Politécnica Nacional, Quito, Ecuador
e-mail: roberto.andrade@epn.edu.ec

I. Ortiz
Universidad de Las Américas, Quito, Ecuador
e-mail: ivan.ortiz@udla.edu.ec

 
A. Joshi et al. (eds.), *Sustainable Intelligent Systems*, Advances in Sustainability Science and Technology, https://doi.org/10.1007/978-981-33-4901-8_11

cloud allow obtaining information from the abundance of data to make decisions in order to promote the proper use of available resources in cities transforming them into smart cities. IoT has been considered a key element to establish a sensorization of the elements that make up the city in the different domains of health, education, transport, energy, and waste management [1]. In contrast, big data techniques and cloud are complements for storing and processing data generated by IoT devices. However, this data could be useless without an appropriate smart way of extracting the knowledge. Thus, AI will play a fundamental role in the optimal management of smart cities transforming data into knowledge.

Under this premise, in recent years there has been an important development in IoT and AI solutions also called smart solutions to meet the needs of the domains of the smart city. Smart solutions allow for instance obtaining, processing, and storing data of the number of cars on a highway, the temperature of a greenhouse, or the amount of electrical energy consumed in a home [2]. According to specialized sites such as Gartner [8], the growth projections of IoT and AI solutions will exceed millions in the coming years.

But when we focus on analyzing city management from an urban planning perspective, we not only consider physical elements such as the number of cars or people, but we also consider human factors related to the pillars of the city (environment, politics, economy, and technology). Urban planning should consider as central element of the citizen, who is a social entity part of the city that has a dynamic and complex characteristic [14]. From a pragmatic point of view, urban planning is more likely to predict the number of cars that will be on a street at a given time, and it has been feasible to predict human behavior and its impact on sustainability goals.

Human behavior, especially those related to sustainability aspects (sustainable behaviors), can be affected by some social factors shown in Fig. 1.

This context motivated to plan the following research questions that mark the development of this work:

1. Are smart solutions focused on sensing the aspects that allow measuring the impact on sustainability?
2. Are smart solutions adaptable to the existing social dynamics in the city?

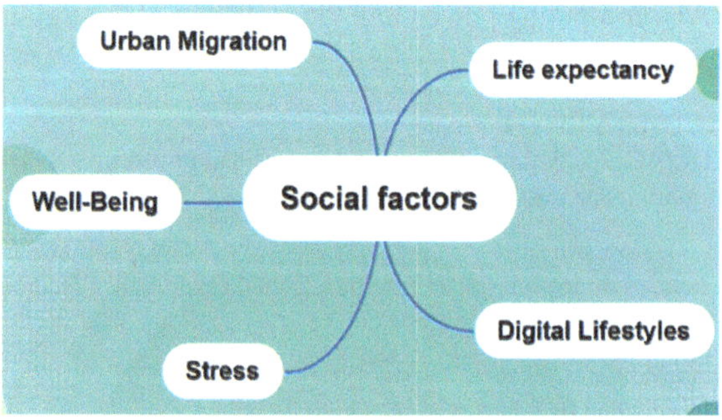


**Fig. 1** Social factors that could affect human behavior

3. What are the behavioral aspects aligned to sustainability that need to be measured?

This study is motivated by this idea "to satisfy our current needs without compromising the capacity of future generations to satisfy theirs" (UN 1987), technological resources such as artificial intelligence and IoT can allow the analysis of the behavior of citizens to identify the dynamics of various factors that must be modified to achieve sustainability.

Understanding the problem of environmental pollution through a holistic vision that integrates artificial intelligence, big data, IoT, and sustainable behavior to identify the relationships that exist between social, economic, and psychological variables, is our main contribution.

The remainder of this paper is structured as follows. Section 2 presents an overview related to people's behavior and sustainability. This section tries to cover the relationship between social factors and people's behavior and how they affect the sustainability of the cities. Section 3 presents the methodology used in this study and the results obtained from the SLR to determine the main aspects of social factors that could affect the sustainability goals of cities. Then, Sect. 4 covers the aspects for modeling social factors related to sustainability. Finally, Sect. 5 shows the conclusions for this study.

## 2 Background

Sustainability not only has an environmental perspective and also includes economic and sociocultural perspectives. Urban sustainability situates citizens as social change agents with an ethical and self-interested [14]. Some researchers have linked the relationship between environmental values and the behavioral intention of people's behavior [3]. Additionally, Barr mentions that people's behavior will also depend on situational factors and psychological variables.

The strategies for sustainability consider two aspects: citizens and responsible environmental behavior. Building a new model of sustainability for the development of cities in the twenty-first century implies taking into account psychological factors such as feelings, values, motives, intentions, and deliberate, planned and systematic behaviors, influenced by the levels of needs, satisfaction, information, capabilities, and resources.

The re-establishment of the "balance" between society nature entails a whole change of mentality and, therefore, of the conception of people regarding development, consumption, progress, and environmental care. People need to face the sustainability crisis that is going through at the beginning of the twenty-first century. Even with the result of successive scientific–technological revolutions, cities have not reached equally for everyone. The actual mode of industrial production could enhance predatory behavior in people in terms of the consumption of material and energy resources, as well as polluting by the generation of waste, which ruins the planetary nature [15].

In recent years, the new version of the city introduces the concept of urban dashboards and platforms that integrates devices, namely Internet of things (IoT). These devices can be sensors, monitoring stations, digital cameras, actuators, tracking systems, etc. IoT devices generate a large number of data. This data could be vastly, and it needs to be processed by techniques, namely big data. Big data techniques are based on gathering massive amounts of data (volume) from different sources (variety) and processing very quickly (velocity) them in order to extract values that support decisions-making, designing solutions, and modifying processes of cities to make them sustainable [16]. In order to extract all the knowledge of the processed data optimally, AI algorithms can be a significant boost. Thus, according to a report of **Pricewaterhouse Coopers** (PWC) related to the real value of AI for business and how to capitalize it, they estimate that the global gross domestic product (GDP) will be up to 14% higher in 2030 as a result of the accelerating development of the use of AI [17].

## 2.1 Social Factors That Influence Sustainability

Developmentalism as a conception of social progress, the denial of the natural, anthropocentrism and ecocentrism beliefs, frugalist, altruism, the pursuit of pleasure in the satisfaction of artificially created needs (hedonism), fatalism and utilitarian behaviors and beliefs of an abundance unlimited resources [5] could influence social behaviors. Natural processes triggered by human action and high speed of social changes need maintain an equality relationship between the availability of resources and the people's needs.

Lifestyles are influenced by instinct and the human psyche that lead us to competence and storage of resources, from an evolutionary perspective [6].

On the other hand, the marketing models that promote greater consumption of products through cultural meanings play an important role as they provide guidelines on acceptable behaviors, marking individual decision-making and that are internalized in the self are combined with the concept as representations of success, that drive behavioral patterns of consumption [9].

## 2.2 Approach of an Ecological Culture

Modification of consumerism lifestyles toward ecological or sustainable styles raises the analysis of trajectories that allow understanding the relationships between behaviors, personality beliefs through intelligent technological solutions that help to minimize the waste of resources, reduce pollution, and approach the search for a community welfare. To achieve these points, the solutions must join forces to build an ecological society that places limits on the use of natural resources, remembering the natural balance, where the common good is above the individual good [5].

Make cities and urban spaces inclusive, safe, resilient, and sustainable, need of an interdisciplinary work between social sciences and engineering that promotes sustainable consumption patterns among citizens. People must be aware of their beliefs and behaviors about the ability to renew natural resources of the planet and the amount of waste that becomes polluting emissions. Build ecological lifestyles to achieve the transition from the urban ecosystem toward sustainability. Ecological footprint measurement is an important concept to identify global human activity and sustainability at the local level. Proposals that are researched and developed globally are framed on the following aspects:

- Environmental awareness.
- Constructive and transformative behaviors, as well as critical thinking.
- Pro-environmental competencies.
- Social representation and community feeling of human development.

Some models and concepts that have been developed:

- Social mind raised by McDougal 1974.
- The model proposed by Baron and Kenny in 1996 seeks the mediation between beliefs, attitudes, and motives about human behavior.
- Conservation beliefs, authenticity, and material beliefs influence recycling behaviors. Obregon 1996.
- Emotional components predict pro-environmental behavior Grob 1995.
- At the personality level, Bustos in 2004 determined that the internal locus of control directly and positively indexes the beliefs of obligation about saving resources.
- Dominant social paradigm.
- Paradigm of human exception, and
- New ecological paradigm.

## 3 Methodology

The methodology proposal in this study is based on the literature review using the PRISMA guideline. We select the following scientific databases: Scopus, IEEE Xplorer, and Science direct. Based on the three research questions, we made a literature review of 252 articles obtained from the following search strings:

- "Citizen" AND "behavior" AND "sustainability".
- "Citizen" AND "sustainability" AND IoT".
- "Citizen" AND "sustainability".
- "Citizen" AND "sustainability" AND "artificial intelligence".

Based on the qualitative analysis carried out using the systematic review tool Rayyan, we identified the following relevant aspects:

| Publication types | |
|---|---|
| Journal Article | 218 |
| SER | 12 |
| CONF | 12 |
| CHAP | 11 |
| BOOK | 4 |
| INPR | 2 |

**Fig. 2** Number of scientific sources in journal, book series, and conferences

| Journal | |
|---|---|
| Sustainability (Switzerland) | 39 |
| World Sustainability Series | 7 |
| Environmental Education Research | 5 |
| Journal of Cleaner Production | 5 |
| Corporate Social Responsibility and Environmental Management | 4 |
| Business Strategy and the Environment | 4 |
| International Journal of Sustainable Development and World Ecology | 4 |
| Frontiers in Psychology | 3 |
| Education Sciences | 2 |
| Journal of Business Ethics | 2 |

**Fig. 3** Journals that consider of topic "sustainable citizen behaviors"

The most of scientific sources that analyze aspect of sustainable citizen behaviors are journal articles, then book series and next conferences (see, Fig. 2.

Based on the analysis of scientific papers, the most relevant journals focus on the following topics: production, business, psychology, environmental, sustainability, and education (see Fig. 3. We observed that exist a lack of journals specialized in technology. This finding is aligned with our first research question and expand our interest to be identified if smart solutions support the measures of sustainability.

Follow the qualitative analysis in this study, we identify the characteristics of citizens based on sustainability (see Fig. 4.

- Environmental or ecological citizen has been suggested within the field of political theory as an approach to realize personal responsibility for the environment.
- Active citizen focused on the construction of democratic values.
- Global citizen is committed to helping and cooperating with others.

## *3.1 Social Factors in Smart Home*

Jagers [10] mentions that there is willingness to change behavior in citizen associated with reducing the negative effects on the environment and mentions that exists in the responders to believe that adapting social practices to nature's limits, for instance through lifestyle changes, is available solution. If this predisposition for change

**Fig. 4** Relationship between sustainability and citizen

exists, how it is coherent with the actual reality or "new normality" and with the consumer lifestyles that have been predominant for several years. Mass consumption was one of the realities that could be evidenced with the onset of the COVID-19 pandemic, which may at first glance turn out to be far from the sustainability goals sought worldwide.

As we had mentioned previously, one of the social factors that affect the development of smart cities is digital lifestyles, which one has seen a huge impact as a result of the COVID-19 pandemic. Teleworking, tele-education, and online shopping became daily activities where the use of electronic devices had a significant increase. This unexpected increase of production and buy of electronic devices could be no aligned with sustainability goals (see Fig. 5.

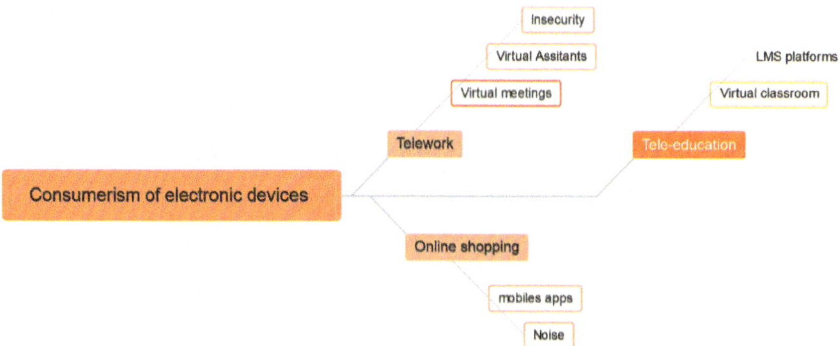


**Fig. 5** Consumerism of electronic devices during COVID-19

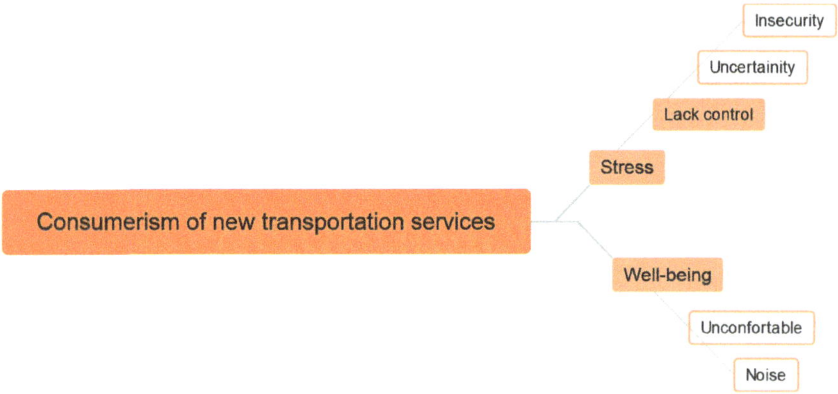


**Fig. 6** Consumerism of transportation services during COVID-19

## *3.2 Social Factors in Transportation*

Urban mobility is another aspect in the urban planning of the city. A study related to New Jersey transit found the relation between time trip and stress [20], and the research also found that some routes commuter has more stress level. Another study shows that metrobus drivers in the city of Mexico are stressed by the insecurity at work [12]. On the other hand, related to passenger, a study to exploring passenger anxiety indicates that the following service items cause passenger anxiety during trains travel: crowding, delays, accessibility to a railway station, searching for the right train on a platform, and transferring trains [4]. Additionally, other study which focuses on foreign people analyzed the stress perceived by foreigners that use public transportation in Bogotá, and this study found the following control variables: lack of control, crime, accidents, cleanness, noise, temperature, and space [13]. Moving around in public transport causes anxiety and stress in cities, which drive people to select other mobility alternatives such as Uber or Lyft. People fell that when they use these services, aspects related to well-being improve because they do not need worries about traffic or parking (see Fig. 6.

# 4 Cognitive Maps for Modeling Consumerism Behaviors in Smart City

So, social behavior is dynamic and has a dependency of situational factors, as Barr mentioned. Some social behaviors generate consumerism approaches, which one could affect to the sustainability goals (see Fig. 7. On the other hand, smart solutions that integrate IoT, big data, and artificial intelligence can reduce traffic jams or improve cars and citizens' mobility. Additionally, the data collected by smartphones

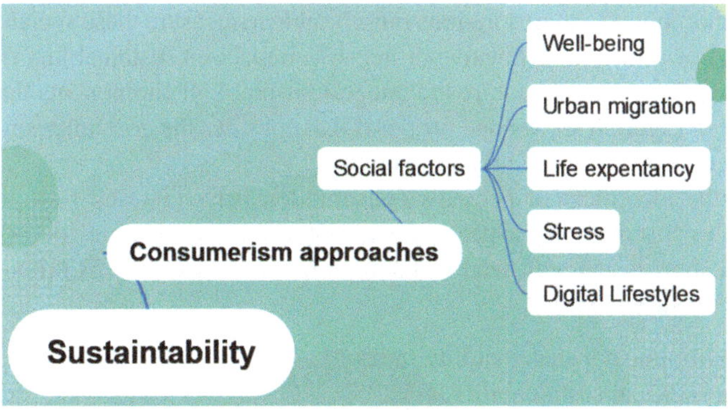


**Fig. 7** Social factors that affect consumerism behavior

and cameras can show citizens information about the traffic situation and suggest optimal routes avoiding stressful situations. [17]. In [18], the authors show that motorized transportation is the highest in air pollution. Thus, they have designed and developed an air quality monitoring instrument (AQMI) using solid-state gas sensors and GPS module. Results can be used for patients with air pollution-related health problems to find less polluted routes. In contrast, [11] propose a study that deals with the complexity of assessing transit service quality by identifying attributes affecting passenger's satisfaction. The authors use a fuzzy model composed of 26 variables as factors. Results can help in improving existing transit facilities and devising strategies for ensuring sustainability. In [19], a novel context air quality prediction model is presented. In this model, the authors include context-aware computing concepts to merge an accurate air pollution prediction algorithm based on artificial intelligence with information from both surrounding pollution sources and user's health profile. Results show 90–96% of precision.

Social factors can positively or negatively affect the sustainability of cities. Table 1 shows the relation between the affectation of social factors to vertical domains in smart city which forms consumerism perspective.

**Table 1** Relationship among smart city vertical domains and social factors

| Social factors/vertical domain | Smart health | Smart transport | Smart home | Smart waste | Public safety |
|---|---|---|---|---|---|
| Urban migration | X | X | | X | X |
| Life expectancy | X | | X | | |
| Well-being | X | | X | X | X |
| Digital lifestyles | | | X | X | |
| Stress | X | X | | | |

Under this context, smart solutions must be able to measure these social dynamics and assess how much is its positive or negative impact on sustainability objectives. Urban planning processes require the integral vision of all components that interact in smart city. Figure 8 shows one proposal that consider the consumerism behavior how element that affects the urban planning.

Model the identifiers of people's consumption behaviors and how much they contribute to sustainability aspects, for example how many times people take an Uber for a short distance instead of taking a bicycle. Machin [15] proposed some indicators for sustainability:

- AA; Environmental sustainability index.
- S.S; Sociological sustainability index.
- EE; Economic sustainability index.
- AS; Sociological–environmental sustainability index (Supportability).
- AE; Economic–ecological sustainability index (viability).
- SE; Economic–sociological sustainability index (Equity).

Assunção [7] mentions that new urban sustainability assessment systems based on land senses ecology are needed, which should combine natural elements, physical senses, and psychological perceptions, and assist decision-makers to develop

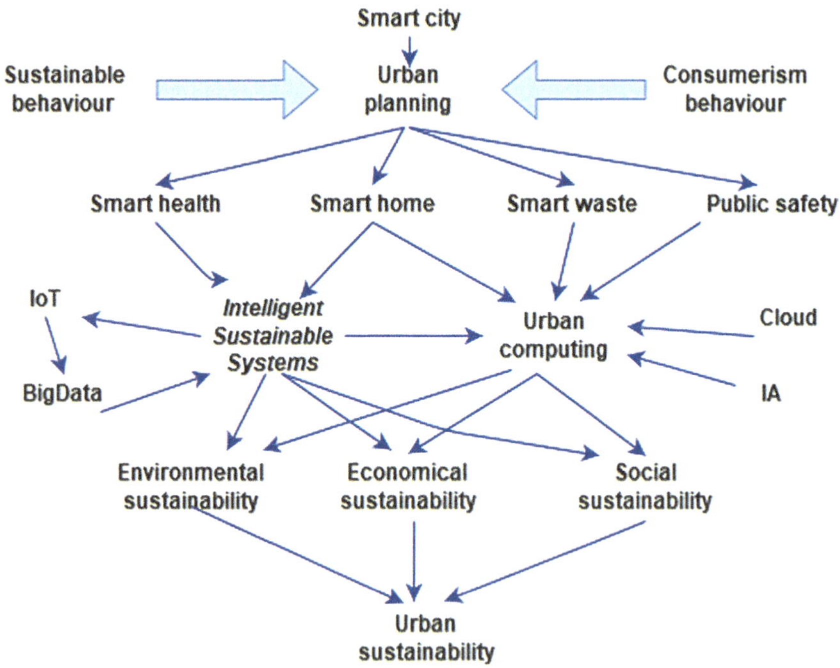


**Fig. 8** Cognitive map for sustainability urban planning

successful management policies. Under this context, Assunção proposes a fuzzy cognitive mapping and system dynamics.

## 5 Discussion

Results of the study suggest that it is necessary to consider the dependence of sustainability on environmental, economic, and social factors in order to establish technological solutions at the local level that allow smart cities to develop a sustainable model of development where citizens are aware of the limits to the exploitation of natural resources and the amount of waste they generate, and there is interest in the construction of models that allow work in this line. For example, in smart transportation, analyzing data from IoT devices such as cameras, traffic sensors, signals, and emissions sensors distributed throughout the city can optimize the traffic flow smartly. As a result, citizens could also be informed of the traffic situation in real time and select the optimal alternative routes avoiding stressful situations and improving their quality of life. Additionally, optimal urban mobility can help use better energy management to save money and improve the environment. So, environmental, social, and economic sustainability can be achieved.

On the other hand, from the smart home approach, the combination of social inclusion and smart health care can underline the importance of social integration, education, and healthy lifestyle changes to bring the people and society to create a participatory and tolerant environment. As a result, sustainable behavior is reached. IoT, big data, cloud computing, and AI can contribute to making possible the transformation of cities to smart cities and sustainable environment.

Promoting equity, justice, solidarity, and respect for the ethnic and cultural aspects of the population, as well as the defense of biodiversity, are basic axes of sustainability, and technology must approach these aspects to build more effective solutions that respond to the complexity of human development problems. Engineers should also create and apply technology to minimize waste of resources, reduce pollution, and protect human health, well-being, and the ecological environment.

The goal is that smart solutions allow us to know if our behaviors are contributing to the environmental care of the planet. Cognitive maps are one interesting alternative to model social behaviors for decision-making. The inclusion of AI models could improve the effectiveness and time response to analyze huge volume of data generated for smart cities. From our literature review, the modeling of cognitive maps using fuzzy techniques could be applied. Figure 9 shows some learning techniques that could improve the accuracy of fuzzy cognitive maps.

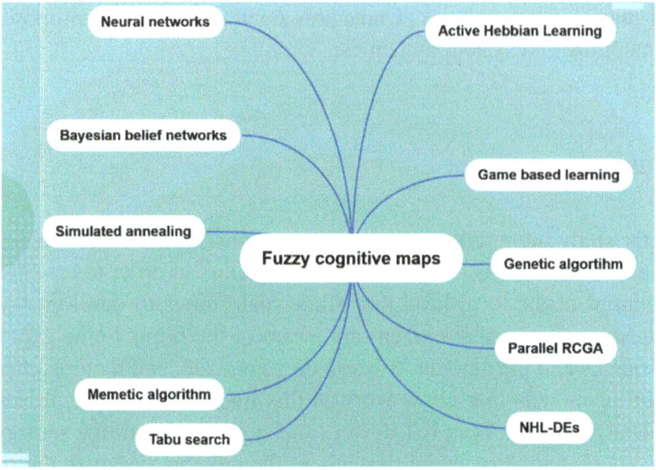


**Fig. 9** Learning techniques for fuzzy cognitive maps

# 6 Conclusions Limitations and Future Research

In this work, we have focused on gathering some relevant evidence about the use of IoT, artificial intelligence, big data, cloud in contrast to promoting equity, justice, solidarity, and respect for the ethnic and cultural aspects without forgetting the defense of biodiversity. All of them are primary axes of sustainability, so technology must approach these aspects to build more effective solutions that respond to the complexity of human development problems.

However, from technological point of view, the lack of politics related to public data access is a limitation for developing experimental and holistic studies. On the other hand, the behavior of citizens is very sensitive to changing their response patterns due to the influence of context variables (persuading to generate new needs or increase the search for satisfaction) and internal variables (thoughts or experiences that stimulate consumption). Uncertainty is a challenge and limitation when seeking to model social factors such as human behavior. The development of artificial cognitive networks must have permanent feedback and dynamism through inductive analysis that improves generalizations since the human experience is built from each human being's individuality. As future research, we expect academia to create and apply technology to minimize waste of resources, reduce pollution, and protect human health, well-being, and the ecological environment. So, the goal is that smart solutions allow us to know if our behaviors contribute to the environmental care of the planet.

## References


1. R.O. Andrade, S.G. Yoo, A comprehensive study of the use of LoRa in the development of smart cities. Appl. Sci. **9**, 4753 (2019)
2. R.O. Andrade, S.G. Yoo, M.F. Cazares, *A comprehensive Study of IOT for Alzheimer's Disease.* in Multi Conference on Computer Science and Information Systems, MCCSIS 2019—Proceedings of the International Conference on e-Health 2019, IADIS Press, pp. 175–182, 11th International Conference on e-Health 2019, EH 2019, Porto, Portugal, 17/07/19.
3. S. Barr, Strategies for sustainability: citizens and responsible environmental behaviour. Area. **35**, 227–240 (2003). https://doi.org/10.1111/1475-4762.00172
4. Y.-H. Cheng, Exploring passenger anxiety associated with train travel. Transportation **37**, 875–896 (2010). https://doi.org/10.1007/s11116-010-9267-z
5. G.J. Carreón, V.J. Hernández, L.C. García, A.J.M. Bustos, F. Morales, L. María de, F.A. Alfonso, *LA PSICOLOGÍA DE LA SUSTENTABILIDAD HÍDRICA. POLÍTICAS PÚBLICAS YMODELOS DE CONSUMO* Aposta. Revista de Ciencias Sociales, núm. 63, España https://www.redalyc.org/pdf/153/15329873005.pdf
6. C.B. Crawford, The theory of evolution: Of what value to psychology? J. Comp. Psychol. **103**(1), 4–22 (1989). https://doi.org/10.1037/0735-7036.103.1.4
7. E.R.G.T.R. Assunção, F.A.F. Ferreira, I. Meidutė-Kavaliauskienė, C. Zopounidis, L.F. Pereira, R.J.C. Correia, Rethinking urban sustainability using fuzzy cognitive mapping and system dynamics. Int. J. Sustainable Develop. World Ecology **27**(3), 261–275 (2020). https://doi.org/10.1080/13504509.2020.1722973
8. Gartner, Gartner *Says Worldwide IoT Security Spending Will Reach $1.5 Billion in 2018.*. Available online: https://www.gartner.com/en/newsroom/press-releases/2018-03-21-gartner-says-worldwideiot-security-spending-will-reach-1-point-5-billion-in-2018. Accessed on 15 Aug 2019
9. M.K. Hogg, P.C.N. Michell, Identity, self and consumption: A conceptual framework. J. Market. Manag. **12**(7), 629–644 (1996). https://doi.org/10.1080/0267257x.1996.9964441
10. S.C. Jagers, S. Matti, Ecological citizens: identifying values and beliefs that support individual environmental responsibility among Swedes. Sustainability **2**, 1055–1079 (2010)
11. S. Jena, H. Dholawala, M. Panda, P.K. Bhuyan, *Assessment of Induced Fuzziness in Passenger's Perspective of Transit Service Quality: A Sustainable Approach for Indian Transit Scenario.* in Transportation Research (Springer, Singapore, 2020), pp. 51–63
12. F. Lámbarry, M.M. Trujillo, C.G. Cumbres, Stress from an administrative perspective in public transport drivers in Mexico City: Minibus and metrobus. Estudios Gerenciales **32**(139), 112–119 (2016)
13. G Luna-Cortés, Stress perceived by foreigners that use public transportation in Bogotá (Colombia). Res. Transp. Econ. 100811 https://doi.org/10.1016/j.retrec.2019.100811
14. M.D. Lytras, A. Visvizi, M. Torres-Ruiz, E. Damiani, P. Jin, IEEE access special section editorial: urban computing and well-being in smart cities: services, applications, policymaking considerations. IEEE Access **8**, 72340–72346 (2020). https://doi.org/10.1109/access.2020.2988125
15. A. Machín, F. Octavio, A. Mena, N. Riverón *Sostenibilidad del desarrollo y formación de ingenieros.* Editorial Universitaria. ProQuest Ebook Central (2013) https://ebookcentral.proquest.com/lib/upsal/detail.action?docID=3216380.
16. V. Sanchez-Anguix, K.M. Chao, P. Novais, O. Boissier, V. Julian, Social and intelligent applications for future cities: Current advances (2020)
17. A. Lavalle, M.A. Teruel, A. Maté, J. Trujillo, Improving sustainability of smart cities through visualization techniques for big data from IoT devices. Sustainability **12**(14), 5595 (2020)
18. P. Partheeban, H. P. Raju, R.R. Hemamalini, B. Shanthini, *Real-Time Vehicular Air Quality Monitoring Using Sensing Technology for Chennai.* in Transportation Research (Springer, Singapore, 2020), pp. 19–28

19. D. Schürholz, S. Kubler, A. Zaslavsky, Artificial intelligence-enabled context-aware air quality prediction for smart cities. J. Cleaner Prod. 121941 (2020)
20. R. Wener, G. Evans, D. Phillips, N. Nadler, Running for the 7:45: The effects of public transit improvements on commuter stress. Transportation **30**, 203–220 (2003). https://doi.org/10.1023/A:1022516221808

# Using Deterministic Theatre for Energy Management in Smart Environments

**Franco Cicirelli and Libero Nigro**

**Abstract** This chapter introduces a deterministic version of the Theatre actor system, suited to modelling, analysis and implementation of time-critical cyber-physical systems (CPS). A Theatre model can be composed of discrete-time actors and continuous-time hybrid modes which act as an interface to continuous evolving variables of an external environment. A Theatre model can be reduced onto timed automata (TA) and hybrid TA of the Uppaal toolbox, to be property checked by the statistical model checker. The chapter develops a significant case study concerning energy management in a smart environment, like a smart home or a smart industrial plant. The goal is to orchestrate a collection of electrical appliances, which can be suspended and re-activated, so as to exploit to a largest extent the power generated by a local source of energy, like a photovoltaic panel, a wind turbine, etc. The system, though, is also able to use energy from an external provider, with a control policy to avoid extra costs, when the local power generator sharply reduces its production. The chapter describes the case study and presents an experimental analysis aimed at comparing different power load scheduling strategies.

## 1 Introduction

Nowadays, there is a growing interest in the development of cyber-physical systems (CPS) [1], which operate with real-time constraints and provide critical services in such application domains like industry 4.0, automotive, healthcare, smart environments and avionics. A CPS integrates three basic components: a physical part (e.g. an industrial plant), an interconnection network and associated protocols and

---

F. Cicirelli
CNR—National Research Council of Italy, Institute for High Performance Computing and Networking—(ICAR), 87036 Rende, Italy
e-mail: f.cicirelli@icar.cnr.it

L. Nigro (✉)
DIMES—Department of Informatics Modelling Electronics and Systems Science, University of Calabria, 87036 Rende, Italy
e-mail: l.nigro@unical.it

 
A. Joshi et al. (eds.), *Sustainable Intelligent Systems*, Advances in Sustainability Science and Technology, https://doi.org/10.1007/978-981-33-4901-8_12

a cyber (software)-part which is in real-time control of the physical part. The physical part, interfaced with sensors and actuators, has its behaviour which evolves in Newtonian physical time. The cyber-part, instead, operates in logical discrete time. A deterministic communication layer is often required between the cyber and physical part of the system, e.g. in the automotive domain, one or more Controller Area Networks (CAN) can be used to ensure a predictable exchange of information among core components.

The development of CPS is challenging because it requires to manage: (a) concurrency/distribution aspects; (b) multiple timing models to be reconciliated; (c) continuous behaviour of external variables which, during modelling and analysis, is abstracted by ordinary differential equations (ODEs); (d) component heterogeneity; (e) determinism in system behaviour; (f) multiple models fostering continuity in the system lifecycle; (g) the fulfilment of timing constraints.

Many research efforts are currently devoted to the development of CPS. A notable example is the Ptolemy community with the recent modelling meta-language Lingua Franca (e.g. [2]), which favours a deterministic approach to the design of models with repeatable behaviour. In this chapter, the Theatre actor system [3, 4] is used for developing CPS by following similar design guidelines as in the Lingua Franca. Despite its asynchronous message passing framework, the control-based character of Theatre can be exploited to ensure deterministic behaviours. Theatre is currently hosted in Java [5] which can be used for modelling, analysis and final implementation. A Theatre model, though, can also be reduced into the terms of the timed automata of the Uppaal toolbox [6], thus opening to the use of the exhaustive model checker for qualitative analysis and/or to the statistical model checker for quantitative property checking [3, 7, 8].

In the last few years, many works in the literature have been devoted to coping with issues tied to the scheduling of electrical appliances and devices in smart environments like, e.g. a smart home. One of the most important objectives of these studies is to improve the efficiency in using electrical energy so as to promote environmental and economic sustainability at the same time. In order to pursue the above-mentioned objective, Energy Management Systems (EMSs) are exploited with the goal of shaping the consumption profiles for energy by properly monitoring and control the electric loads. As an example, EMSs can be used for avoiding peak consumption [9] and reducing the energy bill by lowering energy consumption when the energy cost is high while satisfying, in the meanwhile, some expressed users' preferences in using the available equipment.

The development of effective EMSs, though, is not a trivial issue. In fact, in order to reduce energy cost, several factors require to be considered all together like the presence of renewable energy sources [10] which rises the problem of using as much as possible the self-produced energy with the goal of minimizing the cost of buying energy from the electrical grid [11]. In another case, instead, it is required to consider that energy can be supplied with a variable price [12]. In addition, EMSs challenge for the adoption of methodological approaches helping in revealing some undesirable effects which could emerge when the cyber and the physical parts of the system interact with each other, and to ensure determinism between the behaviour observed during analysis, with that of the system put into execution.

The original contribution of this chapter rests on the use of Theatre deterministic actors for formal modelling and analysis of energy management aspects in a smart grid. The approach is novel because it enables a development methodology for CPS centred on model continuity [5, 13], that is, Theatre actors can be transformed without distortions from early modelling to design and synthesis, e.g. in Java. The contribution significantly enhances previous authors' work. With respect to [13, 14], an EMS model now includes hybrid aspects such as appliances with a continuous behaviour established by some ODEs. The preliminary ideas of deterministic Theatre introduced in [4] are extended and applied in this work to cope also with hybrid behaviour and uncertainty of the controlled external environment, with property checking which is based on statistical model checking [7] and simulations [15, 16].

The remaining of this chapter is structured as follows. First some related works are discussed. Then basic concepts of Theatre actors and continuous modes are presented, together with the determinism concerns. Theatre modelling is exemplified by introducing a case study based on energy management in smart grid (EMSG). The paper goes on by discussing a reduction of Theatre onto the timed automata of the Uppaal toolbox, by stressing the achievement of model determinism. After that, a detailed experimental analysis of the EMSG reduced model is presented. Finally, conclusions are drawn with an indication of further work.

## 2 Related Work

The CPS model-based development approach advocated in this work centres on the use of actors [17, 18] that are highly modular, concurrent and distributed software components, which share no data and communicate with each other by asynchronous message passing. In its basic model, though, actors are best suited for untimed applications which tolerate nondeterminism in the message delivery. In the last few years, some semantic variations of actors have been proposed, with the goal of addressing real-time constraints by an explicit support of timing features, and by enabling determinism and model behaviour repeatability by enforcing an ordered message delivery.

An influential work about deterministic actors is described in [19, 20], where dependences from previous semantics models of *networks of asynchronous processes*, *dataflow systems*, *synchronous-reactive* and *discrete-event* systems are clarified. The modified actor model is based on "*reactors*" which favour system composability by inter-connecting typed input/output *ports* which carry events (messages). The behaviour of reactors is specified by *reactions* that are message handler methods which can modify local data status and act as input/output event transformers, and *actions* which interface a CPS model with asynchronous external events (e.g. interrupts) which schedule corresponding internal events. Timing aspects can be expressed by *timers* to trigger reactions at selected times, and by *delays* and *deadlines* in reactors, which respectively constrain the generation of an output event or specify a time limit on the delivery of an event. Reactors are based on *timestamped messages*. Deterministic behaviour is ensured by messages which are delivered in timestamp order. Timestamps are in *logical times*, that is, *model* time. However, a fundamental issue is keeping a relationship between logical time (LT) and physical (wall-clock) time (PT). LT tends to approximate PT as the worst-case execution time (WCET) of reactions, the network communication latency, etc. are known and modelled, e.g. by delays.

Reactors are part of the current meta-modelling polyglot language Lingua Franca [2], which permits to specify the execution body of reactions in C, Python, Java, etc. A distributed implementation framework of reactors is envisioned on top of proved concepts of Ptides [21] and Google Spanner [22] systems based on global time and distributed timestamped messages.

Basic semantic concepts of reactors are also at the basis of other timed actor-based modelling and implementation languages like Timed Rebeca [23] and Theatre [3]. Timed Rebeca relies on classical actors, where each actor (rebec) owns an internal thread. To each message send two timing attributes, *after* and *deadline*, can be attached to specify respectively how many time units have to pass (*after*) before the message can be consigned, and the maximum allowed amount of time (*deadline*) which can pass to deliver the message. Beyond the deadline, the message becomes discarded. Timed Rebeca rests on a suspensive execution model for reactions (called message servers), where a *delay* statement can be used to specify the duration of a code fragment. In [2, 24], a transformation of a Lingua Franca reactor model into Timed Rebeca is discussed which permits, starting from given assumptions about the timing of the external controlled environment, to model checking the reactor model using the Afra tool [25]. Determinism is pursed in Timed Rebeca by attaching a priority number (unique ID) to actors and message servers.

Theatre, which will be detailed later in this chapter, syntactically resembles Timed Rebeca. However, Theatre models (1) are composed of lightweight actors (they have no internal thread); (2) actors have message reactions with a non-suspensive semantics; and (3) message scheduling and ordered delivery can be regulated by a customizable control strategy. The possibility of exhaustive model checking CPS models based on deterministic Theatre, when arbitrary continuous behaviour in peripheral entities is relaxed, is shown in [4, 26].

In this chapter, energy management in a smart grid is used as a significant CPS modelling case study using deterministic Theatre, with some electrical appliances which have a power consumption curve specified by ODEs. Therefore, statistical model checking [7, 9] and simulation are used for property checking.

The growing demand of electricity and the availability of renewable power sources (photovoltaic panels, wind turbines, etc.) close to the utilization site, put a great challenge in optimizing the consume of electricity in smart environments like a residential smart home [9] or a smart grid industrial plant. An energy management system (EMS) is devoted to monitoring and controlling the available power and the power required by "intelligent" appliances, which can be activated and deactivated dynamically through sensors and actuators, e.g. through a "smart plug", so as to impose suitable scheduling strategies to appliance power loads, with a desirable overall power consumption curve motivated by sustainability requirements.

Devising and implementing effective scheduling algorithms for electric loads is a multifaceted problem [27–30]. Often, a trade-off between energy consumption and the comfort level of the occupants in a smart environment should be established. An adaptive algorithm for diminishing energy consumption without affecting the comfort of home occupants is proposed in [29]. An approach based on deep reinforcement learning for on-line optimization of the scheduling of loads in a smart grid context is described in [30]. In order to favour a more efficient use of electricity, the on-line scheduling algorithm also provides suitable real-time feedbacks to load users. With the goal of helping customers in modifying their consumption patterns, the use of different centralized and decentralized scheduling algorithms is provided in [27]. The paper compares some algorithms on the base of the amount of needed resources and of the complexity of the required implementing infrastructure. The problem of preserving privacy while managing people profiles in an EMS is instead taken into account in [28]. A scheduling algorithm to reduce the electricity bill is proposed in [10]. The approach considers renewable energy sources and uses reinforcement learning techniques. As a test case, the scheduling of six loads with a time granularity of one hour is considered. A real implementation of the system, though, is not considered. An EMS named CAES is proposed in [32]. CAES exploits Q-learning for optimizing energy consumption and to adapt to preferences in consumers' habits that can change over time and to variable energy prices. The approach does not take into account variable profiles for load consumption and self-production of energy.

## 3 An Overview to Theatre

Theatre is both a modelling language and an implementation framework based on timed actors [3]. Theatre actors are thread-less. Messages are (transparently) captured by a *reflective control layer* which defines the delivery order. A parallel/distributed system [5, 33] consists of a collection of Theatre nodes. Each Theatre node is a composition of actors. Actor are supposed to have universal names (e.g. global strings in a distributed system [5], Java object references on a multicore machine [33], etc.). Differently from reactors [20] and similarly, e.g. to Timed Rebeca [23], instead of input/output ports, each actor is assumed to know the identity of partner actors (*acquaintances*) to which messages can be sent. For proactivity, an actor can also send messages to itself. Theatre actors can possibly migrate (through a move() operation) from a node to another, e.g. for load-balancing issues. Two concurrency levels exist. Within a same Theatre, only one actor at a time can be processing a message (*cooperative concurrency*). An actor is at rest until a message arrives. A message reaction method, which is named as a *message server* (msgsrv) as in Timed Rebeca [23], is atomic and cannot be pre-empted nor suspended. However, two or more actors can simultaneously be executing message servers, provided they belong to distinct Theatres (e.g. *true parallelism* on a multicore machine).

Theatre is based on *global time* and *timestamped* messages. As in Timed Rebeca [23], an *after* and a *deadline* attribute can be attached to each message send. They are relative times measured from the sending time. A message cannot be delivered before *after* time units are elapsed. However, a message should be delivered before *deadline* time units elapse. Going beyond a message deadline causes the message to become invalid and to be discarded. When omitted, an *after* defaults to 0 and a *deadline* to $\infty$. It is important to stress that the timeline of a Theatre model, as in Reactors [19, 20] and Timed Rebeca [23], is *logical time*. It is possible to establish a relationship between logical time and *physical time* by using *delay(d)* instructions, which permit to express the duration of a code segment, e.g. an entire message server body. To comply with the non-suspensive semantics of a message server, delay can only be used as the *last* operation of a message server.

During modelling and analysis, a Theatre system can be abstracted as a collection of *processing units* (PU). The effect of a delay is then to occupy a PU for the duration of the delay. It is not possible to dispatch a message to a busy PU. A PU comes back to its free state, when a delay expires.

In [26, 34] Theatre modelling was extended with *hybrid* actors [35]. More in particular, *continuous modes* were introduced. A mode is a special (physical) actor which is intended to interface a discrete actor model with a continuous changing variable of the external environment. As in hybrid automata [36], a continuous mode is defined by (a) an *invariant*, (b) a set of *flows* (ordinary differential equations or ODEs), (c) a *guard* and (d) a final *action*. In a case, a mode can reproduce a dynamical law in a given time period. When the mode terminates (the guard evaluates to true), the final action is executed.

In Theatre, the final action is exploited as another point where physical time *meets* logical time. The effects of the mode are propagated to the rest of a model by the action which sends a message to a normal actor (the so-called *accessor* actor [37]). The net effect for the discrete (cyber) model is to *sample* the value of a continuous variable, which is provided at the mode termination. When a Theatre model is finalized into a concrete implementation, a mode can remain in the synthesized system [9] just as an interface for accessing the value of an external variable, e.g. through an envGateway component [13].

### *3.1 Modelling Energy Management in a Smart Grid*

The shape of Theatre actors will be shown through an EMSG modelling example, which is drawn in an abstract syntax [3, 34] close to Java. An actor encapsulates an internal state of data variables and exposes an interface of message server methods to react to incoming messages. Processing a message can change the local data variables, send messages to known acquaintances (or to itself) and possibly create new actors. Message servers are always executed one at a time and run to completion.

A smart microgrid made up of a certain number of electrical appliances is considered. Each appliance is characterized by its consumption power curve, which can be piece-wise linear (level-based) or continuous. Level-based appliances can have their consumption curve furnished in a tabular way, by providing the power value and the time duration of each consumption segment. A continuous power appliance, instead, is defined by giving a list of ODEs which model each continuous power segment. The microgrid is supposed to have two sources of power: e.g. a local photovoltaic generator and the external energy furnished by a provider. The local generator is another example of a piece-wise level power signal. The model behaviour is assumed to evolve with a given time granularity hereafter called as period. At each period (e.g. 1 min or 1 s), the goal is to orchestrate (activate/deactivate) the electrical appliances so as to make use, to the largest extent, of the local generated power. If the generated power can sustain the requirement of at least one power load, the appliances are regulated so as to reduce the total consumption curve accordingly. Otherwise, when the local power generator is insufficient (e.g. due to a cloud), then the energy is withdrawn from the external provider. In this case, though, a threshold level is supposed to have been negotiated with the provider, so that to not exceeding the threshold in order to keep low the cost of the KWh.

The model is built around the following actors: Signal, Hvac, Controller, Main, plus the modes SignalMode, Rising, Up, Falling and Down. Each signal instance is associated with a level-based electrical appliance. From time to time, its current power value is acquired through the SignalMode instance which samples the corresponding appliance power curve. A particular signal instance corresponds to the local power generator. Hvac is an example of a load whose consumption curve is inspired to that of a physical Hvac. Four modes are associated with the Hvac: Rising (during the first phase of the Hvac operation), Up, when the maximum power is requested by the Hvac; Falling when, due to the inverter intervention, the required power diminishes towards a final value; Down, which corresponds to a final, minimum, required power value. Each Hvac mode operates in a given time interval. All the power loads, including the Hvac, can be suspended. For simplicity, it is assumed that every appliance is able to resume its activities (i.e. its power consumption) from the point it was last left off. The Controller is in charge of exercising a scheduling strategy for activating/shedding of the power loads. The Main component configures the model by creating and initializing the instances of actors and modes, and by assigning actors to Theatres according to a chosen partitioning schema, which can range from the maximum parallelism (each actor is allocated to a distinct theatre), to the minimum parallelism (all the actors are allocated to one only theatre), to intermediate solutions where groups of actors are co-located to selected theatres.

An excerpt of the EMSG model is shown in Figs. 1, 2 and 3. A full version of the model, formalized according to the hybrid timed automata (HTA) of the Uppaal toolbox [8], will be detailed later in this chapter. Some global declarations express scenario parameters, which influence the overall model behaviour.

Each actor/mode is denoted by *self* in its own behaviour. A mode is a special actor whose msgsrvs (activate/suspend) trigger a continuous behaviour. When activated, a SignalMode waits for a RESPONSE time and then sends a sample message to

**Fig. 1** The Signal actor

```
//scenario parameters
env double PERIOD=1.0;
env double RESPONSE=0.05;
env double threshold=2000.0;
env double Tr=1.0, Tu=5.0, Tf=4.0, Td=800.0;
env double K1=0.13, K2=0.4, h=10.0;

actor Signal{ //accessor
  //acquaintances
  Controller ctrl; SignalMode mode;
  //local data
  int[OFF,ON] status=OFF;
  double sampleValue;
  msgsrv init( Controller c,SignalMode m ){
    ctrl=c; mode=m;
  }//init
  msgsrv turn_on(){ status=ON; }
  msgsrv turn_off(){ status=OFF; }
  msgsrv proposal(){ mode.activate(self); }
  msgsrv sample(double value){
    sampleValue=value;
    if(value>=0.0) ctrl.ack(value,self);
    else ctrl.end(self);
  }//sample
}//Signal
```

**Fig. 2** The SignalMode and Rising modes

```
mode SignalMode{
    Signal dest;
    double t;
    msgsrv activate(Signal d ){
        dest=d; t=0;
        inv( t<=RESPONSE ){
          t'==1; //flow
        }
        guard( t>=RESPONSE ){
          dest.sample( getValue(dest) );//action
        }
    }//activate
}//mode
mode Rising{
    Hvac hv;
    double t, p;
    msgsrv activate( Hvac h ){
        hv=h; p=0; t=0;
        inv( t<=Tr ){ t'==1; p'==K1*(h-p); }
        guard( t>=Tr ){ hv.sample( p ); }
    }//activate
    msgsrv suspend(){
        inv( true ){ t'==0; p'==0; }
    }//suspend
}//Rising
```

```
actor Controller{
  actor load;
int cnt, replies;
msgsrv init(){
  self.control() after(PERIOD);
}//init
msgsrv control(){
  cnt=0; replies=activeLoads();
  while( cnt<replies ){
    load=nextLoad();
    load.proposal(); cnt++;
  }
  self.control() after(PERIOD); cnt=0;
}//control
msgsrv ack( actor load, double req ){
  request[id(load)]=req; cnt++;
  if( cnt==replies ) self.actuate();
}//ack
  msgsrv end( actor load ){
    request[id(load)]=0;
  terminated[id(load)]=true; cnt++;
  load.turn_off();
  if( cnt==replies ) self.actuate();
}//end
msgsrv actuate(){
  app_schedule(OPT); replies=activeLoads(),
  cnt=0
  while( cnt<replies ){
    load=nextLoad();
    if( isOn[id(load)] ) load.turn_on();
    else load.turn_off();
    cnt++;
  }
  totConsumption(); decrement();
}//actuate
}//Controller
```

**Fig. 3** The Controller actor

its accessor actor (Signal) with the current value of the power load. Such a value is sampled from the associated power curve (see the getValue() function). Despite the object-oriented notation, sending a message to an actor or to a mode has the asynchronous, non-blocking semantics. Many messages can be sent to be received at current time.

The Rising mode, which is associated with the Hvac actor, relies on an ODE (flow) for the continuous evolution of the electrical power p. When suspended, both the time t and the power p are frozen. In a similar way are modelled the active and suspended states of Up, Falling and Down modes of the Hvac. In the active status, Up has the invariant $t <=$ Tu and the flow $p' == 0$ (which means that the power is kept constant to the value obtained by the Rising mode); Falling has the invariant $t$ $<=$ Tf and the flow $p' == -K2*p$; finally, Down has the invariant $t <=$ Td and the flow $p' == 0$ (i.e. the power is frozen to the last value defined by Falling).

As witnessed in Fig. 1, Theatre actors are initialized by an explicit init() message which carries all the necessary information.

Figure 3 depicts the Controller actor which orchestrates the appliances through a scheduling strategy.

For model compactness, the Controller is supposed to manage all the information related to the appliances through global data. The control loop is established by having that the control() message is received by the controller at the beginning of each period. The first control() is asked by the init() message server. The next control() is self-sent at the end of processing the current control(). At each control iteration, first the candidate set of activatable loads is determined. Then, to each of such loads, a proposal() message is sent. Power loads will reply to proposal() either with an ack() or an end() message. An ack() message carries the identity of the appliance and its instantaneous power request. An end() message informs that a load has exhausted its power curve and it must be flagged as terminated. After all the replies have been received, the controller will execute the control actions by self-sending an actuate() message which implies the *app_schedule(strategy)* function is called which sets the values of the isOn[] boolean array, depending if a load is admitted (true) or not (false) to consume part of the available source power. Correspondingly, a turn_on() or turn_off() message is sent to loads. A turn_off() message sent to a continuous load like an Hvac implies its current mode is suspended. Similarly, a turn_on() causes the current mode to be re-activated. The control iteration is completed by computing, for monitoring purposes, the total power consumption and by decrementing the time interval of the current level of the active power loads of the PERIOD amount.

Different appliance scheduling strategies can be evaluated. The optimal one (OPT), which can be time demanding for its execution, explores exhaustively the solution space by using backtracking. It always finds, at each time point, the best solution of the appliances to activate, in such a way that the total power consumption fits to the largest extent the available power. Two approximate solutions can be considered as well. The first one (AP1) coincides with the first solution found by backtracking. The second one (AP2) is greedy-based and preliminary ranks in ascending order the various power load values and then the candidate appliances are chosen so as to fit as close as possible the current value of the generated power or the threshold of the external provider.

## 3.2 Ensuring Determinism in a Theatre Model

The EMSG model rests on timestamped messages. Timestamps are defined by (logically) absolutizing the *after* value of the messages to *now* + *after*, where *now* is the current global time. For example, each control() message is time tagged with *now* + *PERIOD*. The control layer of Theatre consigns messages in timestamp order. Many messages exist in the EMSG which have the same timestamp and then are simultaneous.

A deterministic model [19] is one that, starting from a set of inputs (e.g. the load power curves), and an initial state (e.g. a chosen scheduling strategy), always produces the same output. In other terms, a deterministic model, for the same input and initial state, admits one only possible behaviour (e.g. the total power consumption curve of the activated loads, at each period, is always the same).

First of all, determinism is a feature of model design. For example, at each period, the controller of the EMSG will expect the power load replies and will take its decision when all the replies arrived. Independently from the model design, the control layer of Theatre contributes to determinism [4] by ensuring that all the simultaneous messages are eventually delivered according to a predefined precedence constraint rule.

As proposed in [2, 4, 24], deterministic Theatre associates a unique integer ID (priority: lower ID implies greater priority) to each actor instance. Moreover, a unique ID (priority) is attached to each message server of an actor class, which normally mirrors the textual order of appearance of the messages. Subsequently, simultaneous messages are first delivered according to the priority of the receiver actor. Simultaneous messages directed to the same actor are ordered according to the message IDs. Requests of distinct actors of a same message of a given receiver actor are ordered according to the priority of the sender actors.

Despite the precedence constraint on the delivery order of simultaneous messages, in a Theatre model, some nondeterminism remains due to the continuous modes and physical time of mode completion. When a mode terminates, its action sends an immediate message to the accessor actor. Such a message will be ordered according to the precedence constraint rule, but all depends on the physical time when the mode terminates its behaviour. The next section will provide further details about Theatre determinism.

## 4 Reducing Deterministic Theatre onto Uppaal Hybrid Timed Automata

The EMSG model can be transformed to Uppaal for verification. Actors can be reduced to timed automata (TA). Modes can be mapped onto hybrid TA [8], due to the use of ODEs. Model transformation relies on a preliminary naming phase which defines a unique integer ID to each entity in the model, that is to each actor and mode instance, and to each message. Then suitable subrange types can be introduced (by typedefs) for each category type, which will automatize TA instance creation at model configuration time. Towards this, it is sufficient to associate each TA/HTA with one only parameter of the corresponding typedef. For readability concerns and to keep a relationship with the abstract model, such a parameter is named self.

Some simple rules (see also [3, 34] can be drawn to derive automatically the automaton for each category type. An actor TA is built around two main locations: Receive and Select. In the Receive location, the next incoming message is awaited. In the Select location, the ID of the message is checked and the corresponding message server body is executed. Message reception is expressed by an input operation (?) on the channel msgsrv[self]. These channels, one per actor ID, are broadcast and non-blocking, to comply with the requirements of the Uppaal statistical model checker (U-SMC) [8]. Sending a message is assisted by the following global variables: S (contains the identity of the sender), R (denotes the receiver), M (holds the ID of the message), A (stores the after time), D (stores the deadline time). By default, $A = 0$, $D =$ INF (a constant denoting infinity). Message arguments, if there any, are held in a global array arg[] of doubles. A message server body is modelled as a cascade of committed locations [6], which guarantees the msgsrv completes its execution *before* a next message can be delivered (msgsrv atomicity). Message scheduling and delivery are responsibility of a scheduler automaton (see Fig. 13) which ensures determinism in the message delivery order. The broadcast channels send and delay are used to signal a message send and a delay setting. Continuous modes understand two (broadcast) channels: activate[self] and suspend[self], which have an element for each mode ID. This organization was adopted to comply with the specification of the associated ODEs. A continuous varying quantity (e.g. the power p of the Hvac in the EMSG model) must be modelled by a stopwatch (clock). It is important to recall that all the clocks of an Uppaal model automatically increase with the same rate. A clock can be reset. To froze the value of a clock p, it is sufficient to specify 0 as its first derivative ($p' == 0$).

Figure 4 portrays the automaton of the signal actor of the EMSG model. There is a one-to-one correspondence with the abstract model in Fig. 1. Figure 5 shows the SignalMode automaton (see also Fig. 2). The action channel is signalled in Fig. 5 to testify mode termination. It is an urgent channel whose communication must be taken as soon as the response time elapsed.

Figure 6 reports the Hvac actor automaton. At initialization time, the Hvac receives and stores internally the identity of the controller and of its four modes (Rising, Up, Falling, Down). The automaton uses the cm internal variable to store the current mode, which can also be NONE when the load was not yet started, or all its temporal behaviour finished. At the arrival of a SAMPLE message, which corresponds to current mode termination, the actor receives and memorizes in the sampleValue, the last updated power value, then switches to the next mode, if there are any, in the order: rising- > up- > falling- > down. A TURN_OFF message causes the current mode to be suspended (and the power variable p to be frozen). At a subsequent TURN_ON message, the current suspended mode is re-activated.

The dynamic of the four modes accompanying Hvac is shown in Figs. 7, 8, 9 and 10. The mode flows are modelled in the invariant of the Hold location. In the

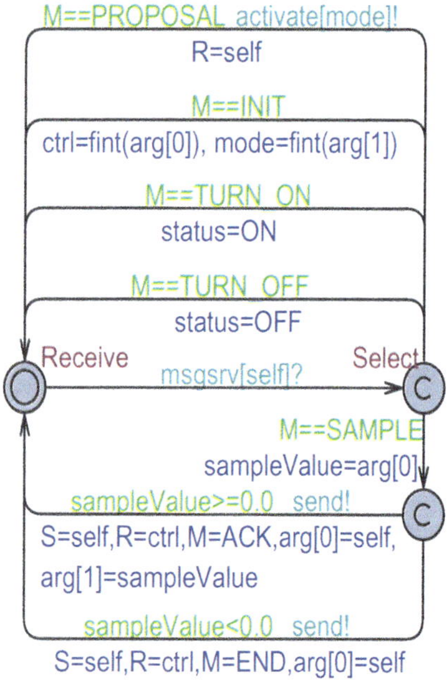


**Fig. 4** The Signal actor automaton

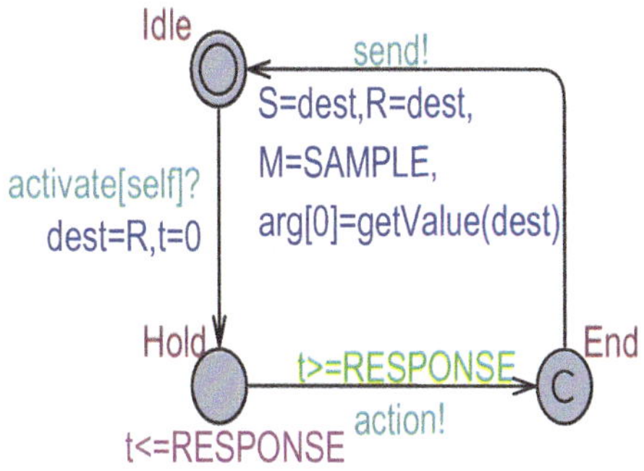


**Fig. 5** The SignalMode mode automaton

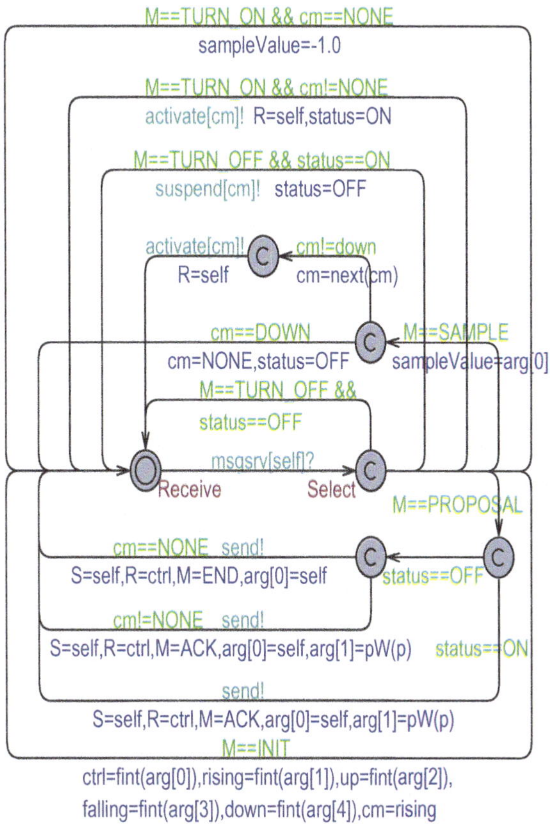


**Fig. 6** The Hvac actor automaton

Suspend location, the mode freezes the value of the two clocks $t$ (used to measure time advancement) and $p$ (used to reproduce the power load dynamic). The function $pW(p)$ converts the $p$ values in Watts.

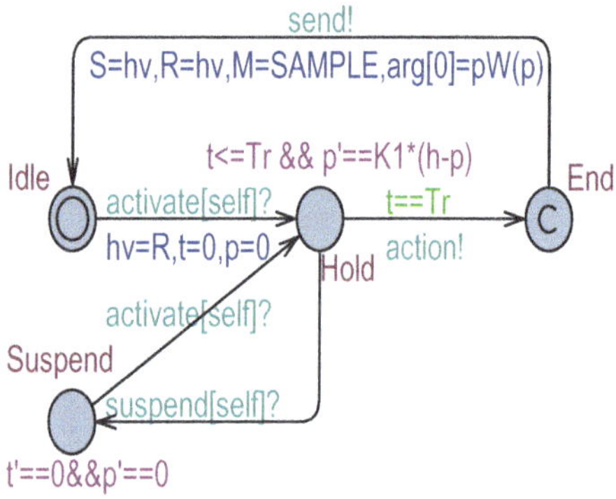


**Fig. 7** The Rising mode automaton

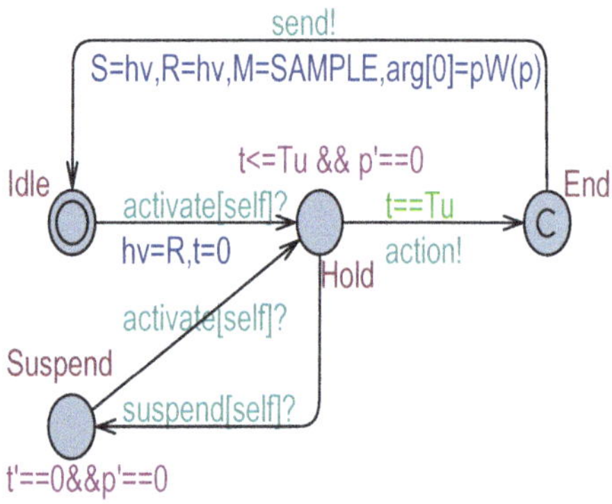


**Fig. 8** The Up mode automaton

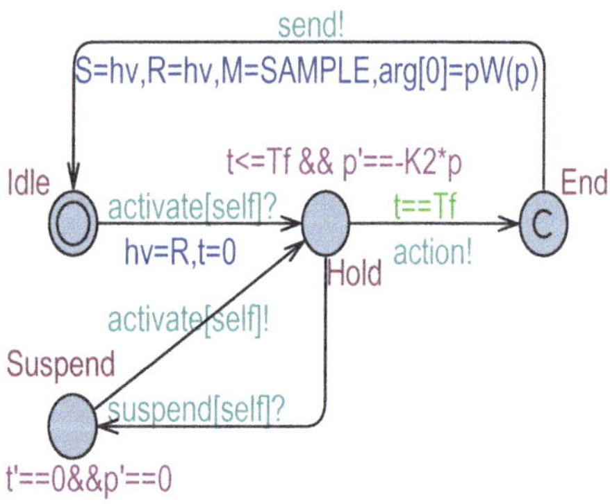


**Fig. 9** The Falling mode automaton

Figure 11 portrays the Controller actor automaton which directly derives from the abstract model in Fig. 3. In Fig. 12 is drawn the automaton of the Main actor which configures and starts the model execution. The Main actor first moves each actor to a distinct theatre. This is done for demonstration purposes. Other partitioning could be adopted as well. After that, the init() message is sent to the controller, then to

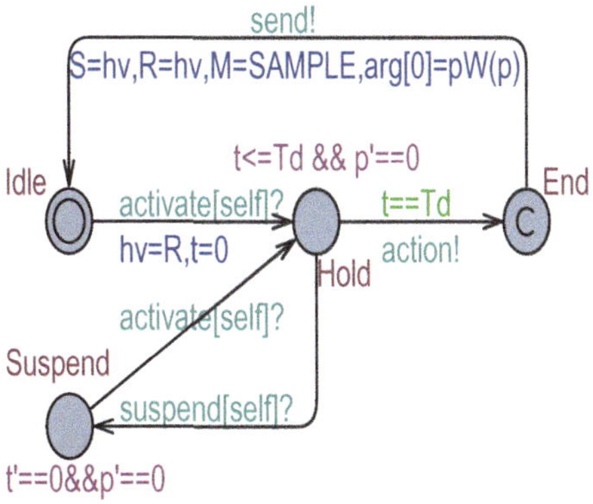


Fig. 10 The Down mode automaton

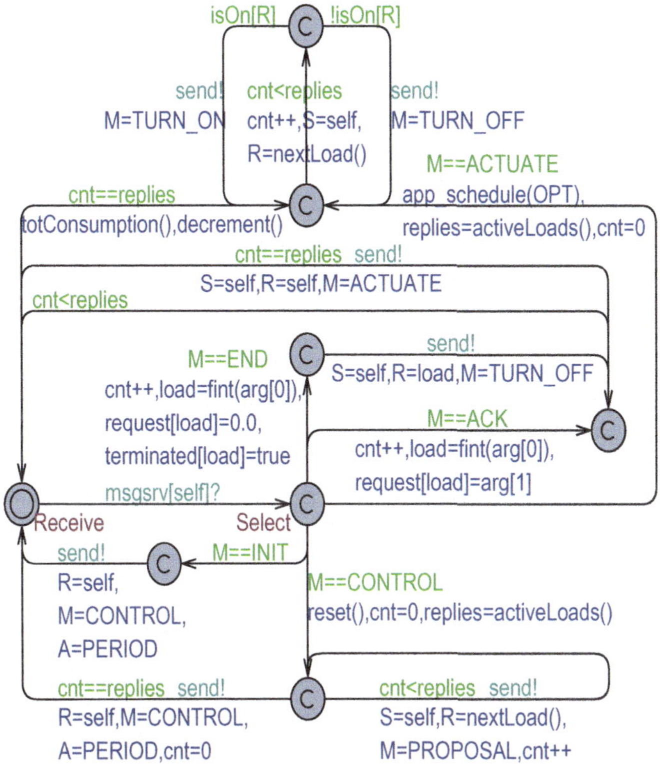


**Fig. 11** The Controller actor automaton

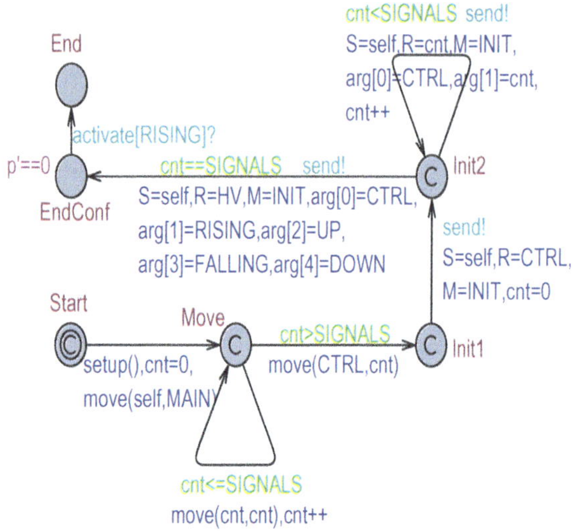


**Fig. 12** The Main actor automaton

each power load (Signal) including the Hvac. The constant SIGNALS denotes the number of power loads. The IDs from 0 to SIGNALS-1 refer to level-based power loads. Signal 0 corresponds to the local generator. The value SIGNALS was used as the ID of the Hvac, which is also denoted with the HV ID.

After sending the init() message to the Hvac, which transmits through the argument buffer arg[], the identity of the controller and that of the four related modes, the location EndConf is reached in Fig. 12 which means the configuration process finished. In EndConf, the clock variable p which mirrors the continuous power variable of the Hvac is thus frozen. This way, it is impeded to p to automatically grow when the first control was not yet started. As a benefit of the broadcast nature of all the used channels, which have exactly one only non-blocking sender and 0, 1 or multiple receivers, as soon as the Hvac activates its first mode Rising, EndConf is

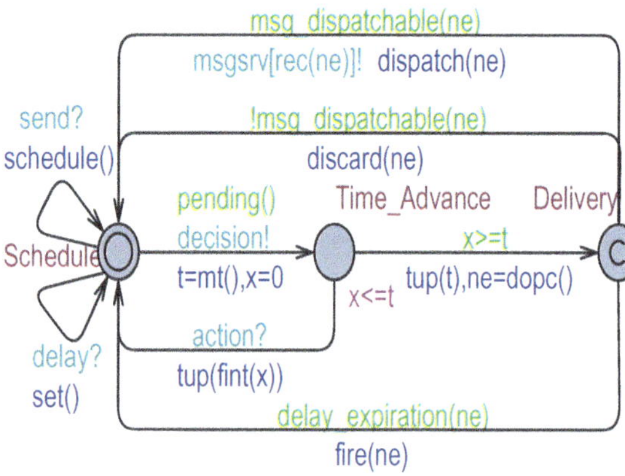


**Fig. 13** The Scheduler automaton

abandoned and the final, harmless, End location is entered where the Main will have no more active role in the model. Most importantly, since the location with $p' == 0$ is exited, the clock $p$ reassumes the default $p' == 1$ and it is enabled to increase its value according to the ODE of the Rising mode (Fig. 7).

**Model determinism and the Scheduler Automaton**
A fundamental role is played by the Scheduler automaton (see Fig. 13) which is an enhanced version of that developed in [4]. The scheduler works with a relative time model. In other terms, the values of *after* and *deadline* of a message are stored in their original relative value.

During the atomic execution of a msgsrv, the Scheduler remains in the Schedule location. Any scheduling operation (message send or delay setup) raised from within the msgsrv is collected by the scheduler through the send and delay synchronizations. As a consequence, any sent message, together with its attributes (sender, receiver, message ID, arguments, after and deadline fields), gets scheduled in a data structure of the Scheduler. Similarly, any delay setting gets scheduled on a second data structure of the Scheduler.

At the end of the msgsrv, the urgent (and broadcast) channel decision is fictitiously signalled (although it is received by no other automaton in the model) and causes the *most imminent time* (mt) of occurrence events (message delivery or delay expiration) to be determined and the Time_Advance location entered where mt is allowed to pass.

During this waiting phase, the non-deterministic firing of an external mode could inject in the model a *straggler* event, that is one with a *lower* occurrence time. In this case, following the trigger of the action urgent channel, the Scheduler comes back to the Schedule location, by taking into account the partial elapsed time in Time_Advance. In Schedule, all the scheduling consequences of processing the final message of the mode are taken, after that the normal behaviour of the scheduler is resumed. If no straggler occurs during its stay in Time_Advance, then the Scheduler first updates all the pending events occurrence time by subtracting from the fields after and deadline the mt value. Finally, the precedence constraint on the delivery of simultaneous events is applied. The scheduler gives precedence to a delay expiration with respect to a message delivery, because it frees a given processing unit (PU or theatre). If a message delivery is instead involved, it is effectively dispatched provided its deadline was not yet expired.

It is the operation of the Scheduler automaton which provides determinism to a Theatre model.

## 5 Experimental Analysis

The EMSG model detailed in the first subsection of the third section of this chapter, formally reduced to the hybrid timed automata of Uppaal [8] as shown in the fourth section, constitutes the experimental setup which was analysed through the

Uppaal statistical model checker (U-SMC). Experiments were carried out on an Asus ZenBook Win10 machine, Intel Core i7-8565U CPU@1.80 GHz, 16 GB Ram.

The experimental model is composed of seven level-based power loads, one power signal generator and one Hvac appliance.

Figure 14 shows the generator power curve (in Watt) which was achieved through the U-SMC query:

simulate [< = 1160] { Signal(0).sampleValue}.

Figure 15 depicts the level-based consumption power curve of load#1. Figure 16 shows the power curve of the Hvac. Due to the activating/suspending operation of the load scheduler, an effective delayed dynamic Hvac power curve is reported in Fig. 17. The time span of simulation experiments was set to 1160 time units as dictated by the time duration of the generator power signal. The generator signal admits an interval of about 100 time units where the produced power is very low (100 W). In this interval, there is the need for the model to acquire power from the external energy provider.

Figure 18 shows the total load power consumption versus time, when the optimal scheduling strategy (OPT) is chosen. The picture was achieved by first launching the query:

simulate [<=1160] {Signal(0).sampleValue, totCons, threshold}.

where the variable totCons holds, from time to time, the total consumed power of the loads. As one can see in Fig. 18, the total consumption curve fits very well the available power, especially in the first part of the behaviour. In the case of a low generated power signal, the scheduler orchestrates the power loads so as to not exceed the provider negotiated power threshold. In the second part of Fig. 18, the total consumption curve is lower than the generator power simply because many appliances have reduced power requirements; in other terms, it is diminished the power demand of the appliances. Figure 19 depicts the number of terminated loads

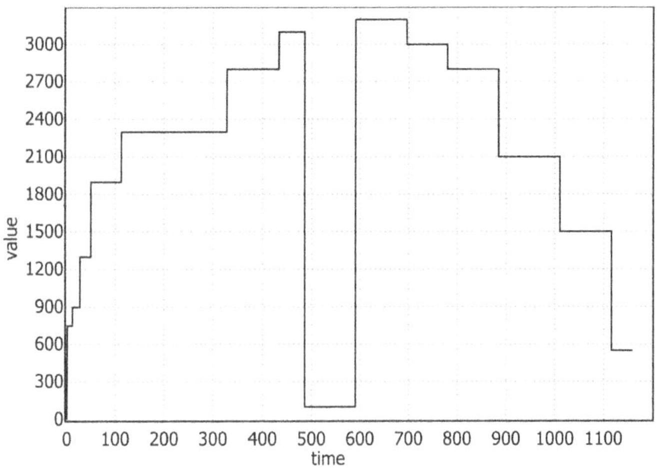


**Fig. 14** The Generator power curve

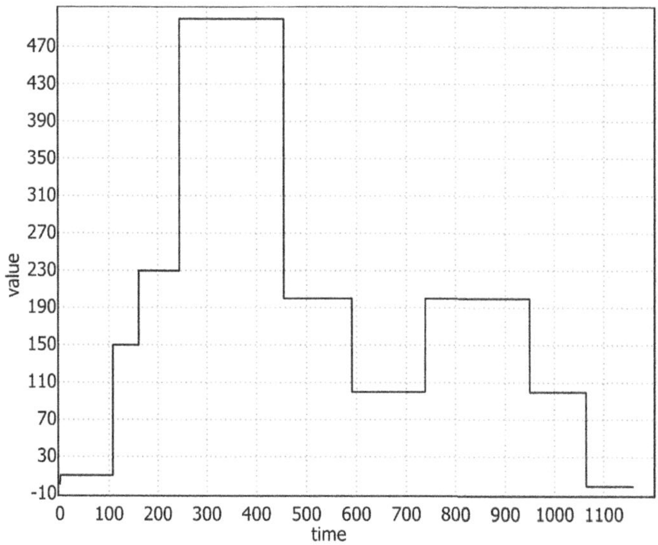


**Fig. 15** The Load#1 power consumption curve

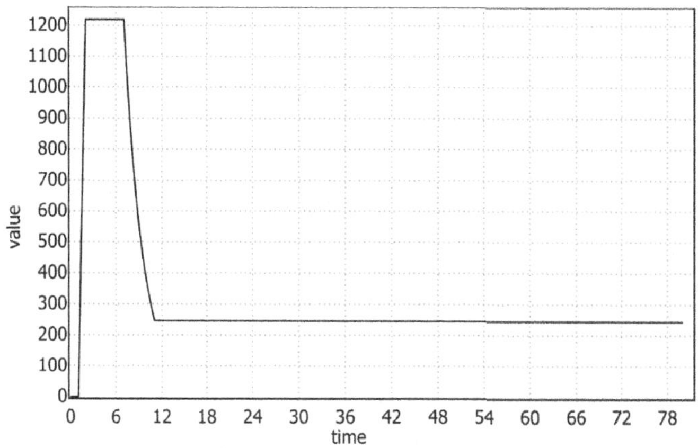


**Fig. 16** The Hvac power curve

versus time. Checking the number of terminated appliances can be important in an industrial smart grid where the more are the loads which finish their work, the better is from a production point of view.

The MITL [U-SMC] query:

Pr[<=1160](<> power_ref == threshold).

was launched to check if effectively there are times for the model when the power is requested to the external provider. After 29 runs, a confidence interval (CI) of [0.901855,1] with confidence 0.95 is proposed. With the default settings (e.g. the

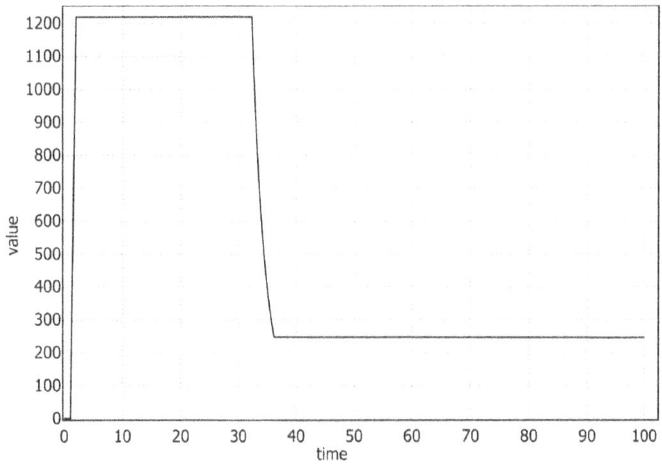


**Fig. 17** Suspension/re-activation shift effects on the Hvac power curve

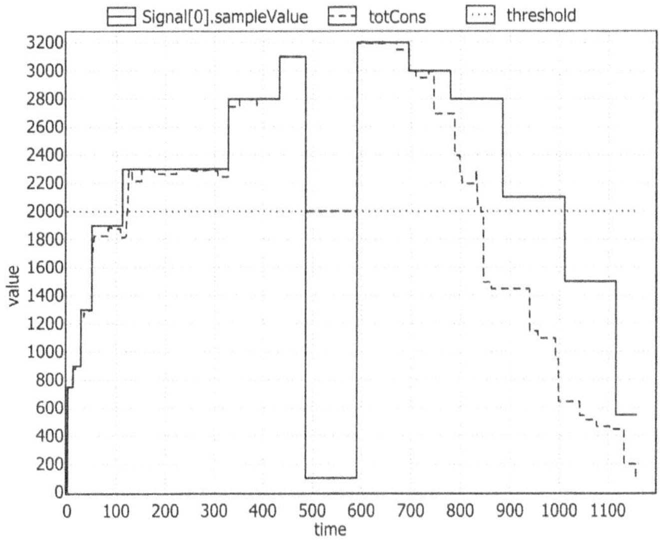


**Fig. 18** Total power consumption versus time—OPT

error in a confidence interval is $\varepsilon = 0.05$), this response indicates an almost certain event.

The following query checked specifically, if the Hvac power load, under OPT, completes its behaviour:

Pr[<=1160] ([] now < 1160||terminated[HV]).

Also in this case, U-SMC provides a probabilistic response of an almost certain event.

**Fig. 19** #Terminated loads versus time—OPT

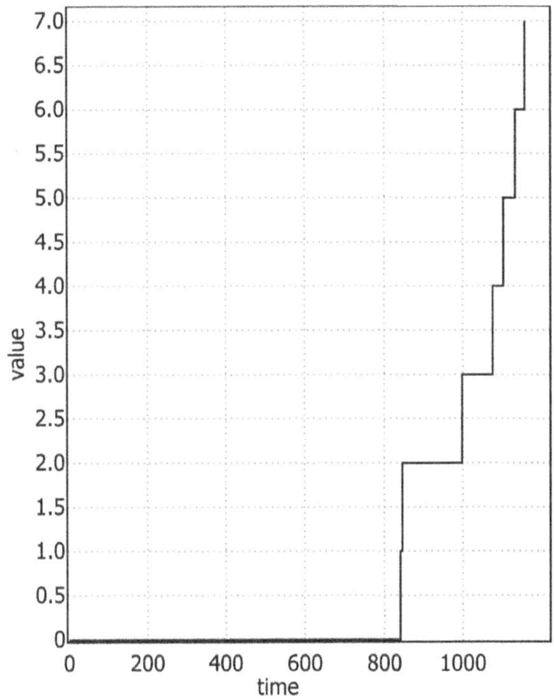


Figure 20 shows the total power consumption fitting curve, and Fig. 21 the number of terminated loads, when the AP1 strategy is used (the first solution found by backtracking).

Figures 22 and 23 portray respectively the total power consumption fitting curve and number of terminated loads versus time; in the case, the AP2 strategy is used (greedy-based solution, with loads preliminarily ranked by ascending power requests).

Figures 20 and 22 confirm the sub-optimality of the strategies AP1 and AP2 with respect to OPT (Fig. 18). Not only the total power consumption fits with a lesser degree the available power but also (and in consequence) the number of terminated loads (Figs. 21 and 23) is smaller than that for OPT. Of course, these results must be taken as indicative, because what really happens depends exactly on the assumed power loads and consumption power curves.

For both the strategies AP1 and AP2, the query about the probability of termination of the Hvac, ends (after 29 runs) with a confidence interval of [0, 0.0981446], with confidence 0.95, testifying the Hvac now is likely to not terminate.

In the light of the experimental results, the EMSG model was found correct from both the logical and, most importantly, the temporal point of view. The behaviour was found perfectly repeatable due to the use of deterministic Theatre.

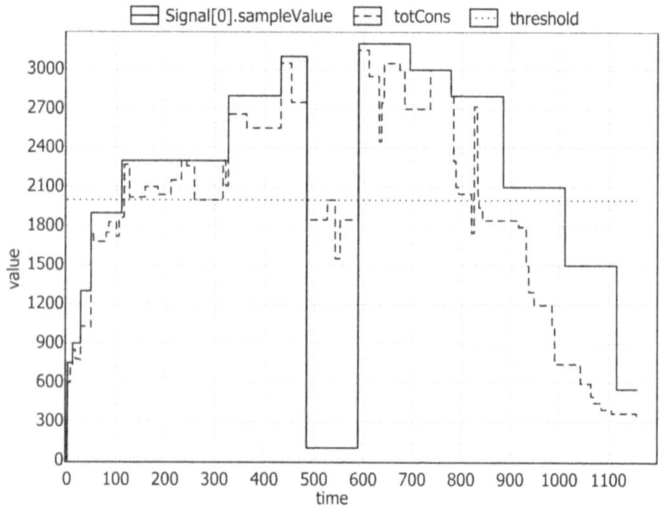


**Fig. 20** Total power consumption versus time—AP1

**Fig. 21** #Terminated loads versus time—AP1

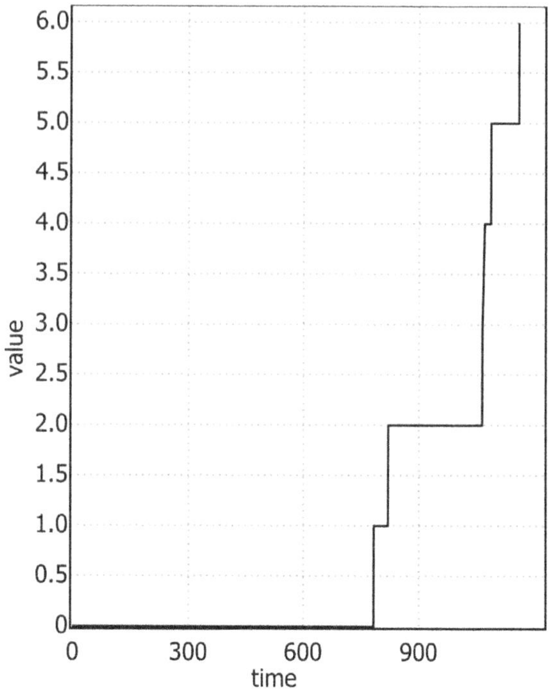

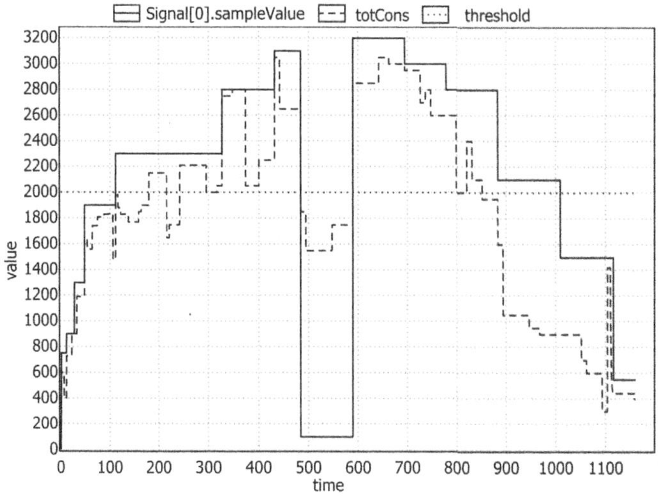


**Fig. 22** Total power consumption versus time—AP2

**Fig. 23** #Terminated loads versus time—AP2

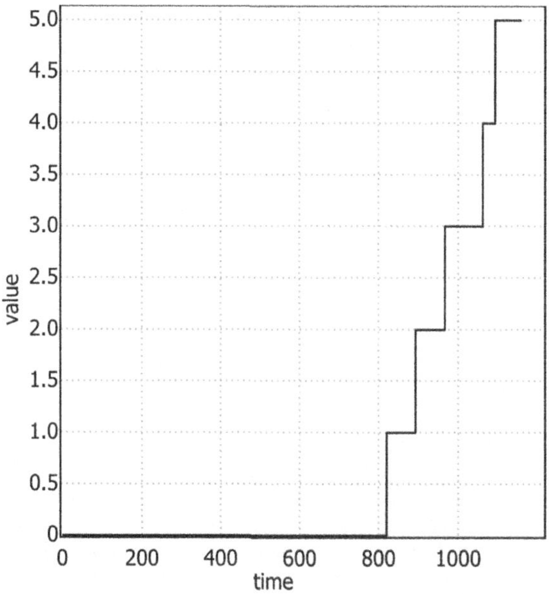


# 6 Conclusions

This paper argues that Theatre [3, 5, 34] extended with deterministic control of simultaneous messages [4] can be effective in the development of cyber-physical systems (CPS), which must fulfil timing constraints and have the problem of reconciliating the external environment physical time with the discrete, logical time of a model. The paper shows how the actors and continuous modes of a Theatre CPS model can be reduced into the terms of timed automata (TA) and hybrid TA of the Uppaal [6, 8] toolbox, for property checking using the statistical model checker. A significant case study concerned with energy management in a smart microgrid, where a collection of appliances is dynamically orchestrated (by activating/deactivating power loads) in order to exploit to the "large extent" an available local source of energy (e.g. a photovoltaic panel), is modelled and thoroughly analysed. It is shown how the model example can benefit from a control layer which ensures a deterministic delivery order of contemporary events, which can guarantee one single model behaviour when started from a given input and state.

Prosecution of the research is geared at:

- continuing the experience with the use of deterministic Theatre in CPS modelling, analysis and implementation, particularly in the automotive [26] and the healthcare [38] application domains;
- applying Theatre modelling and analysis to complex energy management systems in smart environments [15];
- evaluating problems related to the concrete implementation of deterministic Theatre. Although an effective, centralized solution is immediately practicable, namely using, in a parallel/distributed context, one (master) Theatre devoted to scheduling and ordered delivery of messages, a decentralized solution as that advocated for reactors [19, 20], and inspired to the Ptides [21] and Google Spanner [22] systems, could be of much interest.

## References


1. E.A. Lee. *Cyber Physical Systems: Design Challenges*. in International Symposium on Object Component Service-Oriented Real-Time Distributed Computing (ISORC) (IEEE, 2008), pp. 363–369
2. M. Sirjani, E.A. Lee, E. Khamespanah. Verification of Cyber-Physical Systems. Submitted to MDPI (2020)
3. L. Nigro, P.F. Sciammarella, Qualitative and quantitative model checking of distributed probabilistic timed actors. Simul. Model. Pract. Theory **87**, 343–368 (2018)
4. L. Nigro. *Modelling and Analysis of Cyber-Physical Systems Using Deterministic Theatre*. in 4th IEEE World Conference on Smart Trends in Systems, Security and Sustainability (WS4 2020), London, 27–28 July, 2020
5. F. Cicirelli, L. Nigro, P.F. Sciammarella. Seamless development in Java of distributed real-time systems using actors. Int. J. Simulation and Process Modelling. **15**(½), 13–29 (2020)
6. G. Behrmann, A. David, K. Larsen. *A Tutorial on UPPAAL*. in Formal Methods for the Design of Real-Time Systems, LNCS 3185, ed. by M. Bernardo, F. Corradini (Springer, 2004), pp. 200–236
7. G. Agha, K. Palmskog. A survey of statistical model checking. ACM Trans. Model. Comput. Simulat. **28**(1, article 6), 16–39 (2018)

8. A. David, K.G. Larsen, A. Legay, M. Mikucionis, D.B. Poulsen, Uppaal SMC tutorial. Int. J. Softw. Tools Technol. Transf. **17**(4), 397–415 (2015). https://doi.org/10.1007/s10009-014-0361-y
9. F. Cicirelli and L. Nigro. *Home Energy Management Using Theatre with Hybrid Actors*. in 2019 IEEE/ACM 23rd International Symposium on Distributed Simulation and Real Time Applications (DS-RT 2019) (Cosenza, Italy, 2019) pp. 1–8
10. T. Remani, E. Jasmin, T.I. Ahamed, Residential load scheduling with renewable generation in the smart grid: A reinforcement learning approach. IEEE Syst. J. **13**(3), 3283–3294 (2018)
11. G. Belli, A. Giordano, C. Mastroianni, D. Menniti, A. Pinnarelli, L. Scarcello, N. Sorrentino, M. Stillo, A unified model for the optimal management of electrical and thermal equipment of a prosumer in a DR environment. IEEE Trans. Smart Grid **10**(2), 1791–1800 (2017)
12. N. Amjady, M. Hemmati, Energy price forecasting—problems and proposals for such predictions. IEEE Power Energ. Mag. **4**(2), 20–29 (2006)
13. F. Cicirelli, L. Nigro, P.F. Sciammarella, Model continuity in cyber-physical systems: A control-centered methodology based on agents. Simul. Model. Pract. Theory **83**, 93–107 (2018)
14. L. Nigro, P. Sciammarella, *Verification of a Smart Power Control System Using Hybrid Actors*. (IEEE WorldS4, London 2019)
15. R. Egert, C.G. Cordero, A. Tundis, M. Muhlhauser. HOLEG: *A Simulator for Evaluating Resilient Energy Networks Based on the Holon Analogy*. in 21th IEEE/ACM Symposium on Distributed Simulation and Real Time Applications (DSRT 2017), pp. 1–8
16. R. Castro, E.P. Marcosig, J.I. Giribet. *Simulation Model Continuity for Efficient Development of Embedded Controllers in Cyber-Physical Systems*. in Complexity Challenges in Cyber Physical Systems, Using Modelling and Simulation (M&S) to support Intelligence, Adaptation and Autonomy, 1st edn, ed. by S. Mittal, A. Tolk (Wiley, 2020)
17. C. Hewitt, P.B. Bishop, R. Steiger. A *Universal Modular ACTOR Formalism for Artificial Intelligence*. in Proceedings of of the 3rd International Joint Conference on Artificial Intelligence (Standford, CA, USA, August 20–23, 1973), pp. 235–245
18. G. Agha, *Actors: A Model of Concurrent Computation in Distributed Systems* (MIT Press, Cambridge, MA, USA, 1986).
19. M. Lohstroh, E.A. Lee, *Deterministic Actors* (Forum on Specification and Design Languages, Southampton, UK, 2019).
20. M. Lohstroh, M. Schoeberl, A. Goens, A. Wasicek, C. Gill, M. Sirjani, E.A. Lee. Invited: *Actors Revisited for Time-Critical Systems*. in *Design Automation Conference (DAC)* (2019)
21. Y., Zhao, E.A. Lee, J. Liu. *A Programming Model for Time-Synchronized Distributed Real-Time Systems*.in Real-Time and Embedded Technology and Applications Symposium (RTAS) (IEEE, 2007), pp. 259–268
22. J.C. Corbett et al., *Spanner: Google's Globally-Distributed Database* (OSDI, 2012)
23. A. Jafari, E. Khamespanah, M. Sirjani, H. Hermanns, M. Cimini, PTREbeca: modeling and analysis of distributed and asynchronous systems. Sci. Comput. Progr. **128**, 22–50 (2016)
24. M. Sirjani, E. Khamespanah, EA. Lee. *Model Checking Software in Cyber-physical Systems* (COMPSAC, 2020)
25. *Rebeca Modelling Language*, on-line, https://rebeca-lang.org/. Accessed Sept 2020
26. F. Cicirelli, L. Nigro. Model-Checking Actor-Based Cyber-Physical Systems. in 24th IEEE/ACM International Symposium on Distributed Simulation and Real Time Applications (DSRT 2020) (Prague, 14–16 September, 2020)
27. L.C. Siebert, L.R. Ferreira, E.K. Yamakawa, E.S. Custdio, A.R. Aoki, T.S.P. Fernandes, K.H. *Cardoso. Centralized and Decentralized Approaches to Demand Response Using Smart Plugs*. in 2014 IEEE PES T D Conference and Exposition, April 2014, pp. 1–5
28. S.M. Errapotu, J. Wang, Y. Gong, J. Cho, M. Pan, Z. Han, Safe: Secure appliance scheduling for flexible and efficient energy consumption for smart home IoT. IEEE Int. Things J. **5**(6), 4380–4391 (2018)
29. H. Chaouch, J. Ben Hadj Slama. *Modeling and Simulation of Appliances Scheduling in the Smart Home for Managing Energy*. in 2014 International Conference on Electrical Sciences and Technologies in Maghreb (CISTEM), November, pp. 1–5

30. E. Mocanu, D.C. Mocanu, P.H. Nguyen, A. Liotta, M.E. Webber, M. Gibescu, J.G. Slootweg. On-line Building Energy Optimization Using Deep Reinforcement Learning. IEEE Trans Smart Grid 1–1 (2018)
31. A. Tundis, A. Faizan, M. Mhlhuser. A Feature-Based Model for the Identification of Electrical Devices in Smart Environments. Sensors **19**(11) (2019)
32. D.O'Neill, M. Levorato, A. Goldsmith, U. Mitra. *Residential Demand Response Using Reinforcement Learning*. in 2010 First IEEE International Conference on Smart Grid Communications (IEEE, 2010), pp. 409–414
33. C. Nigro, L. Nigro. Lock-Free Parallel Computing Using Theatre. in International Congress on Information and Communication Technology (ICICT 2020); London (UK), 20–21 February (Springer, Advances in Intelligent Systems and Computing, 2020)
34. L. Nigro, P.F. Sciammarella. *Statistical Model Checking of Cyber-Physical Systems Using Hybrid Theatre*. in Proceedings of Intelligent Systems Conference (IntelliSys) 2019, London (UK), 5–6 September (Springer, Advances in Intelligent Systems and Computing, 2019)
35. I. Jahandideh, F. Ghassemi, M. Sirjani, *Hybrid Rebeca: Modeling and Analyzing of Cyber-Physical Systems* (Model-based Design of Cyber-Physical Systems, CyPhy, 2019).
36. T.A. Hensinger, *The Theory of Hybrid Automata* (In Verification of Digital and Hybrid Systems. Springer, Berlin, Heidelberg, 2000), pp. 265–292
37. M. Lohstroh, E.A. Lee, *An Interface Theory for the Internet of Things* (Software Engineering and Formal Methods, Springer, Cham, 2015), pp. 20–34
38. N. Dey, A.S. Ashour, F. Shi, S.J. Fong, J.M.R.S. Tavares, Medical cyber-physical systems: A survey. J. Med. Syst. **42**(4), 74 (2018)

# Design and Synthesis of Filter Bank Structures Based on Constrained Least Square Method Using Hybrid Computing for Audiogram Matching in Digital Hearing Aids

**V. V. Mahesh and T. K. Shahana**

**Abstract** A sustainable world should meet human needs fairly and efficiently. Smart systems with maximum power efficiency and minimum hardware requirements are the product of sustainable technology. Hardware and power efficient filter bank structures suitable for audiogram matching in digital hearing aids are proposed in this work. World Health Organization studies show that millions of people are suffering from hearing loss and majority of them are in the under developed countries. The healthcare facility should also reach the poor to maintain social equity in the society. Compact digital hearing aids with longer battery life can be designed using the proposed filter banks at a reduced cost. This paper proposes three methods to reduce the hardware complexity of filter banks. A constrained least square finite impulse response filter bank using floating point arithmetic is proposed in the first method. The constrained least square method is more suitable for situations, where the spectra of signals and noises are overlapped. The second method proposed here is a hybrid approach in which stochastic computation is used to perform multiplication operation, and floating point arithmetic is used for addition operation. Thus, the hardware and power requirement for implementation of filter bank are reduced considerably. Area and power consumption are further reduced in the third method by using a multiplexer-based stochastic computation.

**Keywords** Constrained least square filter · Finite impulse response filter · Stochastic computing · Audiogram · Hearing loss · Matching error

---

V. V. Mahesh (✉) · T. K. Shahana
Division of Electronics Engineering, School of Engineering, Cochin University of Science and Technology, Kochi, India
e-mail: maheshvv@cusat.ac.in

T. K. Shahana
e-mail: shahanatk@cusat.ac.in



A. Joshi et al. (eds.), *Sustainable Intelligent Systems*, Advances in Sustainability Science and Technology, https://doi.org/10.1007/978-981-33-4901-8_13

## 1 Introduction

Sustainable technologies use less resource and produce minimum waste. It should be resource efficient, energy efficient, and economically suitable. Thus, appropriate technology plays a critical role in sustainability. Smart systems with maximum power efficiency and hardware efficiency are the products of appropriate technology. Hardware, power and delay efficient filter bank structures suitable for audiogram matching in digital hearing aids are proposed in this work. Millions of people are suffering from hearing loss, and around 10% of the affected are children from economically backward underdeveloped countries. So sustainable, energy efficient hearing aid with longer battery life at lower cost is still need of the day. Digital hearing aids with reduced hardware complexity and power requirement can be designed using the proposed filter banks, and thus, cost of the system can be reduced considerably. Hearing aids enhance the speech intelligibility by rectifying hearing loss at specific audio spectrum of the affected person. A filter bank splits the input audio signal in to different frequency bands which may be then processed separately to compensate for the hearing loss. The sensitivity of a hearing impaired person is low at certain frequencies, and the design consideration of filter bank is to adjust sound levels at arbitrary frequencies within a given spectrum. This is achieved by passing audio signal from microphone through a set of filters that divide audio signals in to different frequency bands. The sub-band filter gains are adjustable to match the audiogram of the hearing impaired person. The design considerations of a compact hearing aid involve minimum hardware, low power consumption, low delay, and higher flexibility. Long delays more than 10 ms may cause annoying effect in hearing aid users. Due to this discomfort, a considerable percentage of the hearing impaired persons is not ready to use any type of hearing aids. So a filter bank structure with low delay and reduced hardware complexity are the requirements for the design of a compact hearing aid with longer battery life. So optimization of hardware complexity, power requirement, and delay are still act as a volatile research problem.

Infinite impulse response (IIR) filter banks have low computation complexity but Finite impulse response (FIR) filters are preferred due to the linear phase characteristics, stability, and regular structure. A low-power design and realization of ANSI S1.11 filter bank for digital hearing aids is reported in [1]. A designer has to minimize filter bank delay as delays more than 10 ms may result disturbing perception for hearing aid users. The computational complexity is reduced by using interpolation, decimation, and frequency response masking methods in the reconfigurable FIR filter bank reported in [2]. But the price paid is longer delay, greater than 20 ms, which is not suitable for a practical hearing aid. The long delays will introduce many issues such as the disparity between aid generated and airborne vented sound, etc. A nine band uniform filter bank is presented in [3] by a recursive low pass filter and discrete cosine modulation method. Group delay and complexity of the derived 1/3 octave filter bank are low but recursive nature of prototype filter affects stability and linear phase characteristics of the system. Sample rate conversion technique based 16 band uniform and non-uniform filter bank is proposed in [4], but hardware complexity is high for the system. Variable bandwidth filter based on Farrow sub-filters is presented in [5]. Canonic signed digit representation of coefficients is reported in [6], in which multipliers are realized by adders and shifters. Hybrid evolutionary algorithms are also used in the system to improve filter characteristics. A relaxed ANSI S1.11 specification-based multirate filter bank structure is reported in [7], in which complexity oriented cases, delay-oriented cases, and patient-oriented cases are presented in detail. Huang et al. proposed a non-linear transformation-based reconfigurable filter bank in [8]. Farrow structure-based variable bandwidth digital filter for hearing aid application is proposed in [9] in which the hardware complexity is reduced by compromising the precision of Canonic Signed Digit (CSD) arithmetic. Additional circuitry involving accumulator is required to apply tunable factor value to the system which will increase the overall system complexity. A reconfigurable filter bank using fractional interpolation and symmetry property of linear phase filter is proposed in [10]. In [11], a non-uniform modified Discrete Fourier Transform (MDFT) filter bank is proposed, in which order of the filter is varied for different bands. A constrained least square (CLS) FIR filter bank-based structure is reported in [12], in which more number of multipliers is required as the order of filter is high.

Fault-tolerant capacity and less area requirement are the important features of stochastic computing (SC) which makes it suitable for many computation intensive applications. Many applications of SC are reported in the fields of neural networks [13] and digital image processing [14]. The large number of pixels involved in the image processing systems makes the overall transformation process extremely complex. Image processing systems with reduced hardware and low cost are possible if these functions are realized using SC. Limited accuracy and long processing latency are the issues associated with SC applications [15]. Decoding of low-density parity check (LDPC) codes and many new stochastic LDPC decoders are reported in [16]. Stochastic computing-based compact digital filters is presented in [17] by Ichihara et al. In [18], Manohar et al. give a comparative study of stochastic and deterministic computing. Three methods are proposed in this work to reduce hardware complexity and power requirement of filter banks for hearing aid applications. Low-order CLS FIR filter bank is proposed in the first method for audiogram matching. Stochastic computation is used to realize CLS FIR filter bank in the second and third method. So hardware and power requirement for the implementation of filter bank is reduced considerably compared to floating point arithmetic used in the first method. In the second method, stochastic numbers corresponding to input samples are computed instantaneously. The area and power requirement are further reduced in method 3 compared to method 2 because in this system, stochastic numbers representing input samples are selected from memory by using a multiplexer.

The chapter is organized as follows. Audiogram chart is discussed in Sect. 2. An overview of constrained least square FIR filter is given in Sect. 3. CLS FIR filter bank implementation and audiogram matching are given in Sect. 4. Basics of stochastic multiplication are discussed in Sect. 5. In Sect. 6, synthesis of CLS FIR filter banks is presented. Performance evaluation of filter banks is given in Sect. 7. Finally, Sect. 8 concludes the chapter.

## 2 Audiogram and Hearing Loss

Audiogram charts hearing ability of a person and is produced by audiometric testing. Hearing ability or threshold is defined as the softest sound a person can hear at different audio frequencies. Figure 1 represents the audiogram of a person with normal hearing. Human ear response measured in decibels (dB) is plotted in $Y$ axis, and the corresponding frequency in Hertz (Hz) is represented in $X$ axis. Audiogram generally plot the individual hearing threshold compared to normal hearing average. All thresholds upto 20 dB are treated as normal hearing. The symbol "$O$" is used to represent the response of right ear, and "$X$" is used to represent the response of left ear. Sensitivity of a hearing impaired person is low at certain frequency bands, and a particular type of hearing loss is rectified by the selective amplification of sound signals at specific band of frequencies.

Audiograms used as reference are selected from the website of independent hearing aid information source (https://www.earinfo.com). Hearing level measurements are carried out at octave frequencies (250/500/1 k/2 k/4 k/8 k) in a standard audiogram. So uniform filter bank approach requires large number of frequency bands to match the audiograms; otherwise, matching becomes difficult at low frequencies. Thus, non-uniformly spaced digital FIR filter bank becomes a good choice for hearing aid applications. Non-uniform digital FIR filter bank structures are used in the proposed designs. Response of the filter bank should match audiogram of the patient for selectively amplifying audio signals to compensate the respective hearing loss. When abrupt changes in slope occur, matching of audiogram can be obtained by using large number of sub-bands with narrow bandwidths. Audiograms of three different hearing loss cases are used in this paper for the analysis of performance parameters such as hardware complexity, delay, and matching error, etc. Matching error represents the maximum overall error between filter bank output and audiogram. The maximum tolerable matching error is considered as 3 dB, because psycho

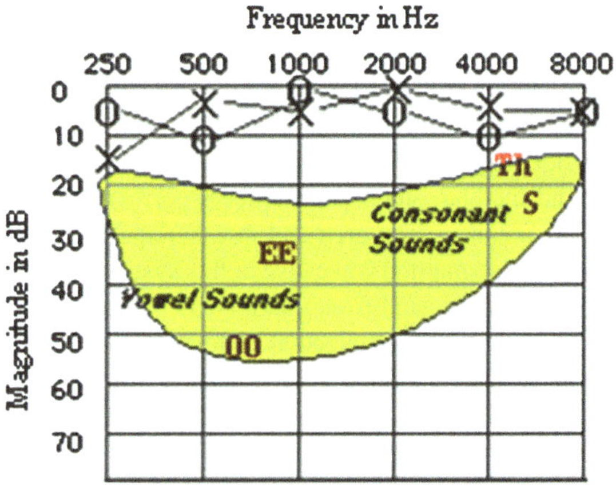


**Fig. 1** Audiogram for normal hearing

acoustical studies showed that [19]; generally, people are not sensitive to errors less than 3 dB.

## 3 Constrained Least Square FIR Filter

The method defines upper and lower thresholds that include the maximum allowable ripple in the magnitude response of a filter. With consideration of this constraint, least square error minimization technique is applied over the whole frequency range of the filter response, instead of over specific bands [20–22]. The areas of discontinuity in ideal filter response are also considered while the error minimization process. The region around cut-off frequency is also included in integral square error calculation and altogether overcomes Gibb's phenomenon without windowing or "smoothing out" the discontinuity of the ideal low pass filter. The frequency response of an FIR filter can be represented by the equation

$$H(\omega) = \sum_{n=0}^{N-1} h(n)\mathrm{e}^{-j\omega n} \tag{1}$$

If $h(n) = h(N-1-n)$, then $H(\omega)$ has linear phase and can be written as

$$H(\omega) = C(\omega)\mathrm{e}^{-jM\omega} \tag{2}$$

where $C(\omega)$ is the real valued amplitude, and $M = \frac{(N-1)}{2}$ for filter of length $N$ [23]. For symmetric odd length filters, $C(\omega)$ can be represented as

$$C(\omega) = \frac{1}{\sqrt{2}}c(0) + \sum_{n=1}^{M} c(n)\cos n\omega \tag{3}$$

where the impulse response coefficients $h(n)$, in terms of the cosine coefficients $c(n)$ are

$$h(n) = \begin{cases} \frac{1}{2}c(M-n) & \text{for } 0 \le n \le M-1 \\ \frac{1}{\sqrt{2}}c(0) & \text{for } n = M \\ \frac{1}{2}c(n-M) & \text{for } M+1 \le n \le N-1 \\ 0 & \text{otherwise} \end{cases} \tag{4}$$

Let $D(\omega)$ be the desired amplitude of an ideal low pass filter. Approximating this discontinuous function by the cosine polynomial $C(\omega)$ is the fundamental filter design problem. The approximation error can be written as

$$E(\omega) = D(\omega) - C(\omega) \tag{5}$$

The primary measures of approximation used in filter design are.

(i) The weighted integral square error is given by

$$E(\omega)_2 = \left(\frac{1}{\pi}\int_0^{\pi} W(\omega)(C(\omega) - D(\omega))^2 d\omega\right)^{\frac{1}{2}} \tag{6}$$

(ii) The weighted Chebyshev error is given by

$$E(\omega)_\infty = \max_{\omega\varepsilon 0,\pi} W(\omega)|(C(\omega) - D(\omega))| \tag{7}$$

where $W(\omega)$ is a non-negative error weighting function. Optimal FIR filters can be designed in a simple way by minimizing the value of $C(\omega) - D(\omega)_2$.

Unweighted integral square error is minimized in the CLS FIR filter design with constraint that the local minima and maxima of $C(\omega)$ lie within the specified lower and upper bound functions $B_L(\omega)$ and $B_U(\omega)$. The expressions for lower and upper bound functions are given by

$$B_L(\omega) = \begin{cases} 1 - \delta_p & \text{for all } \omega \in [0, \omega_p] \\ -\delta_s & \text{for all } \omega \in [\omega_p, \pi] \end{cases} \tag{8}$$

$$B_U(\omega) = \begin{cases} 1 + \delta_p & \text{for all } \omega \in [0, \omega_p] \\ \delta_s & \text{for all } \omega \in [\omega_p, \pi] \end{cases} \tag{9}$$

where $\delta_p$ and $\delta_s$ are the maximum allowed deviations in the passband and stopband. This filter design method is more suitable for situation, where the maximum peak error size is to be controlled, and there is no reason to treat that the input signals to be filtered have no energy in the transition band.

## 4 Filter Bank Implementation and Audiogram Matching

In the proposed designs, CLS FIR filter of order $n = 36$, $\delta_P = 10^{-4}$, $\delta_s = 10^{-4}$ is used as the prototype filter for the filter bank structure. Due to the smooth transition band characteristic of CLS FIR filter, it is possible to reduce filter order and number of bands required for matching the audiogram. The individual filter band widths used for filter bank design of three different hearing loss cases are given in Table 1. Frequency response of low pass CLS FIR filter with cut-off frequency 2 kHz for different values of $n$ is shown in Fig. 2. MATLAB software is used to simulate the filter response. Sampling frequency used for the design is 16 kHz, and pass band group delay of the filter is 1.125 ms, which is obtained by the product of half of the filter order ($n$) and sampling time. Number of coefficient multiplications required for the filter is only 37 as the order of the filter is 36. Pass band group delay and

**Table 1** Frequency bands used for CLS FIR filter bank

| Hearing loss type | No. of bands | Bandwidth (Hz) |
|---|---|---|
| Mild hearing loss at high frequencies | 7 | 700, 2000, 1500, 1100, 1600, 2400, 1800 |
| Mild hearing loss at all frequencies | 7 | 300, 1100, 2000, 2200, 1800, 1400, 1000 |
| Mild to moderate hearing loss at low frequencies | 7 | 500, 800, 1000, 3200, 2100, 2100, 800 |

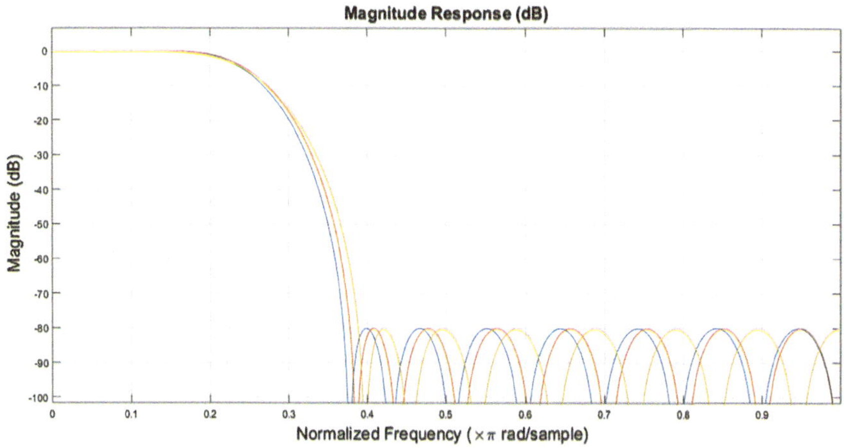


**Fig. 2** Frequency response of CLS FIR low pass filter

number of multipliers required for the proposed CLS FIR filter are lower than the reported methods. Audiogram matching for the filter order $n = 38, 36, 34$ are carried out. Minimum number of multipliers is required in the design with order, $n = 34$ but matching error obtained in this case is greater than the tolerable limit of 3 dB. So CLS FIR filter with order $n = 36$ is selected for the design of proposed filter banks. Transition width and matching error for different values of $n$ for audiogram matching of mild to moderate hearing loss at low frequencies are given in Table 2.

High computational cost and long passband group delay are the main drawbacks of reported FIR filter designs, and this is due to the involvement of large number of multipliers proportional to the order of the filter. Here, a constrained least square FIR

**Table 2** Transition width and matching error for different values of n for mild to moderate hearing loss at low frequencies

| Filter order($n$) | Transition width (Radians) | Matching error (dB) |
|---|---|---|
| 38 | $0.153\pi$ | 1.76 |
| 36 | $0.163\pi$ | 2.18 |
| 34 | $0.171\pi$ | 3.04 |

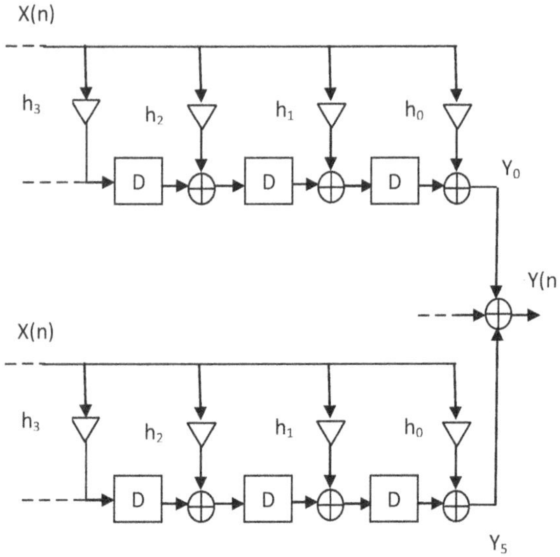


**Fig. 3** CLS FIR filter bank structure

filter bank is proposed with simple structure and low computational complexity. The filter bank structure is shown in Fig. 3. Initially, midpoint value of a specific band of audiogram of a particular hearing loss case is taken as sub-band gain of the filter bank. Then, it is varied to get an optimum value which results in a minimum overall matching error. Data broadcast structure is used to realize the filter bank, in which data are transmitted to all the multipliers at the same time. By this structure, critical path can be reduced without using pipelining latches, and it will improve clock speed or sample speed of the system [23]. The minimum time required to process a new sample is termed as critical path.

## *4.1 Audiogram Matching*

A bank of digital filters is designed to obtain audiogram matching of a particular type of hearing loss. CLS FIR filter of order 36 with $\delta_p = 10^{-4}$ and $\delta_s = 10^{-4}$ is used as the prototype filter. The stop band attenuation of the CLS FIR filter is 80 dB, and the transition bandwidth is 0.163 $\pi$ radians. Optimal filter bandwidths for audiogram matching are selected by simulating them individually for a minimum matching error. Performances of filter bank structures are evaluated by MATLAB software. Filter banks for matching audiogram patterns of mild to moderate hearing loss at low frequencies, mild hearing loss at high frequencies, and mild hearing loss at all frequencies are designed. The optimum gains and frequency bands required for filter bank for each audiogram matching are obtained by a number of simulation trials.

### 4.1.1 Audiogram Matching of Mild to Moderate Hearing Loss at Low Frequencies

Loss of overall loudness is the primary effect of mild to moderate hearing loss at low frequencies. The audiogram shown in Fig. 4a. Normal hearing is possible at high frequencies in this case. The frequency response of 7 bands generated by CLS FIR filter bank is shown in Fig. 4b, the matching is shown in Fig. 4c, and the maximum matching error obtained is 2.18 dB.

### 4.1.2 Audiogram Matching of Mild Hearing Loss at High Frequencies

Person suffering from this type of hearing loss cannot hear *s*'s, *z*'s, th's, *v*'s and other soft, high frequency consonants. Seven bands are used in the bank structure for getting best results in this case. The audiogram is shown in Fig. 5a. The frequency response of the CLS FIR filter bank is shown in Fig. 5b. Frequency response of the filter bank after separate gain adjustment to each band in order to match the audiogram is shown in Fig. 5c. In this case, maximum matching error obtained is 2.41 dB.

### 4.1.3 Audiogram Matching of Mild Hearing Loss at All Frequencies

Normal communication distance is up to 12 m, whereas this is reduced to 2 m for a person affected by mild hearing loss at all frequencies. The audiogram is plotted in Fig. 6a. Figure 6b shows frequency response of the filter bank. Best matching results are obtained by carefully varying the magnitude gains of 7 bands as shown in Fig. 6c. Maximum matching error obtained is 1.8 dB. Matching errors for all the three audiogram matching cases are plotted in Fig. 7.

## 5 Multiplication by Stochastic Logic

In SC, real numbers are represented by either unipolar or bipolar format. In unipolar format, the input $x$ is a real number with in the unit interval $(0 \leq x \leq 1)$. A binary input {110} can be represented as stochastic bit stream X, with 6 one bits, out of 10 bits. The stochastic bit streams {1,101,010,101}, {1,111,110,000} …, etc. represent the input stream {110}. This stochastic bit streams X is also interpreted as $x = P(X = 1) = 6/10$. $P(X)$ represents the probability for a bit to be one in the stream of bits. Multiplication operation can be performed by AND gate logic in SC, whereas it is computational intensive in conventional signed binary representation. A single AND gate is used in unipolar format to perform multiplication operation and is shown in Fig. 8. In this system, the input and output values are probabilities of number of bits with value one over the total number of bits in the representation. The accuracy

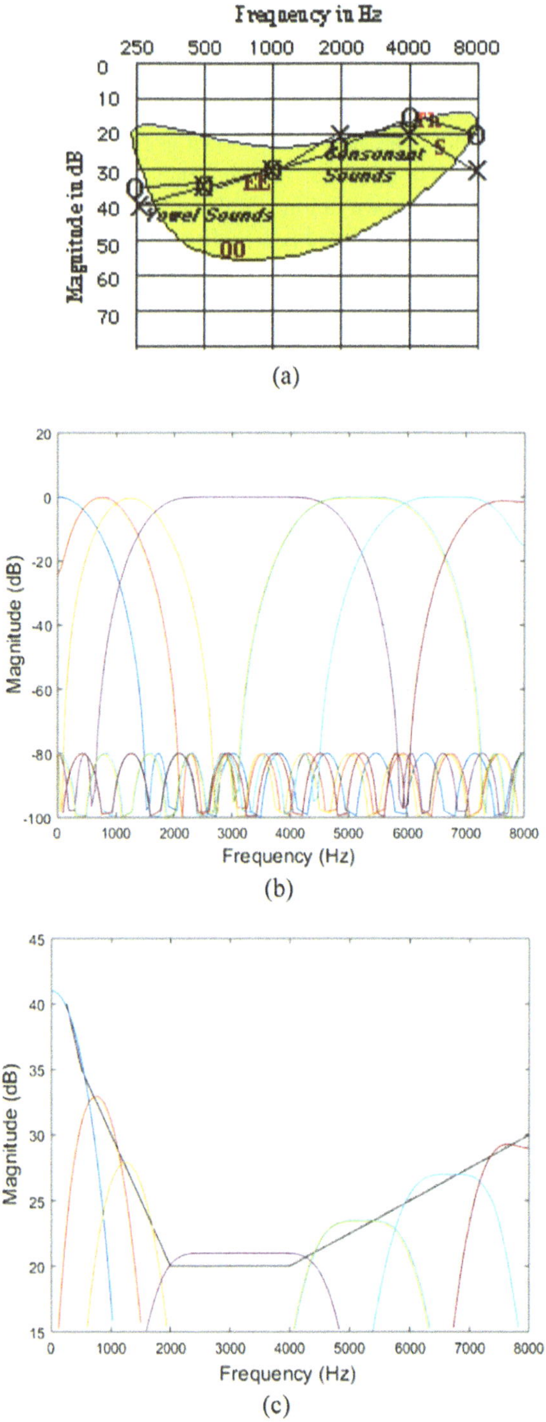


**Fig. 4** Audiogram and experiment results. **a** Audiogram for mild to moderate hearing loss at low frequencies, **b** Frequency response of the filter bank, **c** Frequency response of the filter bank after separate gain to each band

**Fig. 5** Audiogram and experiment results. **a** Audiogram for mild hearing loss at high frequencies, **b** Frequency response of the filter bank, **c** Frequency response of the filter bank after separate gain to each band

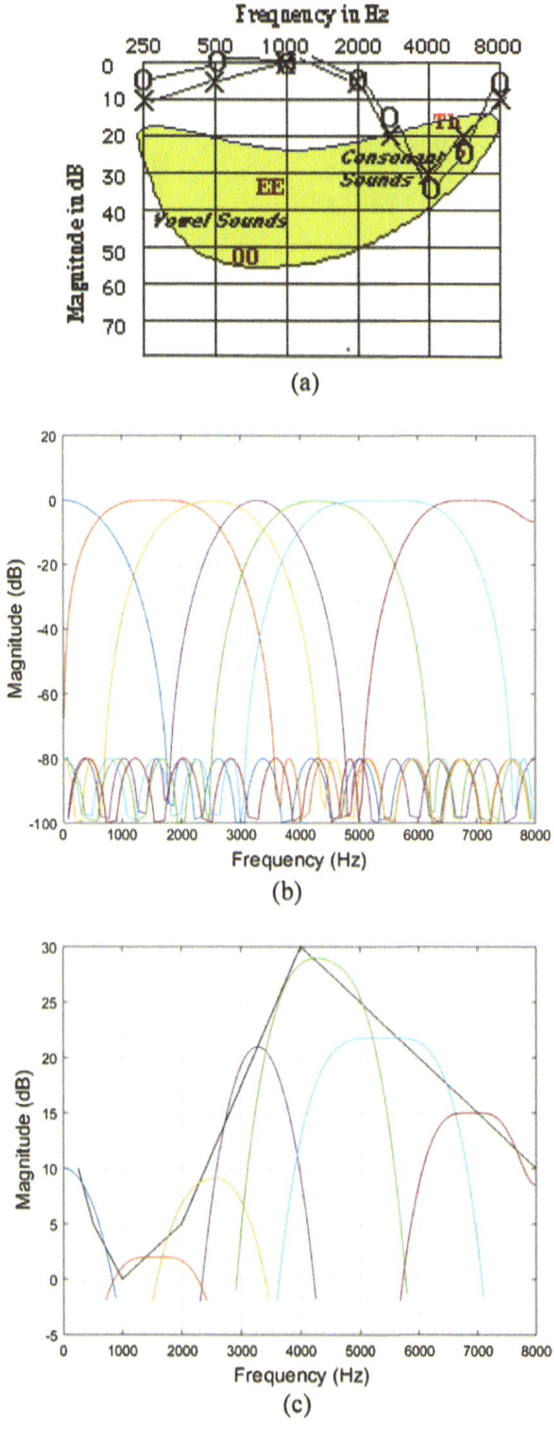

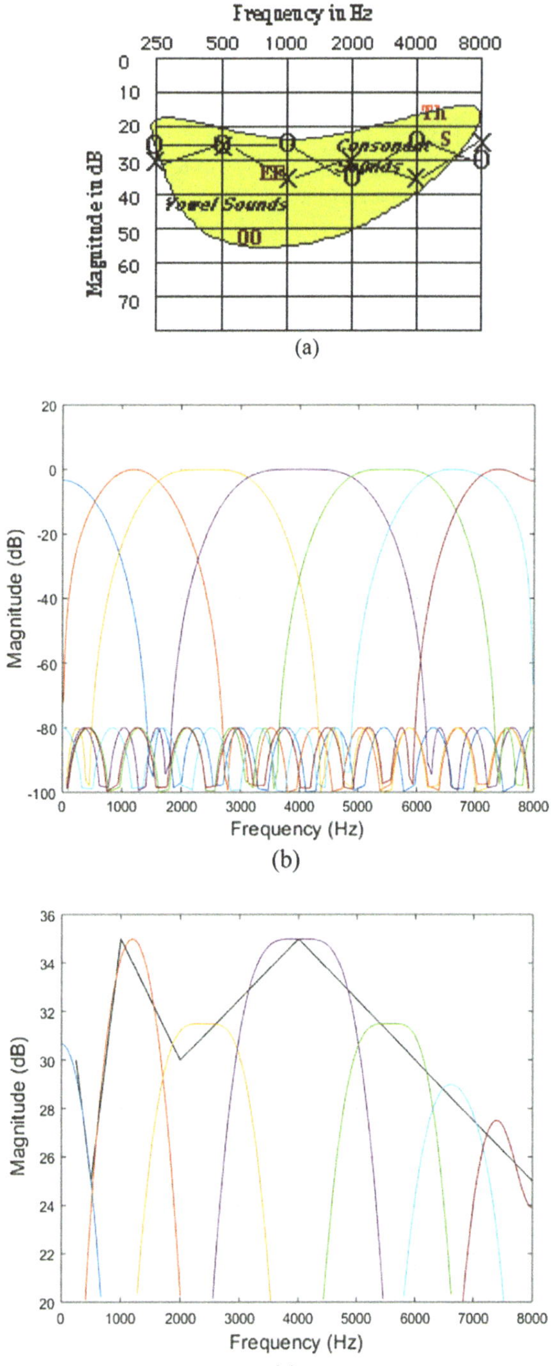


**Fig. 6** Audiogram and experiment results. **a** Audiogram for mild hearing loss at all frequencies, **b** Frequency response of the filter bank, **c** Frequency response of the filter bank after separate gain to each band

**Fig. 7** Matching error. **a** Mild to moderate hearing loss at low frequencies, **b** Mild hearing loss at high frequencies, **c** Mild hearing loss at all frequencies

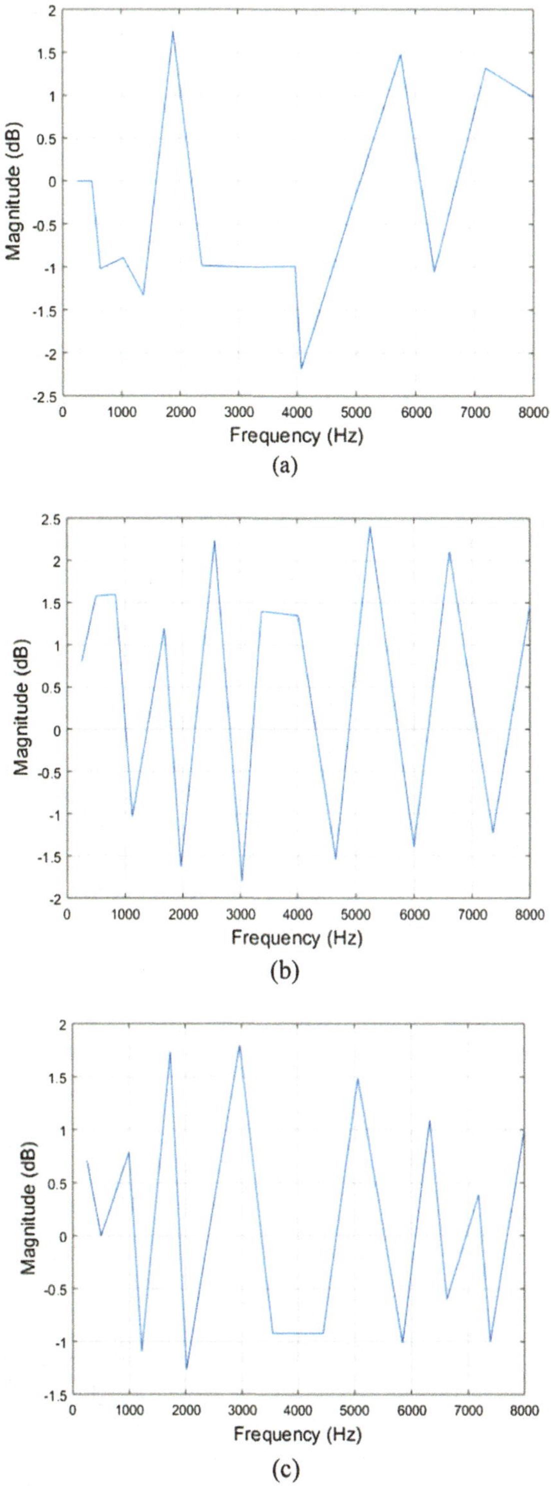

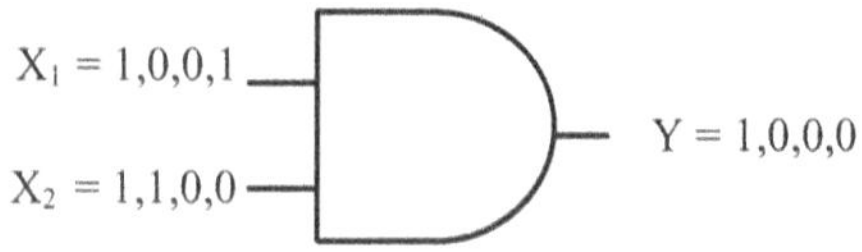


$$x_1 \times x_2 = y$$

$$\frac{1}{2} \times \frac{1}{2} = \frac{1}{4}$$

**Fig. 8** Stochastic multiplication

of computation is function of input bit stream correlation and it decreases with the increase in correlation. Increasing number of bits in the stream of representation will improve accuracy, because in long sequences, minor variations in bits will not affect much the probability of ones in the sequence.

Realization of addition operation is little bit complex in stochastic domain, so only multiplication is performed in stochastic logic in the proposed filter bank designs. In the proposed design, input samples and filter coefficients are represented in 255 bit long stochastic bit streams in unipolar format. The two important functions in the system are binary to stochastic conversion at input section and stochastic to binary conversion at output section of the multiplier.

## 5.1 *Stochastic Number Generator*

Linear feedback shift register (LFSR) is used to generate random sequences to make the stochastic bit streams. LFSR is a finite state machine (FSM) and is sequential circuits that generate a finite number of states in a deterministic fashion. The LFSR used for the proposed design is shown in Fig. 9. Output of the 8 bit LFSR is compared with the binary input data. If LFSR output data is greater than binary input data, then the comparator output become one otherwise zero. Then, LFSR data is shifted by one bit position and compared again with the binary input data [24]. This process is repeated for 255 times to generate a 255 bit long stochastic number. The block diagram of the binary to stochastic converter is shown in Fig. 10. In this work, stored

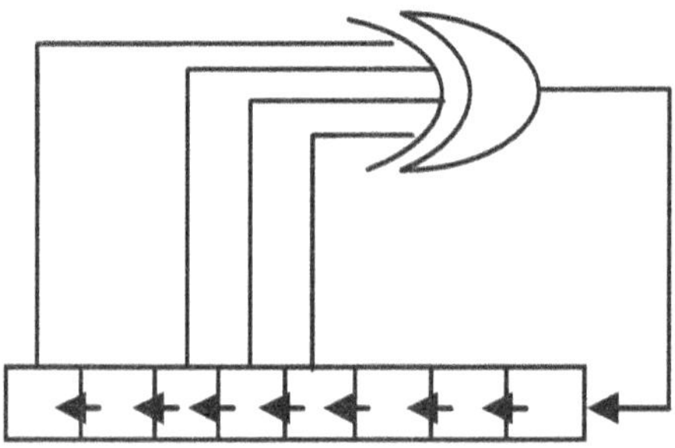

**Fig. 9** Linear feedback shift registers (LFSR)

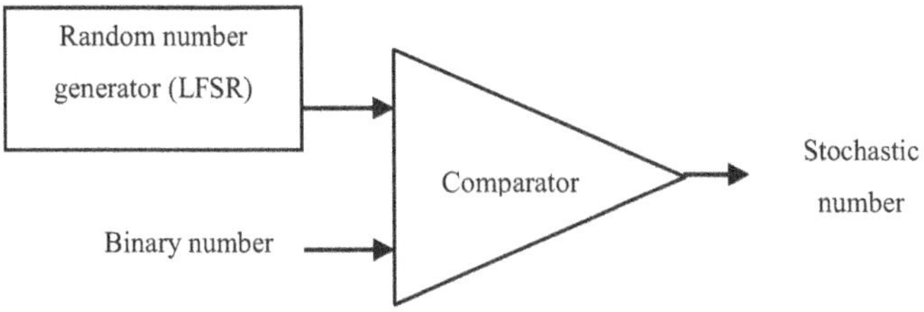


**Fig. 10** Binary to stochastic converter

values of filter coefficients in stochastic bit stream format are used stochastic bit stream format and are used. Stochastic numbers corresponding to input samples are calculated instantaneously, so that the required area can be reduced considerably.

### *5.2 Stochastic to Binary Converter*

Stochastic to binary conversion is very easy to realize compared to generator. An 8 bit binary counter is used to convert stochastic number back to binary format. Binary counter counts the number of one bits present in the input stochastic bit stream. Block diagram of stochastic to binary converter is shown in Fig. 11.

## 6 Synthesis of CLS FIR Filter Banks

Filter banks are synthesized in three different ways in the proposed work. Floating point arithmetic is used in the first method to synthesize CLS FIR filter banks. The logic of stochastic computing is used in the second method, and a modified stochastic computing using multiplexer is adopted in the third method. The long computation delay in stochastic computing is reduced in the proposed methods by using an array of 255 AND gates for multiplication operation rather than a single AND gate used in conventional methods.

**Fig. 11** Stochastic to binary converter

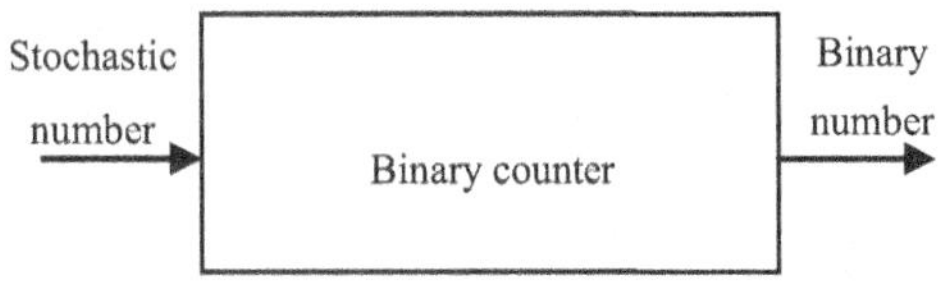

**Table 3** Synthesis results of filter bank for mild to moderate hearing loss at low frequencies

| | Mahesh [12] (floating point) | Proposed 1 (floating point) | Proposed 2 (stochastic-1) | Proposed 3 (stochastic-2) |
|---|---|---|---|---|
| Area ($\mu m^2$) | 2,499,210.81 | 1,738,968.74 | 1,040,837.21 | 859,744.67 |
| Power (nW) | 513,682,049 | 312,225,615 | 101,646,417 | 82,678,702 |
| Delay (ps) | 4919.40 | 4452 | 4931.5 | 4452 |

## 6.1 Synthesis of Filter Banks with Floating Point Arithmetic

CLS FIR filter of order $n = 36$, $\delta_P = 10^{-4}$, $\delta_s = 10^{-4}$ is used as the prototype filter for the filter bank structure and is synthesized using floating point arithmetic. Cadence Encounter(R) RTL compiler v11.10 software is used to synthesize the filter banks, and area, power, and delay reports of filter bank structure for audiogram matching of mild to moderate hearing loss at low frequencies are given in Table 3.

## 6.2 Synthesis of Filter Banks with Stochastic Logic

In this proposed method, multiplication operations involved in the implementation of CLS FIR bank are performed in stochastic domain in unipolar format. Here, a hybrid method is used; only computational intensive multiplication is performed in stochastic domain, and addition is performed in binary domain itself. Synthesis results of filter bank for audiogram matching of mild to moderate hearing loss at low frequencies are given in Table 3.

## 6.3 Synthesis of Filter Banks with Modified Stochastic Logic Using Multiplexer

Here, a multiplexer is used to select a stochastic number stored in the memory corresponding to the input sample. Block diagram of the method is shown in Fig. 12. Thus, hardware and computation time required for stochastic sequence generation are saved in this method. Synthesis results of CLS FIR filter bank for audiogram matching of mild to moderate hearing loss at low frequencies are shown in Table 3. Comparison of area utilization and power requirement is shown in Fig. 13.

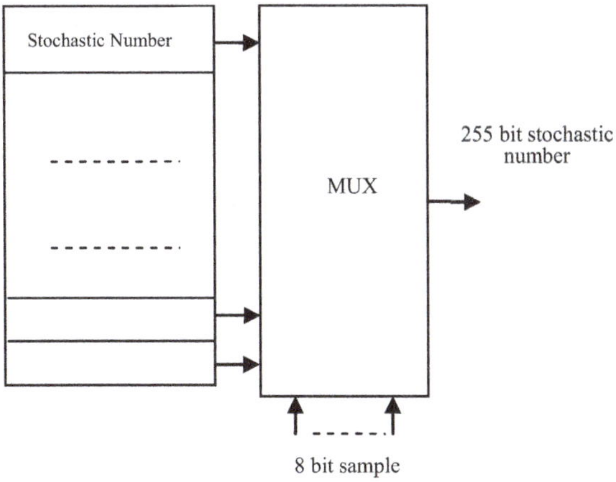


**Fig. 12** Binary to stochastic converter using multiplexer

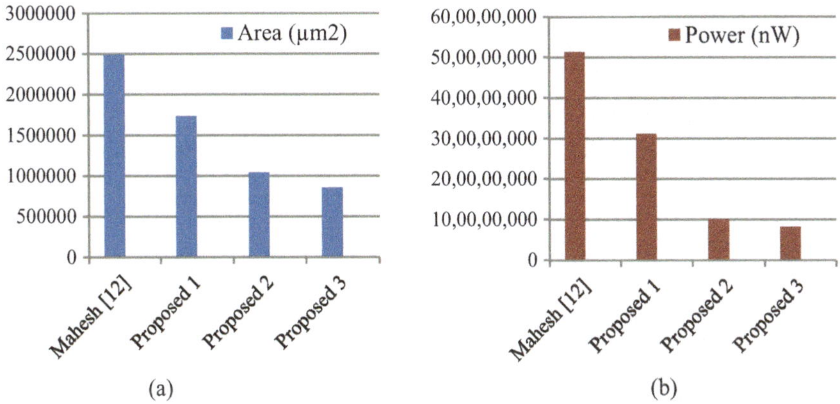


**Fig. 13** Comparison of **a** Area **b** Power

## 7 Performance Evaluation

Most of the reported FIR filter methods have the drawback of high computational cost due to the involvement of large number of multipliers in the design. In the proposed methods, hardware requirement for the multiplication process is reduced drastically. The battery life and size of hearing aid depend on the power requirement and area utilization of the filter bank structure used in the device. Filter bank structures for matching of audiogram for mild to moderate hearing loss at low frequencies are synthesized using Cadence Encounter(R) RTL compiler v11.10 software, and the results are shown in Table 3. From results, the percentage saving in area utilization is 44.14 and 50.36% in proposed method 2 and proposed method 3, respectively, compared to proposed method 1. Similarly, percentage saving in power is 67.44 and

73.52% in proposed method 2 and proposed method 3 , respectively compared to method 1. Stochastic computing suffers accuracy compared to floating point arithmetic, and accuracy can be improved by increasing the stochastic sequence length. Percentage saving in area utilization and power requirement in the proposed method 1 compared to earlier work [12] are 30.42% and 39.22%, respectively. So the proposed filter bank structures consumes less area and low power and thus reduces size of hearing aid and increases its battery life.

The order and pass band group delay of the prototype CLS FIR filter used in the proposed methods are lower than reported methods. The area utilization and power consumption increase with the increase in number of multipliers of filter bank. Performance comparison of proposed structures with the recently reported filter bank structures used for digital hearing aid design is given in Table 4. The filter bank proposed by Huang et al.[8] is based on non-linear transformation in which number of multipliers/band required is 187, and the delay is 7.77 ms. Filter bank based on fractional interpolation reported in [10] uses 84 multipliers/band, and the delay is 13.5 ms. In the non-uniform MDFT filter bank proposed by Sakthivel et al. [11], complexity is expressed in terms of order of filter, and the minimum number of multipliers required for $n$th order filter is generally $(n + 1)$. In this work, a CLS

**Table 4** Performance comparison

| Hearing loss | Method | Multipliers/band | No. of bands | Matching error (dB) |
|---|---|---|---|---|
| Mild hearing loss at all frequencies | Huang et al.[8] | 187 | 7 | 1.35 |
| | Sakthivel Elias [11] | 56 | 9 | 1.69 |
| | Amir, Bindiya [10] | 84 | – | 1.49 |
| | Mahesh, Shahana [12] | 74 | 8 | 1.61 |
| | Proposed | 37 | 7 | 1.8 |
| Mild to moderate hearing loss at low frequencies | Huang et al. [8] | 187 | 7 | 3.75 |
| | Sakthivel and Elias [11] | 29 | 6 | 1.87 |
| | Amir and Bindiya [10] | 84 | – | 1.49 |
| | Mahesh and Shahana [12] | 74 | 8 | 2.6 |
| | Proposed | 37 | 7 | 2.18 |
| Mild hearing loss at high frequencies | Huang et al.[8] | 187 | 7 | 1.65 |
| | Sakthivel and Elias [11] | 102 | 10 | 2.2 |
| | Amir and Bindiya [10] | 84 | 13 | 2.54 |
| | Mahesh and Shahana [12] | 74 | 10 | 1.82 |
| | Proposed | 37 | 7 | 2.41 |

FIR filter bank is proposed as the prototype filter for realizing filter bank structures. The number of multipliers per filter is only 37, and the pass band group delay is 1.125 ms. Stop band attenuation of the proposed CLS FIR filter is 80 dB. Matching errors of the proposed structures are within the tolerable limit of 3 dB. Comparisons show that the proposed filter bank structures have better performance than reported methods. So hearing aid devices using the proposed filter bank structures consume less area and low power and thus improves battery life and reduces the size of device. The proposed methods are applicable to other common hearing loss patterns also. Low hardware and power requirement are the characteristics of a sustainable product from an appropriate technology. Hearing aid devices at reduced cost affordable by poor people can be manufactured by using the proposed filter bank structures.

## 8 Conclusion

Sustainable technology produces smart systems with maximum energy efficiency and minimum resource requirement. Small size and power efficiency are the two important features of an optimal digital hearing aid. Filter bank structures used in this work consume less area and low power than the reported methods and thus suitable for the design of compact hearing aids. Due to significant reduction in multiplications per sample, hardware requirement is reduced drastically in the proposed designs. So digital hearing aids can be produced at a lower cost affordable by the low income community also by using the proposed filter bank structure. The proposed filter bank structures based on constrained least square finite impulse response filter show very low pass band group delay and matching errors of filter banks are within the tolerable limit.

## References


1. Y.T. Kuo, T.J. Lin, Y.T. Li, C.W. Liu, Design and implementation of low-power ANSI S1.11 filter bank for digital hearing aids. IEEE Trans Circuits Syst. **57**(7) (2010)
2. Y. Wei, D. Liu, Reconfigurable digital filter bank for hearing-aid systems with a variety of sound wave decomposition plans. IEEE Trans. Biomed. Eng. **60**(6) (2013)
3. S.-C. Lai, C.-H. Liu, L.-Y. Wang, S.-F. Lei, *14-ms-Group-Delay and Low Complexity Algorithm Design of 18-band Quasi-ANSI S1.11 1/3 Octave Filter-Bank for Digital Hearing Aids*. in International Conference on Intelligent Information Hiding and Multimedia Signal Processing (2014), pp. 81–84
4. J.T. George, E. Elias, A 16—band reconfigurable hearing aid using variable bandwidth filters. Global J. Res. Eng. (F) Electri. Electron. Eng. **14** (2014)
5. N. Haridas, E. Elias, Efficient variable bandwidth filters for digital hearing aid using farrow structure. J. Adv. Res. (2015). https://doi.org/10.1016/j.jare.2015.06.002
6. N. Haridas, E. Elias, Design of reconfigurable low complexity digital hearing aid using farrow structure based variable bandwidth filters. J. Appl. Res. Technol. **14**(2), 154–165 (2016)

7. C.-Y. Yang, C.-W. Liu, S.-J. Jou, A systematic ansi s1.11 filter bank specification relaxation and its efficient multirate architecture for hearing-aid systems. IEEE/ACM Trans. Audio Speech Language Process. **24**(8), 1380–1392 (2016)
8. S. Hang, I. Tian, X. Ma, Y.A. Wei, A reconfigurable sound wave decomposition filter bank for hearing aids based on nonlinear transformation. IEEE Biomed. Circ. Syst. **10**(2), 487–96 (2016). https://doi.org/10.1109/TBCAS.2015.2436916
9. I. Raghu, H. Nisha, E. Elias, High performance continuous variable bandwidth digital filter design for hearing aid application. Int. J. Electron. Commun. **92**, 36–53 (2018). https://doi.org/10.1016/j.aeue.2018.05.006
10. A. Amir, T.S. Bindiya, E. Elias, Design and implementation of reconfigurable filter bank structure for low complexity hearing aids using 2-level sound wave decomposition. Biomed. Signal Process. Control **43**, 96–109 (2018). https://doi.org/10.1016/j.bspc.2018.02.020
11. S. Vellaisamy, E. Elias, Design of hardware-efficient digital hearing aids using non-uniform MDFT filter banks. Signal Image Video Process. **12**, 1429–1436. (2018). https://doi.org/10.1007/s11760-017-1225-1
12. V.V. Mahesh, T.K. Shahana, Constrained least square non uniform dynamic filterbank for delay and hardware efficient digital hearing aids. Health Technol. **9**, 355–363 (2019). https://doi.org/10.1007/s12553-018-0268-9
13. J.A. Dickson, H.C. Card, *Stochastic Arithmetic Implementation of Neural Networks With In Situ Learning*. in Proceedings of International Conference on Neural Networks (1993), pp. 711–716
14. P. Li, D.J. Liija, "Computation on stochastic bit streams, Digital image processing, Case studies. IEEE Trans. VLSI, **22**, 449–462 (2014)
15. Y. Liu, K.K. Parhi, Synthesis of correlated bit streams for stochastic computing. IEEE Trans. Signal Process. **64**(14) (2016)
16. S.S. Tehrani, S. Manner, W.J. Gross, Fully parallel stochastic LDPC decoders. IEEE Trans. Signal Process. **56**, 5692–5703 (2008)
17. H. Ichihara, T. Sugino, S. Ishii, T. Iwagaki, T. Inoue, Compact and Accurate digital filters based on stochastic computing. IEEE Trans. Emerging Topics Comp. **99**, 1–10 (2016)
18. R. Manohar, Comparing stochastic and deterministic computing. IEEE Comput. Archit. Lett. **14**(2), 119–122 (2015)
19. S.P Lipshitz, M. Pocock, J. Vanderkooy, On the audibility of midrange phase distortion in audio systems. JAES**30**(9), 580–595 (1982)
20. Ivan W Sclesnick, Markus Lang and C Sidney Burrus, "Constrained Least Square Design of FIR Filters without specified Transition Bands," *IEEE Tran.on Signal Processing,* vol.44, no.8, August 1996.
21. I.W. Sclesnick, M. Lang, C.S. Burrus, Modified algorithm for constrained least square design of multi band FIR filters without specified transition bands. IEEE Trans. Signal Process. **46**(2), 497–501 (1998)
22. V.V. Mahesh, T.K. Shahana, *Delay and Area Efficient Sound Wave Decomposition by Nonuniform Filter Bank for Digital Hearing Aids*. in Proceedings of SAI Intelligent Systems Conference (IntelliSys). Lecture notes in Networks and Systems (Springer, 2016). ISBN 978-319-56994-9
23. K.K. Parhi, *VLSI Digital Signal processing Systems, Design and Implementation* (Wiley, Delhi, India, 2011)
24. P.K. Gupta, R. Kumaresan, Binary multiplication with PN sequences. IEEE Trans. Acoustics Speech Signal Process. **36**(4), 603–606 (1988)

# Genetic Sequence Alignment Computing for Ensuring Cyber Security of the IoT Systems

**Haejin Cho, Sangwon Lim, Maxim Kalinin, Vasiliy Krundyshev, Viacheslav Belenko, and Valery Chernenko**

**Abstract** Functional integrity and sustainability of cyber systems, coherency, and connectivity of the flexible infrastructures based on the Internet of Things (IoT) concept arise in the cyber security as a stability manifestation. There is a set of polymorphic attacks that have mutations (local differences and time gaps) in the sequences of the operational acts forming an IoT-specific class of the intrusions such as a forced power consumption and a forced topology change. Our research proposes the nature-inspired technologies, namely a genetic sequence alignment computing and a sequence similarity calculation, instead of massive equity checking of multiple packets and signatures traditionally used for the system protection. The sequence alignment computing is used by nature to compare and copy the DNA and nucleotide chains keeping the stability of the bioinformatic structures, which are rather similar to sequences of the security events. The methods based on the Needleman-Wunsch, Needleman-Wunsch with the position weight matrix (PWM), Smith-Waterman, and Mauve techniques have been implemented and experimentally studied with a BoT-IoT dataset on the IoT Raspberry Pi4 platform. The experiments have shown the best effectiveness obtained with the Needleman-Wunsch algorithm with the PWM: accuracy—0.98%, precision—0.99%, and recall—0.98% achieved at detection of the polymorphic intrusions.

**Keywords** Alignment algorithm · Intrusion detection · Needleman-wunsch · Mauve · Pattern recognition · Polymorphism · Raspberry · Security · Sequence · Smith-waterman · DNA · IoT · IIoT · VANET

H. Cho · S. Lim · V. Belenko · V. Chernenko
LG Electronics Inc., Seoul, South Korea

M. Kalinin (✉) · V. Krundyshev
Peter the Great St.Petersburg Polytechnic University, St. Petersburg, Russia
e-mail: max@ibks.spbstu.ru



A. Joshi et al. (eds.), *Sustainable Intelligent Systems*, Advances in Sustainability Science and Technology, https://doi.org/10.1007/978-981-33-4901-8_14

# 1 Introduction

Today, the Internet of Things (IoT) network concept has become an integral part of the modern information infrastructure of society. The IoT is a single network that connects the smart objects of the real world and virtual objects around us. The number and variability of such objects (smart devices) continues to grow, affecting various aspects of human life, from vehicles to mobile electronics and healthcare [26]. In other words, the IoT is not just a set of different devices and sensors connected by wired and wireless communications and the global cyber space of the Internet, but it is a closer integration of the real and virtual worlds, in which communication takes place between people and devices.

The sustainability of IoT is the availability of the system to perform its functions in the context of external influence [6]. Considering that the subsystems and components of the IoT have many horizontal connections, thus forming a heterogeneous cyber infrastructure with a large number of entities, most of which lack built-in protection mechanisms; the problem of network security in such networks is especially actual. Main attack vectors are aimed for overtaking the control by the IoT device and capturing the confidential data, as well as the purposeful node deactivation [1]. Not only objects of the virtual world—data, files, personal information—are under threat, but also physical objects of the real world. According to a report by Kaspersky Laboratory, in the first half of 2019, more than 100 million attacks were detected on the smart IoT devices, which are seven times more than in the H12018 [10]. Network infrastructure and wireless data transmission channels are the most vulnerable objects of the modern infocommunication environment [27]. Attackers using wireless links can remotely invade a target subnet or device (group of devices), intercept traffic, launch denial of service attacks (including distributed attacks), and take control of the IoT devices to create megabotnets.

In 2020, the world is facing with the COVID-2019 coronavirus pandemic, and life has almost completely moved to the network, people around the world work, study, shop and have fun online. This could not affect the goals of the cybercriminals, medical organizations, educational services, delivery, etc., faced massive DDoS attacks. A large network of hospitals in France fell victim to one of these attacks, when attackers tried to disable the infrastructure of medical institutions. As a result of the attack, hospital staff working remotely was unable to use work programs and corporate e-mail for some time. In Germany, on the first day of distance learning, a distance learning platform was attacked. The service through which teachers exchange teaching materials, homework, and tests with students was not available. In the Netherlands, the food delivery service Thuisbezorgd was unable to process orders as a result of a DDoS attack and had to refund customers. In the first quarter of 2020, there was a significant increase in the number and quality of DDoS attacks. In particular, the popularity of smart attacks that exploit vulnerabilities in the network infrastructure is noted. SYN flood is the leader among the types of attacks (92.6%); the share of attacks via ICMP continued to grow and exceeded the share of UDP and TCP floods (Kaspersky [11].

To counter cyber attacks at the IoT, as in traditional networks, various intrusion detection systems (IDS) are applied (e.g., [17]. There are two main approaches on which most the IoT IDSs are based: anomaly detection and misuse detection. The advantages of the former are a global view on the network, the ability to detect new attacks, and the use of fewer rules (compared to signature methods); their disadvantages include a high level of errors of the first and second kind and low adaptation for changing conditions (new profiles, dynamic anomalies). The advantages of an IDS based on search for abuse are reliability, speed, low false positives for known intrusions. Its disadvantages are the strict dependence on the relevance of the signature base and the impossibility of detecting new attacks.

The work of an IDS usually consists of three stages: (1) monitoring, (2) sequence extraction and misuse detection by pattern matching, (3) signaling of attack to provide the attack response. Matching with attack patterns is currently the main method for IDS when detecting malicious activity [13]; however, it has two significant drawbacks. Firstly, the IoT-specific attacks, such as forced power consumption, forced topology change, etc., have a split sequence of operations, time intervals in the chain of actions, and some attacks may also have a non-linear sequence of actions. Secondly, there is a class of polymorphic attacks that have mutations (namely local differences, omissions, delays) in the sequences of actions during the implementation of the attack. These polymorphic attacks adapt to a wide range of conditions, operating systems, and circumstances and try to avoid scanning by defenses to infect endpoints [8]. They are called the polymorphic attacks because they have different signatures that make them difficult to identify using a single signature. For example, in order to disguise ongoing network attacks, attackers can use polymorphic shellcode to create unique attack patterns. This method usually involves encoding the payload, with the decoder placed at the front of the payload before sending it. All known the pattern matching methods do not detect such variations of a single sample.

Our research proposes new, the nature inspired, technologies, namely a genetic sequence alignment computing and a sequence similarity calculation, instead of massive equity checking of multiple packets and signatures traditionally used for the IDS composing. The aim of our research is to assess the prospects of using the biological coded sequence alignment algorithms for solving the task of detection the specific network intrusions. The novelty of our results is the fact that it has been proposed to encode network packets (or the chain of activity acts) into the sequences of the nucleotide codes for their subsequent alignment and matching with patterns. The matching operation is based on the similarity checking, not equity checking, as it was usually applied for the intrusion detection. Therefore, our contribution to the cyber security is the proposed bio-inspired approach that, in contrast to traditional methods, makes it possible to detect new variations of the attacks in the IoT networks, to reduce the signature database, and to increase the detection effectiveness on the low-resource devices.

The following material is organized in the next manner: Sect. 2 reviews successful cases of applying modern nature-inspired methods in solving cyber security issues; Sect. 3 discusses methods for aligning sequences: global, local, and additive approaches; Sect. 4 presents the proposed coding and alignment technique; Sect. 5

presents our results of the experiments on detecting the DDoS attacks using a BoT-IoT dataset on the portative IoT Raspberry Pi4 platform; Sect. 6 presents a discussion of the results of applying the bio-inspired methods; Sect. 7 concludes our contribution.

## 2 The Related Works

Recently, IDS researchers and developers have been experimenting with mechanisms successfully used in other fields, such as artificial neural networks and genetic algorithms [19].

Existing solutions and proposals based on neural networks are described in [14]; however, these solutions have the following drawbacks: neural networks tend to overfit or diverge, and the behavior of the network may not always be unambiguously predictable, which increases the likelihood of false positives or missing events. Decisions on the application of bio-inspired methods for detecting network intrusions are much less [18]. Several successful examples are known:

1. IDS based on DNA using ASP module [23].

The main goal of this project is to optimize the most resource-intensive part of network traffic analysis—matching nucleotide sequences to calculate the offset between them. In this work, a low-level Content Addressable Memory-based (CAM) module called ASP was developed. Associative String Processor (ASP) is capable of matching fixed-length patterns at a rate of one character per processor clock. Thanks to this module, sequence matching occurs at high speed.

2. Combination of IDS with a genetic algorithm [2].

In this project, authors present a genetic approach with an improved initial population and selection operator, to efficiently detect various types of network intrusions. Genetic approach is used to optimize the search of attack scenarios in audit files, thanks to its good balance exploration and exploitation. Experimental results show an increase in intrusion detection efficiency in a reasonable processing time and a reduce the false positive rate.

3. IDS based on the theory of evolution [9].

In this project, authors also present an intrusion detection system, by applying genetic algorithm to efficiently detect various types of network intrusions. This approach uses evolution theory to information evolution in order to filter the traffic data and thus reduce the complexity. To implement and measure the performance of developed system is used the KDD99 benchmark dataset and obtained reasonable detection rate.

4. DNA-based IDS using the Smith-Waterman algorithm [5].

This project presents a host-based intrusion detection system, which encodes each activity in the system into a DNA sequence (signature). Using the Smith-Waterman signature matching algorithm, a reference signature is searched for. This approach is flexible from the point of view of matching two sequences, since in such a system, there are adjustable parameters, each of which can affect the magnitude of the offset and, consequently, the choice of the reference signature. Such parameters are, for example, the value of the penalty for the mismatch of two characters, for the gap, which is put in place of the missing characters in the sequence, etc.

The problem of finding correspondences and differences between sequences has been widely studied, since this allows one to determine the presence of many mutations. The main methods of bioinformatics can be divided into the following groups:

- methods based on RNA and DNA;
- protein algorithms;
- algorithms of genomics and proteomics.

All of the above algorithms work with long strings that have a limited alphabet. In biology, it is said that if two proteins or two nucleotide sequences have great similarity, then they are homologues. For bioinformatics, such homologous sequences are important, since they have a common origin, a similar 3D structure and, to one degree or another, a similar aminoacid sequence. Thus, their alignment is used to establish the correspondence of individual regions of the sequences [24].

## 3 Sequence Alignment Methods

Sequence alignment is a procedure for comparing two or more sequences by searching for rows of individual elements or characteristic combinations of sequence elements that are located in the sequences being aligned in the same order. Alignment algorithms allow you to search the databases of sequences of certain genes or patterns and find phylogenetic relationships using multiple alignments [3]. Thus, an important task of gene matching is to determine the degree of sequence similarity of genetic material. The general task of matching sequences is finding the largest common subsequence, that is, finding a subsequence that is a subsequence of several sequences [25]. A subsequence can be obtained by removing zero or more (optionally adjacent) characters. Variants are possible when the compared sequences have different lengths. Then, the smaller sequence is complemented by breaks that are of a conditional nature and do not carry a functional load. An important component is the quality assessment and comparison. The length of the longest common subsequence is an intuitive measure of the similarity between two sequences. To increase the capabilities as a comparison tool, you can use the scoring function to rank the different alignments so that biologically similar alignments are scored higher. The scoring function assigns positive scores to aligned characters that either match or are known to be like. Negative scores are assigned to both aligned characters that differ and characters that are aligned with spaces. Typically, the alignment score is the sum of the scores for each aligned character pair. A graphical representation of aligned

```
--GTGACATGCGAT--AAGAGG---CCTT--AGATCCGGATCTT
  | ||| ||||||  |||      ||||  ||||| |  ||||
GGGAGAC-TGCGATACAAG---TTACCTTGTAGATCTG–TCTT
```

**Fig. 1** Example of sequence alignment

sequences is a record containing sequences, where the aligned characters are written underneath each other. Thus, one column contains amino acids that are similar or have the same functional load. An example of a graphical representation of aligned sequences is shown on Fig. 1.

There are two basic matching methods: global and local alignment.

## *3.1 Global Alignment*

Global alignment looks for the best alignment of two sequences from start to finish. And best of all in situations where the sequences are sufficiently similar and approximately the same length.

An efficient dynamic programming algorithm for optimal sequence alignment was first introduced by Needleman and Wunsch [16]. The task of this algorithm is to find the maximum occurrence of a sequence in another. Let there be given two lines $A = a_1 a_2 \ldots a_m$ and $B = b_1 b_2 \ldots b_n$ where $m \leq n$. The algorithm searches for the optimal alignment of the sequence data. Matrix $F$ is used to search for alignment. The element of matrices $F_{i,j}$ takes the value of the weight of optimal alignment of subsequences $a_1 a_2 \ldots a_i$ and $b_1 b_2 \ldots b_j$. The matrix $F$ is filled recursively according to the following formulas:

$$F_{i,0} = -i * d,\ F_{0,j} = -j * d,$$

$$F_{i,j} = \max \begin{cases} F_{i-1,j-1} + s(a_i, b_j), \\ F_{i-1,j} - d, \\ F_{i,j-1} - d. \end{cases},$$

where $s(a_i, b_j)$ is the value of the substitution matrix for elements of biological sequences; $d$ is penalty for breaking the sequence. If $F_{i,j} = F_{i-1,j-1} + s(a_i, b_j)$, then the elements $a_i$ and $b_j$ of sequences $A$ and $B$, respectively, are compared with each other. If $F_{i,j} = F_{i-1,j} - d$, then the element $b_j$ is removed. If $F F_{i,j} = F_{i,j-1} - d$, then the element $a_i$ is removed.

The complexity of this algorithm is $O(mn)$. But most often $m$ and $n$ are close, so we can assume that $O(n^2)$.

## 3.2 Local Alignment

Local alignment looks for similar areas within two sequences, which are best used if the sequences are of different lengths or are not similar enough. It is often important to identify more subtle types of similarities. The two strings may not be uniform over their entire length; they may contain smaller substrings that are similar. Moreover, global alignment requires lines to be approximately the same length. To account for this possibility, Smith et al. [22] proposed a modification of the Needleman and Wunsch alignment algorithm to compute local alignment. Given two lines $A = a_1 a_2 \ldots a_m$ and $B = b_1 b_2 \ldots b_n$ where $m \leq n$. Task: find substrings α and β of rows $A$ and $B$, respectively, for which the alignment estimate is maximum for all possible pairs of substrings from $A$ and $B$. To find the optimal alignment, the matrix $F$ is filled according to the following formulas:

$$F_{i,0} = 0,\ F_{0,j} = 0,$$

$$F_{i,j} = \max \begin{cases} 0 \\ F_{i-1,j-1} + s(a_i, b_j) \\ F_{i-1,j} - d \\ F_{i,j-1} - d \end{cases}.$$

It can be noted that the comparison of sequence elements is performed in the same way as in the Needleman-Wunsch algorithm, but when $F_{i,j} = 0$, a new alignment begins; that is, when the alignment estimate is negative, a new alignment of the subsequence is sought. This algorithm has complexity $O(n^2)$. Local alignment allows you to match a substring of a larger input string to a smaller string. Typically, the global alignment algorithm is unable to detect such similarities because most characters from a longer string must be space aligned, resulting in a negative score. A graphical representation of global and local alignment is shown in Fig. 2 [5].

```
Global Alignment:
--AGATCCGGATGGT--GTGACATGCGAT--AAG--AGGCGTT
    ||| |  |  |  ||||| ||||||  |||  |  | ||
GTCCATCTG--TCTTGGGTGAC-TGCGATACAAGTTA--CCTT

Local Alignment:
--AGATCCGGATGGT--GTGACATGCGATA--AG--AGGCGTT
                 ||||| |||||||
GTCCATCTG--TCTTGGGTGAC-TGCGATACAAGTTA--CCTT
```

**Fig. 2** Example for global and local alignment

## 3.3 The Alignment with Additive Matrices

Typically, the alignment score is the sum of the scores of each aligned symbol pair. The task of optimal sequence alignment is to find the highest score alignment for a given score function and a pair of rows. There are three main aspects to consider when scoring:

- the value assigned to the matching characters;
- value assigned to non-matching characters;
- value assigned to spaces.

It should be noted that when aligning the DNA sequence, it is always are penalized the same; that is, when constructing the alignment, no additional information is taken into account. However, the alignment of amino acids, which are not all equally different from each other, exploits their evolutionary conservatism.

The structural features of the compared symbols are taken into account as follows: substitution matrices (Fig. 3) or amino acid similarity matrices (Fig. 4) are used. Amino acids with similar biochemical properties, such as charge, polarity, and others, are more likely to be paired. In order to consider the unequal probability of substitutions, these matrices were developed, which contain estimates of the partial weights for any pair of replacement of an amino acid (or nucleotide) $i$ with an amino acid (or nucleotide) $j$. The larger the number at the intersection of the matching letters, the greater the quantitative indicator of similarity.

| | | | | | | | | | | | | | | | | | | | | |
|---|---|---|---|---|---|---|---|---|---|---|---|---|---|---|---|---|---|---|---|---|
| Ala (A) | 4 | | | | | | | | | | | | | | | | | | | |
| Arg (R) | −1 | 5 | | | | | | | | | | | | | | | | | | |
| Asn (N) | −2 | 0 | 6 | | | | | | | | | | | | | | | | | |
| Asp (D) | −2 | −2 | 1 | 6 | | | | | | | | | | | | | | | | |
| Cys (C) | 0 | −3 | −3 | −3 | 9 | | | | | | | | | | | | | | | |
| Gln (Q) | −1 | 1 | 0 | 0 | −3 | 5 | | | | | | | | | | | | | | |
| Glu (E) | −1 | 0 | 0 | 2 | −4 | 2 | 5 | | | | | | | | | | | | | |
| Gly (G) | 0 | −2 | 0 | −1 | −3 | −2 | −2 | 6 | | | | | | | | | | | | |
| His (H) | −2 | 0 | 1 | −1 | −3 | 0 | 0 | −2 | 8 | | | | | | | | | | | |
| Ile (I) | −1 | −3 | −3 | −3 | −1 | −3 | −3 | −4 | −3 | 4 | | | | | | | | | | |
| Leu (L) | −1 | −2 | −3 | −4 | −1 | −2 | −3 | −4 | −3 | 2 | 4 | | | | | | | | | |
| Lys (K) | −1 | 2 | 0 | −1 | −3 | 1 | 1 | −2 | −1 | −3 | −2 | 5 | | | | | | | | |
| Met (M) | −1 | −1 | −2 | −3 | −1 | 0 | −2 | −3 | −2 | 1 | 2 | −1 | 5 | | | | | | | |
| Phe (F) | −2 | −3 | −3 | −3 | −2 | −3 | −3 | −3 | −1 | 0 | 0 | −3 | 0 | 6 | | | | | | |
| Pro (P) | −1 | −2 | −2 | −1 | −3 | −1 | −1 | −2 | −2 | −3 | −3 | −1 | −2 | −4 | 7 | | | | | |
| Ser (S) | 1 | −1 | 1 | 0 | −1 | 0 | 0 | 0 | −1 | −2 | −2 | 0 | −1 | −2 | −1 | 4 | | | | |
| Thr (T) | 0 | −1 | 0 | −1 | −1 | −1 | −1 | −2 | −2 | −1 | −1 | −1 | −1 | −2 | −1 | 1 | 5 | | | |
| Trp (W) | −3 | −3 | −4 | −4 | −2 | −2 | −3 | −2 | −2 | −3 | −2 | −3 | −1 | 1 | −4 | −3 | −2 | 11 | | |
| Tyr (Y) | −2 | −2 | −2 | −3 | −2 | −1 | −2 | −3 | 2 | −1 | −1 | −2 | −1 | 3 | −3 | −2 | −2 | 2 | 7 | |
| Val (V) | 0 | −3 | −3 | −3 | −1 | −2 | −2 | −3 | −3 | 3 | 1 | −2 | 1 | −1 | −2 | −2 | 0 | −3 | −1 | 4 |
| | A | R | N | D | C | Q | E | G | H | I | L | K | M | F | P | S | T | W | Y | V |

**Fig. 3** Example of a similarity matrix

**Fig. 4** Example of positional weight matrix

$$F = \begin{matrix} A \\ C \\ G \\ T \end{matrix} \begin{bmatrix} 3 & 6 & 1 & 0 & 0 & 6 & 7 & 2 & 1 \\ 2 & 2 & 1 & 0 & 0 & 2 & 1 & 1 & 2 \\ 1 & 1 & 7 & 10 & 0 & 1 & 1 & 5 & 1 \\ 4 & 1 & 1 & 0 & 10 & 1 & 1 & 2 & 6 \end{bmatrix}$$

Similarity matrix data may also be applicable for sequence analysis in cybersecurity. Indeed, many events in the system can be different, but legal and interchangeable, when equalized, one cannot strongly penalize for non-coincidence. To take into account, the similarity of mismatched elements in this work proposed to construct a matrix based on the frequency of occurrence. In bioinformatics, there is also the concept of positional weight matrix (PWM)—a method that is used to search for motives in biological sequences [4]. The weights reflect the frequency of the sequence element at a particular position.

Thus, using the PWM, it is possible to vary the values of the estimates for different elements of the sequences in pairwise alignment, multiplying the base estimates by the weights, that is, by the values from the matrix. Then, at the testing stage, where the actual alignment of the sequences of network traffic packets takes place, PWM constructed at the training stage is used. The size of PWM is $n \times m$, where $n$ is the cardinality of the set of unique elements from the training sample, and $m$ is the length of the sequences. To build a matrix need:

(1) for each unique element $i \in [0; n]$ calculate the number of occurrences at each position $j \in [0; m]$ in all original data sequences;
(2) write the obtained value into the corresponding cell $(i, j)$ of the matrix;
(3) convert the values of the matrix by dividing by the total number of sequences in the alignment.

At this stage, the elements of the resulting matrix show the probability of meeting a specific element at a specific position in the original alignment. To get weights, you need to normalize the values.

Taking into account the matrix, the assessment of the alignment quality at the testing stage is calculated as follows:

- If the elements match, then add 1.
- If the element was skipped, then add 0.
- If the elements are different, then add the corresponding value from the substitution matrix, located at the intersection of the current compared elements.

## 4 The Proposed Bio-Inspired Method

The developed bio-inspired intrusion detection method based on the analysis of network traffic packets consists of two main stages: a training stage and a testing stage. Figure 5 shows the detailed diagram of our method, which uses the global alignment algorithm with a pre-built PWM.

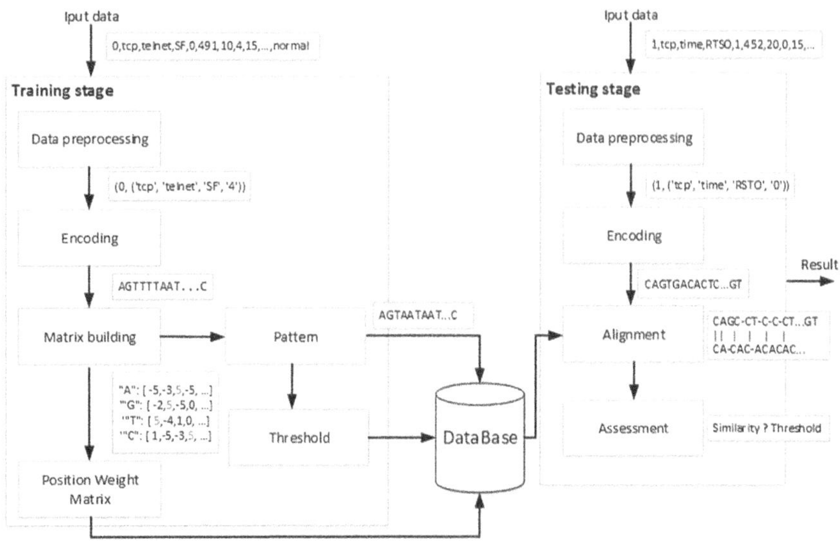


**Fig. 5** Proposed bio-inspired IDS work process

## *4.1 Training Stage*

The training stage is the training of the system, where a behavior profile (pattern) is created, with which the comparison will be made. A pattern is a set of sequences of labeled (normal or abnormal) training network traffic. The training pattern sequences are preliminarily presented in the same form as the test ones used in the analysis; that is, processing and converting are similar to the test mode described in Sects. 4.1.1 and 4.1.2. At this stage, a PWM is constructed based on the obtained pattern. The description of the construction of the matrix is presented below, in Sect. 4.1.3. The pattern and the constructed matrix are placed in the data storage of the information system. In addition, at this stage, a sign is determined; that is, a criterion (threshold) by which the input data is classified as normal or abnormal traffic. In this work, the threshold is determined on the basis of statistical data. For this purpose, the training data sequences are aligned with each other, and for example, the average value of the alignment quality assessment is taken.

### 4.1.1 Data Preprocessing

To carry out preprocessing of "raw" data, significant properties necessary for detecting an attack and reflecting the unique characteristics of an object are selected, and noise removal and normalization are performed. This work poses the problem of choosing meaningful fields in network packages, which are tuples formed into a set of a fixed size of the form:

$$\text{sequence} = \text{packet}_1(\text{field}_1,\ \text{field}_2,\ \ldots,\ \text{field}_n),$$

$$\text{packet}_2(\text{field}_1,\ \text{field}_2,\ \ldots,\ \text{field}_n)\ldots$$

Each sequence is a matching unit for alignment. For a practical experiment, the original training set was transformed as follows:

- Selected dataset $X$ of normal traffic (labeled "normal");
- From the obtained dataset, a dataset $X'$ is formed, which represents a set of structures with the following fields:

```
data=(part[1],      #protocol_typepart[2],      #servicepart[3],
#flagpart[10] #num_failed_logins)
```

- All possible types of packets were selected from the set $X'$, that is, combinations of fields, and thus, a matrix with types of packets $M$ was obtained, where each type of packet has its own serial number:

```
(0, ("tcp," "telnet," "SF," "4"))
(1, ("tcp," "time," "RSTO," "0"))
(2, ("tcp," "ftp," "SF," "0"))
(3, ("tcp," "other," "S1," "0"))
(4, ("tcp," "smtp," "S1," "0"))
(5, ("tcp," "ftp," "RSTO," "0"))
(6, ("udp," "domain_u," "SF," "0"))
(7, ("tcp," "telnet," "S0," "0"))
(8, ("tcp," "finger," "RSTR," "0"))
(9, ("tcp," "ftp," "SF," "1"))
(10, ("tcp," "private," "REJ," "0"))
(11, ("tcp," "finger," "S3," "0"))
(12, ("tcp," "ftp_data," "S1," "0"))...
```

- From the dataset $X'$, a set of sequences of packets $S'$ was formed.
- In future, the resulting set will be used as a standard with which alignment should be performed when detecting attacks.

### 4.1.2 Encoding

To construct a PWM, the biopython was used. This module works with strings of characters $A, C, G, T$, which are sequences of amino acids. For this reason, it is necessary to convert the data obtained in the previous paragraph, for which the following encoding was used:

- From the matrix $M$, the ordinal number of each element, that is, each type of packet, is converted into a quaternary number system.
- Each digit of the obtained numbers is replaced by the corresponding amino acid symbol, that is, 0–to $A$, 1–to $C$, 2 - to $G$, 3–to $T$. Thus, after converting the formed sequences of packages to a form suitable for use in Biopython, the following was obtained:

```
(0,Seq("TTTCCCCTGATGATGATGATGATGATGACCCCCAACTGATGATGATGACG...TGA"))
(1,Seq("CAGTGACACTCGTTTTGATGATGATGATGACCCCTGATGACGCAGCCCCT...CTG"))
(2,Seq("TGATGACAAGTGATGATGACCTCCGCAGCGGTGATGACCCCTGATGATGA...TTT"))
(3,Seq("TGATGACCTGCACACAGTTTCAGCAGTGATGATGATGATTTTGATGATGA...TGA"))
(4,Seq("TTTTGACAGTGACACTCGCCTCTGATGATGATGACAACTGATGATGATGA...TGA"))
(5,Seq("TTTTGACGTGACCTGATGATGATGACCCCTTTTGACCATGATTTTGATGA...TGA"))
(6,Seq("GTGTGACACTCCTGTGATGATGATGATTTCAGTGACCTGTGATGATGATG...CGG").
```

### 4.1.3 Building a PWM

At this stage, a PWM is built. The values of the PWM show the frequencies of occurrence of each symbol in each position.

```
"A": [0, 351, 1261, 62, 431, 928, 270, 433, 804, 392, 469, 674, 449,
491,
606, 504, 499, 589, 503, 497, 543, 564, 487, 550, 545, 496, 559, 551,
512, 529, 495, 520, 554, 507, 539, 526, 543, 561, 543, 504, 531, 559,
522, 513, 545, 532, 518, 549, 494, 517, 579, 511, 542, 543, 542, 533,
514, 528, 548, 507, 573, 521, 507, 511, 523, 546, 541, 500, 528, 539,
513, 506, 524]
```

The resulting values have been normalized to a range of −5 to 0:

```
"A": [ −5.0,
−3.6926605504587156,
5.0,
−5.0,
−3.8375165125495374,
5.0,
−5.0,
−4.294755877034358,
5.0,
−5.0,
−3.354114713216958,
4.999999999999998,
−5.0,
−2.6504297994269344,
4.441340782122909,
−3.643216080402012,
−2.7737226277372273,
1.7622950819672134,
−3.881118881118879,
−2.908163265306123,
−0.9668508287292799,
1.25,
−4.450549450549448,
0.09727626459143934,
0.15789473684210353,
−2.5213675213675235,
0.7943925233644888,
0.6372549019607865,
```

```
-3.4732824427480935,
-1.1111111111111116,
-3.647058823529412...
```

Matrix values are in the range [−5; 0], showing that the larger the number, the more likely it is to meet the given character at the given position. Consequently, during alignment, the higher the score, the larger the corresponding values in the matrix; that is, the sequence is more similar to the pattern. Thus, a PWM was obtained; the values of which are used as coefficients in evaluating the alignment of sequences.

## *4.2 Testing Stage*

At the testing stage, the incoming test data is aligned with the earlier received:

- Sequences of normal and attacked traffic were obtained from the test suites, which were pre-processed.
- For each input test sequence, an alignment was carried out with the "reference," and the score was calculated using a matrix. An example of equalizing normal and attacked traffic and evaluating it using a matrix:

```
CAGC-CT-C-C-CTGTGACGT-C-CGTGACAGTG-ATG-ACAGTTTTG-ATG-AC-C-C-
CTGAC-CTG-C-C-C-CG-CTGATG-ATG-ATG-ACGTGACAGTG-ACAGC-C-C-C-CA-
AG------------------------------------------
|| | | | | |   ||  | |  |||   |  |||     |  || | | | || |  | | |
| | |  |  |  ||  |||  ||| | | | || |                         CA-
CAC-ACACAC----AC--ACAC---ACA---CA--CACA------CA--CACACACAC--
ACAC--ACACACAC-AC--A--CA--CA--CAC---ACA---CACA-CACACACACACA-
CACACACACACACACACACACACACACACACACACACACACA
"MyScore:", 100
```

For each comparison, a quantitative quality assessment is made, which is compared with the accepted threshold. The result goes to the response subsystem, which notifies the information system itself and counteracts the attack if necessary. The results are also fed into the database, which can be used to update it. The proposed bioinspired IDS architecture is shown in Fig. 6.

# 5 The Experimental Study and Results

The BoT-IoT dataset was used for training and testing [12]. It is based on the traffic of an emulated IoT network. This dataset contains 69 GB of traffic, which is divided into normal and infected portions. For the experiments, dataset records containing traffic with a DoS/DDoS attack and traffic without attack were selected.

The methods based on the global alignment: Needleman-Wunsch, Needleman-Wunsch with a PWM, and local alignment: Smith-Waterman and Mauve [7] techniques have been implemented and experimentally studied on the IoT Raspberry

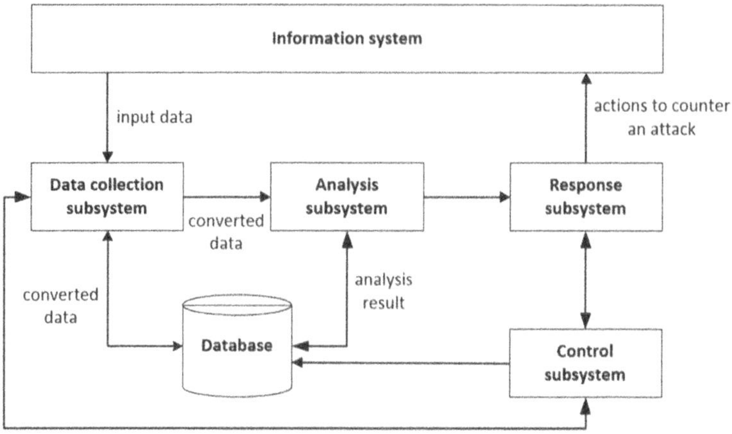


**Fig. 6** IDS architecture

Pi4 platform. The experimental setup had the following configuration: Raspberry Pi 4 Model B, Broadcom BCM2711 processor model, 4 cores, frequency 1500 MHz, LPDDR4 RAM 4096 MB, Raspbian operating system.

A comparison was made between two approaches to choosing a pattern: a white pattern (based on the normal traffic) and a black pattern (based on the abnormal traffic—with a DoS/DDoS attack).

The implementation was carried out using the following software tools and kits:

- Python3.6—an interpreter of the Python programming language;
- Biopython—a set of freely distributed tools for biological computing, written in Python by an international team of developers;
- NumPy—a software library written in Python, which provides a wide range of functions for working with various mathematical algorithms and processing of multidimensional datasets;
- scikit-learn—a software library written in Python that implements various machine learning algorithms.

Figures 7 and 8 show the ROC curves constructed in the course of experiments with white pattern and black pattern. Table 1 shows the accuracy, precision, recall for the algorithms under consideration when using the white pattern (WP) and black pattern (BP).

As it can be resulted out of the above graphics and the summary table, the best results have been obtained for the Needleman-Wunsch algorithm with the PWM with the black pattern. This can be explained by the fact that attack traffic is similar to each other, even despite some mutations. At the same time, pure traffic can differ quite strongly from each other, and the resulting averaged pattern may not take into account the features of individual secure packets, as a result of which the number of false positives will increase during the testing stage.

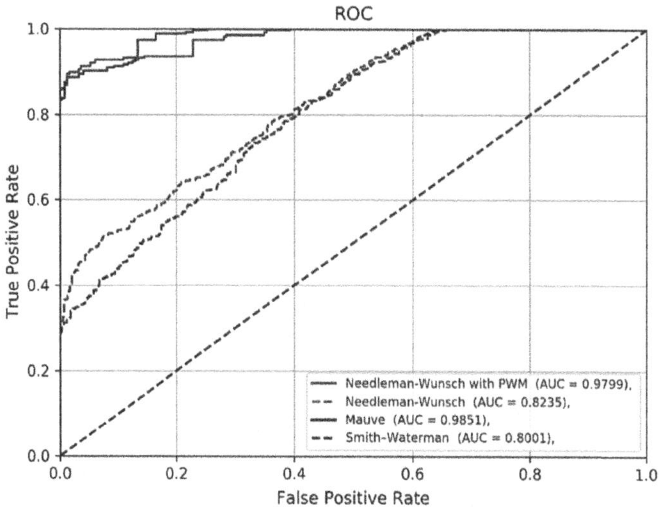


**Fig. 7** ROC curves: the white pattern use case

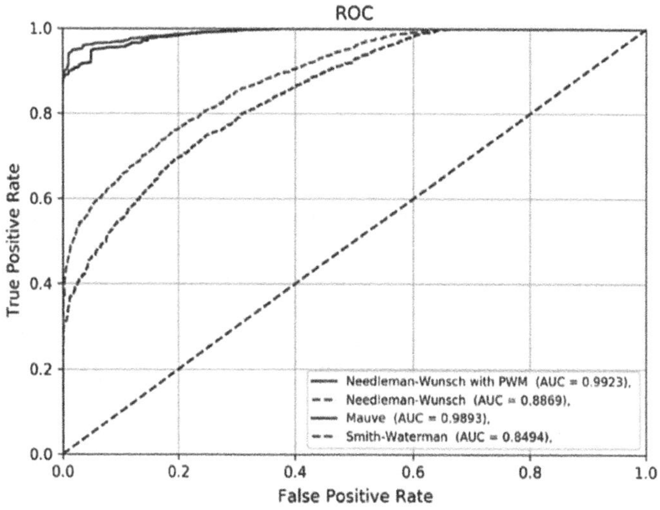


**Fig. 8** ROC curves: the black pattern use case

## 6 Discussion

Analysis of the existing related works devoted to the use of various genetic, nature-inspired, and artificial intelligence algorithms confirms the promising of the proposed bioinformatics approach. For example, in their work, the authors propose the use of chaotic variants of grasshopper optimization for solving protein folding optimization

**Table 1** Quality of traffic classification

| Alignment algorithm | Accuracy | Precision | Recall |
|---|---|---|---|
| Needleman–Wunsch with PWM (WP) | 0.96 | 0.95 | 0.97 |
| Needleman–Wunsch (WP) | 0.82 | 0.80 | 0.87 |
| Smith-Waterman (WP) | 0.85 | 0.85 | 0.88 |
| Mauve (WP) | 0.96 | 0.94 | 0.96 |
| Needleman–Wunsch with PWM (BP) | 0.98 | 0.99 | 0.98 |
| Needleman–Wunsch (BP) | 0.89 | 0.93 | 0.87 |
| Smith-Waterman (BP) | 0.86 | 0.88 | 0.85 |
| Mauve (BP) | 0.96 | 0.97 | 0.93 |

problem [21]. The genetic algorithm, differential evolution, particle swarm optimization, the flower pollination algorithm, and the plant propagation algorithms are widely used for different solutions [15]. A gene expression programming technique can be applied for CBR-based traffic and file transfer applications in the networks [20]. Our paper extends the wide range of the nature-inspired research studies.

Our work deals with the area of cyber security and intrusion detection, namely the classification of the network packets. As the result of the experiments, it has been found that the encoding of network packets into nucleotide chains occurs in a short time, and the use of parallelization allows avoiding time delays in the IDS, which is especially important for the highly mobile and low-resource IoT networks.

Much attention should be paid to the selection and calculation of the pattern and the similarity threshold. Composing several patterns (each type of attack has its own pattern) can improve the accuracy of the traffic classification, but this leads to a decrease in the speed of the IDS. The choice of the similarity threshold is a question about the number of the false positive and false negative. Depending on the type, task, and criticality of the IoT system, the threshold can be either overestimated or underestimated. In case of matching with the black pattern, an overestimation of the threshold will lead to the risk of missing the attack packet. An underestimate of the threshold will lead to the fact that normal packets will be dropped out of the IDS and will not reach the recipient.

## 7 Conclusion

The presented work discusses the task of detecting the variant types of the network intrusions in the IoT-based systems. The research study reviews the successful cases of applying the modern nature-inspired methods for solving the cyber security tasks. To detect the network intrusions in the IoT, it was proposed to apply the genetic sequence alignment calculations.

The developed methods based on the Needleman-Wunsch, Needleman-Wunsch with a PWM, Smith-Waterman, and Mauve techniques have been implemented and experimentally studied with the BoT-IoT dataset on the IoT Raspberry Pi4 platform. The best results for the classifying of the network traffic have been obtained with the global Needleman-Wunsch alignment accomplished with the PWM and the black pattern.

Our approach is adapted for the low-performance IoT devices, which compare favorably with existing analogues that use massive signature databases. The limitation of our technique is that the accuracy of the attack detection decreases with long chain length. But the real-life use case does not meet this limitation, because the network packet's length is limited.

The advanced nature-inspired data processing technology is likely to play a leading role in sustainability of a newly volatile environment of the IoT, thereby decreasing the risk of undermining the secure, stable, and useful application of cyber space to facilitate the economic growth and individual lives.

**Acknowledgments** The research is financially supported by LG Electronics Inc.

## References


1. E. Aleksandrova, Methods of group authentication for low-resource vehicle and flying self-organizing networks. Automat. Control Comput. Sci. **51**(8), 947–958 (2017)
2. S. Benaicha, L. Saoudi, S. Guermeche, O. Lounis, *Intrusion Detection System Using Genetic Algorithm*. in 2014 Science and Information Conference (London, 2014), pp. 564–568
3. H. Carrillo, D. Lipman, The multiple sequence alignment problem in biology. SIAM J. Appl. Mathe. **48**(5), 1073–1082 (1988)
4. X. Chen et al., Learning position weight matrices from sequence and expression data. Comput. Syst. Bioinfo. **6**, 249–260 (2007)
5. S. Coull, J. Branch, B. Szymański, E. Breimer, *Intrusion Detection: A Bioinformatics Approach*. in 19th Annual Computer Security Applications Conference (Las Vegas, NV, USA, 2003), pp. 24–33
6. A. Dakhnovich, D. Moskvin, D. Zeghzda, An approach for providing industrial control system sustainability in the age of digital transformation. IOP Conf. Series Mater. Sci. Eng. **497**, 1–10 (2019)
7. A. Darling et al., Mauve: multiple alignment of conserved genomic sequence with rearrangements. Genome Res. **14**, 1394–1403 (2004)
8. P. Fogla, M. Sharif, R. Perdisci, O. Kolesnikov, W. Lee, *Polymorphic Blending Attacks*. in USENIX (2006), pp. 241–256.
9. M. Hoque, M. Mukit, M. Bikas, An implementation of intrusion detection system using genetic algorithm. Int. J. Netw. Sec. Appl. **4**(2), 109–120 (2012)
10. Kaspersky Lab Report, IoT under fire: Kaspersky detects more than 100 million attacks on smart devices in H1 2019 (2019) URL: https://www.kaspersky.com/about/press-releases/2019_iot-under-fire-kaspersky-detects-more-than-100-million-attacks-on-smart-devices-in-h1-2019
11. Kaspersky Lab, DDoS attacks in Q1 2020. URL: https://securelist.ru/ddos-attacks-in-q1-2020/95949/
12. N. Koroniotis, N. Moustafa, E. Sitnikova, B. Turnbull, *Towards the Development of Realistic Botnet Dataset in the Internet of Things for Network Forensic Analytics: Bot-IoT Dataset* (2018) https://arxiv.org/abs/1811.00701

13. D. Lavrova, M. Poltavtseva, A. Shtyrkina, *Security Analysis of Cyber-Physical Systems Network Infrastructure*. in 2018 IEEE Industrial Cyber-Physical Systems (ICPS) (St. Petersburg, 2008), pp. 818–823.
14. E. Malyshev, D. Moskvin, D. Zegzhda, Application of an artificial neural network for detection of attacks in VANETs. Automat. Control Comput. Sci. **53**(8), 889–894 (2019)
15. M. Mellal, A. Salhi, Parallel-Series System Optimization by Weighting Sum Methods and Nature-Inspired Computing, in *Applied Nature-Inspired Computing: Algorithms and Case Studies*. ed. by N. Dey, A. Ashour, S. Bhattacharyya (Springer Tracts in Nature-Inspired Computing, Springer, Singapore, 2020), pp. 231–251
16. S. Needleman, C. Wunsch, A general method applicable to the search for similarities in the amino acid sequence of two proteins. J. Mol. Biol. **48**(3), 443–453 (1970)
17. E. Pavlenko, D. Zegzhda, M. Poltavtseva, *Ensuring the Sustainability of Cyberphysical Systems Based on Dynamic Reconfiguration*. in 2019 IEEE International Conference on Industrial Cyber Physical Systems (ICPS) (Taipei, Taiwan, 2019), pp. 785–789
18. A. Procopiou, N. Komninos, *Bio/Nature-inspired Algorithms in A.I. for Malicious Activity Detection*. in Nature-Inspired Cyber Security and Resiliency: Fundamentals, techniques and applications, ed. by E.-S. El-Alfy, M. Elroweissy, E.W. Fulp, W. Mazurczyk (2019)
19. A. Shenfield, D. Day, A. Ayesh, Intelligent intrusion detection systems using artificial neural networks. ICT Express **4**(2), 95–99 (2018)
20. J. Sangeetha et al., *Analyzing and Predicting the QoS of Traffic in WiMAX Network Using Gene Expression Programming*. in Advancements in Applied Metaheuristic Computing, ed. by N. Dey (IGI Global, 2018), pp. 34–70
21. A. Saxena, R. Kumar, Chaotic Variants of Grasshopper Optimization Algorithm and Their Application to Protein Structure Prediction, in *Applied Nature-Inspired Computing: Algorithms and Case Studies*. ed. by N. Dey, A. Ashour, S. Bhattacharyya (Springer Tracts in Nature-Inspired Computing, Springer, Singapore, 2020), pp. 151–175
22. T. Smith et al., Identification of common molecular subsequences. J. Mol. Biol. **147**(1), 195–197 (1981)
23. Z. Trabelsi, R. Hamdy, *Implementation of a Dna-Based Anomaly Identification System Utilizing Associative String Processor (ASP)*. in ACS/IEEE International Conference on Computer Systems and Applications—AICCSA (Hammamet, 2010) pp. 1–8
24. L. Volkert, D. Stoffer, *A Comparison Of Sequence Alignment Algorithms for Measuring Secondary Structure Similarity*. in 2004 Symposium on Computational Intelligence in Bioinformatics and Computational Biology (La Jolla, CA, USA, 2004), pp. 182–189
25. Y. Wang, H. Wu, Y. Cai, A benchmark study of sequence alignment methods for protein clustering. BMC Bioinfo. **19**, 529 (2018)
26. D. Zegzhda, D. Lavrova, E. Pavlenko, Management of a dynamic infrastructure of complex systems under conditions of directed cyber attacks. J. Comput. Syst. Sci. Int. **59**(3), 358–370 (2020)
27. P. Zegzhda, M. Poltavtseva, A. Pechenkin et al., A use case analysis of heterogeneous semistructured objects in information security problems. Autom. Control Comput. Sci. **52**(8), 918–930 (2018)

# A Concept of Operations to Embody the Utilization of a Distributed Test, Track and Track System for Epidemic Containment Management

David Bird

**Abstract** The severe acute respiratory syndrome coronavirus 2 pandemic revealed that many countries had insufficient strategies to conduct test, track, and trace of the viral transmission once infected people entered a country's borders. Computer science could be used to understand the seats of infection, and hotspots that may fuel potential outbreaks. As well as the added benefit of steering on-the-ground epidemiological surveillance activities to contain further outbreaks. However, there is more to a computerized solution than an architectural design of an end-to-end distributed test, track, and trace system and its use of machine learning technologies. A successful implementation encompasses a number of key areas that include people, processes, and technology. Comparisons are drawn with cyber security operations center use-cases in support of a strategy and concept of operations to enable: (a) front-end test teams at the border chokepoints to collect test samples; (b) cloud processing of test subject records and laboratory test results; (c) emergency operations center containment monitoring; (d) data analysis of test subject groupings to identify hotspot areas; (e) use of epidemiological trends to direct further testing; and (f) conduct epidemiological monitoring to detect new chains of transmission.

**Keywords** Cloud · Concept of operations · Incident and operations management · Cyber security operations center model

## 1 Introduction

South Korea learned from their Severe Acute Respiratory Syndrome (SARS) experience when it came to SARS coronavirus 2 (SARS-Cov-2). Because of this they were able to develop and deploy operational polymerase chain reaction (PCR) tests within two weeks at scale [23]. Borders were closed at the start of the pandemic, and epidemiological surveys were prioritized [17]. The South Korean government deployed a track and trace system that utilized multi-source data from the government

---

D. Bird (✉)
The Institution of Engineering Technology, London, UK
e-mail: david.bird@bcs.org

 
A. Joshi et al. (eds.), *Sustainable Intelligent Systems*, Advances in Sustainability Science and Technology, https://doi.org/10.1007/978-981-33-4901-8_15

as well as medical records and cellular data; hence, locations of credit card transactions and geo-location tracking were used which has consequently been deemed intrusive by the West [9, 30]. Thus, a "big data" approach was used that utilized testing, tracking, and tracing and proved successful in coordinated containment activities. Localized epidemiological response teams were able to contain outbreaks in the community, and this enabled the country to avoid a country-wide lockdown [30]. More recently, SARS-Cov-2 disease of 2019 (COVID-19) outbreaks were still occurring with 76% imported from abroad and 24% sourced domestically from within the country [3, 4]. For a country that has a population that is three quarters the size of the United Kingdom (U.K.), they have had only 441 deaths [46]. South Korea, as well as Germany and New Zealand, all used regional epidemiological teams to root out SARS-Cov-2 and COVID-19 [4].

From a U.K. perspective, the Scientific Advisory Group for Emergencies (SAGE) relied on modeling that was fine tuned for influenza predictions rather than COVID-19 [15, 18]. Several months into the pandemic, University College London highlighted this deficiency, and models were honed using COVID-19 data; the outcomes indicated that because of the U.K.'s initially flawed control plan, closing the border by March would have postponed the inevitable arrival of the pandemic *en masse* by only 5-days [39]. Subsequently, epidemiology capabilities were overwhelmed and abandoned, and the borders were kept open well into the peak of the pandemic enabling at least 20,000 inflected people to enter the country [19, 20, 27]. Consequently, the government changed tact and began to favor “herd immunity by infection” which received critical disparagement [31, 33, 36]. Hence, a country-wide lockdown was instigated [19] followed by a slow response in the scaling up of PCR swab testing that was also labored by a shortage of test laboratories [1]. This was followed by a late reinstatement of epidemiological surveys some six weeks after the “lockdown”.

Precious time was apportioned to the development and testing of a U.K. National Health Service (NHS) Digital Service (NHSX) contact tracing smartphone application (app) which was intended to be used for centralized track and trace efforts. The original NHSX app worked with the android operating system but was inoperable using Apple's iOS; so, the government changed tact and abandoned this NHSX app variant in preference for the decentralized version that relied on Google and Apple collaborative technology; this capability was only realized in the autumn of 2020 [4]. During the intervening period, the U.K. approach depended upon a call center-orientated trace approach in order to contact people who may have encountered someone suspected of having COVID-19. What has become apparent is better diagnostics are required ideally before the start of a pandemic but certainly as part of a country's recovery from the ensuing health crisis and critically in preparation for the next epidemic.

This chapter continues the justification for a Distributed Test, Track and Trace System (DTTTS) authored by Bird [3]. Motivated by the difficulties encountered by the U.K. to implement a successful test, track and trace system, this chapter presents a novel approach for implementing such a system that is supported by a service wrap. This chapter is dedicated to justifying the approach of a DTTTS and provides an argument that the concept can be developed further into a useable concept of operations (CONOPS). The pertinent points from Bird [3] are discussed further, followed by an expansion of ideas comparing this system concept to the analogy of the Cyber Security Operations Center (CSOC). Both approaches are in principle incident response orientated even though the contexts of use and data sources are different. Incident response principles could be used to contribute to epidemic containment management. The related work section presents a case for a DTTTS. This is followed by a discussion section that provides a comparative analysis of a DTTTS in relation to CSOC models. The further work section articulates dependencies that need to be considered and resolved to ensure a DTTTS approach can be achieved.

## 2 Related Work

The NHS test and trace system had to be instituted without a tracking component and is heavily dependent on the British Public self-notifying the central call center by telephone. In August 2020, the NHS test and trace advisors reached nearly 80 percent people who had tested positive, and 80 percent of these people declared their contacts from which only 66% were further reachable by the service [4]. By October 2020, the U.K. government's scientific community admitted that they had little confidence in the existing test and trace system—even though it had been augmented by the new NHSX COVID-19 app; still only two thirds of close contacts of people who tested positive could effectively be reached [10, 11].

The cloud 2.0 era offers society an opportunity to re-think traditional medical strategies; and adapt our thinking to provide new approaches and methods for epidemic containment in a bid to avert widespread infection and a repeat of this global health crisis. In light of the U.K. Government's test and trace strategy debates were fueled about its inadequacy, and this section presents a case for a DTTTS, which is proposed as a more reliable systematic approach and crucially has an enabling trackable capability.

### *2.1 The Case for a Distributed Test, Track and Trace System*

A cloud-based DTTTS would bring a "big data" approach to the COVID-19 problem and could enable the U.K. to pursue containment measures in a similar vein to those used by South Korea. Testing would be initially conducted at the borders and this would be less resource intensive and more efficient than testing at scale across the country's population; thereby lessening the impact on NHS resources. Test subject details would be collected at transport hubs; swab tests would be taken and then dispatched to laboratories, where test outcomes would be centrally appended to records [3]. Test subjects would then be instructed to quarantine at their declared addresses until the results are received. An Emergency Operations Center (EOC) would then monitor this data store using Machine Learning (ML) algorithms [21] to keep an eye on hotspot clusters of positively tested subjects and utilize trending to notify regional authorities of greater congregations; as well as direct concentrated and targeted epidemiological surveillance activities [3]. Moreover, Smith [41] showed how cyber defense principles could provide insights for the medical profession. Bird [3] provided a comparison of cyber versus epidemic incident response measures, which compares the two disparate domains revealing their relatable similarities.

- Cyber security domain:
  - Active defense: Proactively lessens the attack landscape (e.g., patching, active sensor monitoring, network traffic filtering, protocol blocking, and so forth). Derived and shared threat intelligence reduces the opportunities for attackers to gain a foot hold and persist.
  - Protective monitoring: Reacts to passive sensor alerts, applies technical controls, minimizing impact, and re-occurrence using processes/procedures. The aim is to minimize the opportunities afforded to attackers to deny them from completing their objectives.
- Medical domain:
  - Reactive testing: Implements swab tests and records the details of test subjects. Collated positive results are used to identify clusters for further monitoring by epidemiologists. Data provides a steer to track and trace in the community using epidemiological surveillance teams on the ground.
  - Epidemiological surveillance and monitoring: Monitoring is used to produce statistics of positive infection rates in the community. This can be used to proactively break the chains of transmission by tracking hotspots and the tracing the seats of infections in society.

The use of a centralized system populated by definitive data—rather than the crowd-sourced and decentralized app sourced data—would provide a reliable record of potential hotspots and provide intelligence on pockets of infection from which a virus could spread. Admittedly, this type of system will process both personal and medical data of multiple test subjects and would need to comply with the European Union's general data protection regulations. Pseudo anonymization and full anonymization techniques are permissible under this regulation [3].

Bird [3] proposed that the DTTTS concept could use public cloud platforms to provide an adequate and resilient design that is compliant with NHSX and international standards for cloud and UK government guidelines. Applying secure-by-design principles for secured endpoint to back-end connectivity facilitates secure infrastructure that would underpin the service to monitor test, track, and trace activities at pre-, intra-, and post-pandemic stages of a national health crisis. The system would require multi-factor authentication for user access, front-end application level authentication, and back-end data validation processes. The system could be developed using a cold standby use-case until it is required, alleviating the costs of a hot-standby solution. Service management approaches can be applied to ensure performance, capacity planning, change, and configuration management.

By using a DTTTS approach, an operational model can be articulated on how this distributed solution could be utilized and operated as part of an epidemic containment strategy. In the following section, the DTTTS concept is taken to the next stage; specifically, how it could be used in practice by being enshrined with an outline CONOPS.

# 3 Discussion

Even though Public Health England (PHE) tried to redeploy their epidemiological surveillance resources in response to SARS-Cov-2, it became apparent that they were overwhelmed partly due to the lack of initial testing capabilities. So consequently, the U.K.'s containment strategy was abandoned by mid-March [20, 27, 32]. The U.K. Government has recently decided to disband the PHE blaming the organization for a number of failings early on in the pandemic [5, 44]. In reality, an over reliance on the failed NHSX app compromised on-the-ground epidemiological efforts and subsequently track and trace coordination.

The U.K. Government has instigated the *Joint Biosecurity Center* (JBC). The JBC has taken ownership of the PHE alert monitoring system [37]. In addition, a new service called the national institute for health protection has been stood-up to combine the existing PHE and NHS test and trace capabilities under a single-public health service umbrella [6].

In this section, comparisons are drawn from CSOC use-cases in support of a strategy and CONOPS to enable: (a) front-end test teams at the border chokepoints to collect test samples; (b) cloud processing of test subject records and laboratory test results; (c) EOC containment monitoring; (d) data analysis of test subject groupings to identify hotspot areas; (e) use of epidemiological trends to direct further testing, and (f) conduct epidemiological monitoring to detect new transmission chains.

## *3.1 Distributed Test, Track, and Trace System Comparative Analogy to Cyber Security Operations Center Models*

The JBC has a remit to lead the U.K. disease monitoring efforts using real-time analysis of infection outbreaks. Therefore, the JBC is an EOC derivative tasked to inform SAGE by conducting: (a) data analysis to understand infection rates; and (b) advise on how the government should respond to spikes in infections [12]. But the JBC is significantly constrained by the NHS test and trace system being a significant contributary data source and its inability to effectively track infections in a similar manner to the South Korean example.

To remedy this predicament and continuing the cyber security versus medical domain comparison from the previous section, there is in principle a direct correlation between a CSOC and the required functions of an EOC; they are both based around centralized data collection [34] but use different data sets to provide leading threat intelligence. Generally, CSOCs are implemented by organizations at enterprise scale as part of infrastructure defense strategies, for the purposes of monitoring operational network and system level functions across the enterprise real estate. Following the anatomy of cyber defense monitoring performed by CSOCs, the following benefits are provided [2]:

- System logs and network flow statistics can be reliably processed in a timely manner.
- A combination of specialist tools and deployed agents can be used to drive patching and configuration remediation activities in order to minimize vulnerabilities.
- Ingress and egressing network traffic and application activities can be trended in order to detect abnormal or suspicious activities.
- Events analytics can be used to discern true positives and enable false positives from inadvertent sensor firing to be tuned out.

The functions of the CSOC have been likened to data management principles comprising the following contributory activities as part of a workflow: (a) data collection; (b) parse and normalize; (c) store data from collect/parse and normalize; (d) analyze data; and (e) report information derived from the data to generate an appropriate response [29]. In addition, the Observe, Orient, Decide, and Act (OODA) loop has been adopted from its militaristic origin [26] and is pertinent for CSOC operations to assimilate tactically derived threat intelligence [29]:

1. Observe: Collect, monitor, and store data.
2. Orient: Analyze the data to identify potentially suspicious events.
3. Decide: Based on loop iterations, define a course of action based on the facts.
4. Act: Perform the pre-determined action.

Both the data management workflow and the OODA loop are complementary in this context, where the following are asserted to be true:

1. Observe → Collect/parse and normalize/store alerts, logs, and netflow data.
2. Orient → Analyze data to detect a cyber threat.
3. Decide → Report on an identified cyber security event.
4. Act → Undertake response activities to perform mitigating and remedial actions to minimize the impact from the detected cyber threat.

This chapter required the functional analysis of CSOC tools to discern end-to-end structures pertaining to Security and Incident Event Monitoring (SIEM) technologies; and to learn how to adapt these principles into a DTTTS-based solution. A background study was therefore performed that compared renowned enterprise-level SIEM technologies employed by CSOCs. Vendor technologies comprise Hewlett Packard's Arcsight [28], LogRhythm Nextgen SIEM Platform [25], Splunk [43], and SolarWinds Orion Platform [42]. The output is shown in Table 1.

**Table 1** Comparison of cloud SIEM technology functions

| Vendor | Data source | Data aggregation | Data accumulation | Analytics |
|---|---|---|---|---|
| Arc sight | Smart agents<br>Connectors | Logger | Management center | Third-party tooling |
| LogRhythm | Application Programming Interfaces (API)<br>System monitor agents | Data indexer | Data processor | Platform manager |
| Splunk | Forwarders | Indexer cluster(s) | Search head cluster | Search head<br>Cluster master |
| SolarWinds | Cloud native APIs<br>Connectors | Orion database server | Network traffic or log analyzer(s) | Server and application monitor<br>Log analyzer client |

The SIEM functions shown in Table 1 focus on cloud product deployments using the infrastructure-as-a-service model with platform-as-a-service elements. What became apparent by comparing these different vendor approaches is that in order to fulfill these functions CSOCs rely on the following prerequisite criteria: (a) originating data source; (b) data aggregators; (c) data accumulators; and (d) analytics capabilities. Of note, this study does not include SIEM-as-a-Service cloud technologies provided by some of these vendors because of the similarity to the software-as-a-service model; in these cases, the entire service and underlying nodal deployments are supplied by vendors and therefore does not correlate to the DTTTS IaaS-based analogy.

These four discerned criteria are in fact the essential ingredients for the DTTTS to function adequately in a health incident response context. As a distributed computerized system, it requires an ecosystem that in principle follows the key criteria of a CSOC, the only difference is that it will use different source data, and it operates for a different purpose. In the case of CSOCs, the tooling and processes already exist requiring trained, competent, and qualified staff as an absolute necessity [38]. In the case of the DTTTS concept, salient tools will need to be researched and purchased or developed and provisioned; teams would need to be trained and salient processes designed and implemented.

Henceforth, the wider system of interest, shown diagrammatically in Fig. 1, will be needed to perform the following: (1) compulsory swab testing of selected test subjects at border chokepoints and onward distribution of personal details pertaining to test subjects via endpoint devices to an exposed cloud application programming interface, (2) testing of swab samples at medical laboratories, distribution of the results, and updates appending test subject records; (3) centralized cloud processing/storage of definitive personal/medical data on positively diagnosed test subjects; (4) analytics and trends analysis to be used to centrally monitor hotspots within society; and (5) provide intelligence to trace transmission in the community

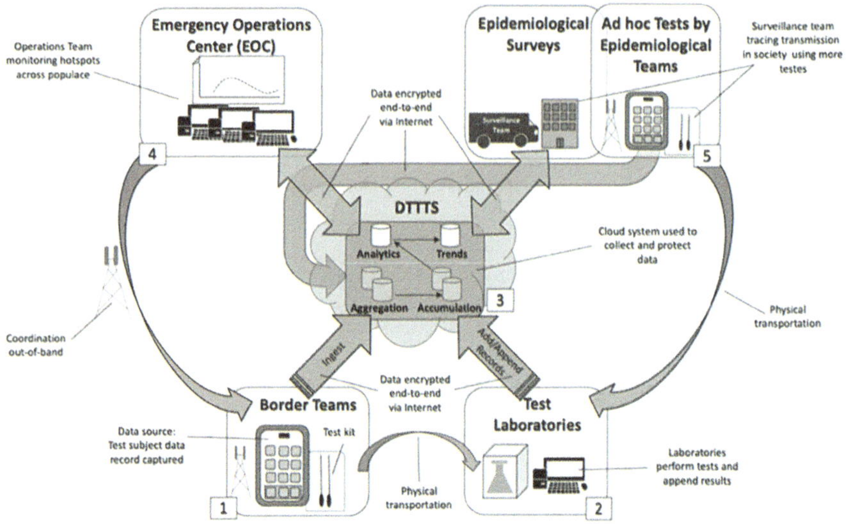


**Fig. 1** Elaborated DTTTS concept

emanating from hotspot clusters and coordinate epidemiological surveillance teams to conduct ad hoc tests that are further recorded by the system [3].

This approach, thereby provides the people, processes and technology to support containment operations during a public health event; by using OODA loop decision-making within a health setting similarly to how CSOCs employ measures to respond to cyber threats. A service wrap is also required in order to use this capability in the real world. Fitzsimmons et al. [13] identified the following pertinent aspects of service operations: (a) services are generally widely dispersed; (b) services are perishable; and (c) services are intangible requiring government regulation. These aspects are pertinent for health security operations, where parliament would need to pass laws to enforce a DTTTS health-based strategy and compel people transiting the U.K.'s borders to comply with the testing regime. From a governance perspective, the Department of Health and Social Care would need to oversee the DTTTS. The service wrap would be the enabler for the coordination and cooperation of decision-making activities in support of health crisis incident response.

In the CSOC analogy coordination can be facilitated by SIEM analysts but a cyber security incident response team [47] is also required to react to a security event in order to apply mitigations and remediations—remotely accessing components of the enterprise information technology infrastructure and on the ground as first responders interacting with assets. In the DTTTS context, an EOC is required to facilitate communication and intelligence, command and control and incident coordination to vector epidemiological teams into areas of potential infection within the community [14].

The use of the CONOPS outline provided in Fig. 2 would fuse and define the inter-relationships that are needed after test samples have been processed by laboratories.

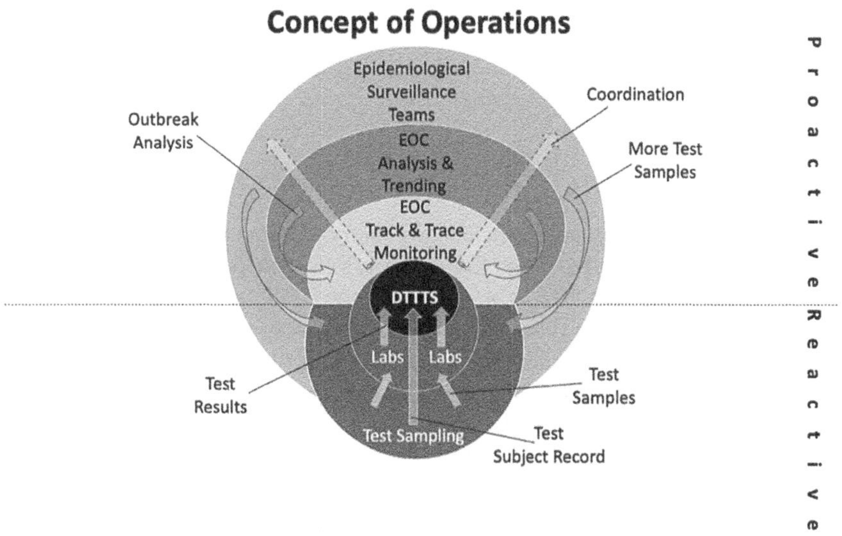


**Fig. 2** DTTTS concept of operations

In the lower area of the diagram, test results from collected test samples provide elaborating data to EOC staff, the EOC activities in the upper area inform tracking and trace activities and undertake coordination efforts of deployed epidemiological surveillance teams to identify any outbreaks from known hotspots. The epidemiological approach might necessitate further swab tests, and these are fed back into the laboratory test process. In a similar vein to a CSOC, the DTTTS could be used as part of the EOC ecosystem for operational management to: (a) correlate and corroborate test data; (b) identify infected/socially isolated residences; (c) monitor and track outbreaks from known hotspot locales; and (d) drive epidemiological surveillance to detect and trace transmission [3].

## 4 Further Research

The following dependencies need to be resolved to achieve the DTTTS concept and outline CONOPS [3]:

- Pass legislation so test subjects comply with the testing, track, and trace regime.
- Ability to rapidly deploy trained testing staff to border crossing points.
- Ability to speedily distribute front-end testing kits to transport hubs.
- Develop new methods to reliably screen passengers reducing inconvenience.
- Ensure sufficient laboratory capacity is maintained to process swab tests.
- Maintain a resourced cadre of epidemiological surveillance teams.
- DTTTS design is to be scalable to monitor, contain, and respond effectively.

- Create EOC processes to monitor, report, and coordinate with local government authorities.

## 5 Conclusion

The U.K. prime minister has conceded the country's overall response was probably too slow. Thus, compared to South Korea, it was a catalyst to produce a comparative U.K. death rate ratio of 1:98.5 [3, 4, 46]. Professor Shridhar, medical advisor to the devolved Scottish Assembly, has always advocated a test, trace, and isolate strategy from the outset of the pandemic [40]. The incumbent chief scientific advisor regrets that testing could not have been implemented sooner at the start of the pandemic [45]. Thus, it is becoming widely accepted that the failure of an early testing program and slow contact tracing process was counterproductive as a means of understanding the seats of infection.

Local councils have subsequently received new powers to implement stricter local lockdown measures [7], and the U.K. Government has also retrospectively implemented travel restrictions and quarantine measures since late July in an attempt to avoid the inevitable "second wave" being brought in from overseas [4]. However, Sir King [24], the former U.K. Chief Scientific Advisor, stated that containment has always been required from the start rather than at end of the pandemic. Consequently, there is proof that for the "first wave" and "second wave" countries who attacked the virus early reaped the benefit of a lower death toll. This certainly appears to be the case for Germany, New Zealand, and South Korea who successfully used regional epidemiological teams [4]. Meanwhile, other nations have surmised that testing at the borders is part of an effective strategy to identify the ingress of SARS-Cov-2 or COVID-19; these countries and autonomous regions include at least Iceland, Greece, Austria, Jersey, Madeira who test passengers on entry. France and Germany also started testing passengers entering their countries from high-risk destinations [22].

There are two trains of thought regarding the pandemic: The World Health Organization believes that it will take up to two years to defeat COVID-19 with a vaccination program, whilst a preeminent SAGE scientist thinks that SARS-Cov-2 may never be eradicated [8]. This is a particularly important consideration when continual in-country outbreaks provide a warning of the resulting risks from COVID-19; especially considering the recently identified coronavirus mutation D614G that potentially facilitates faster propagation across the populace [16]—albeit it in a weakened state [35].

Resultantly, computer science could greatly contribute to the betterment and sustainability of human health by contributing to the DTTTS concept and outline CONOPS; thereby providing a capability to perform better epidemic containment management. Therefore, the case for a DTTTS is even more justifiable in the sense that pre-emptive tracking and tracing provides a means to detect and track epidemics from the border inwards. The utilization of service operations can conjoin the test, track, and trace facets under a common umbrella. The DTTTS concept and supporting CONOPS would facilitate coordinated monitoring and containment intelligence, as well as inform better epidemic containment management. Albeit a limitation, and dependency for success, is changes would be necessary to legislation in order to make test subjects comply with the test, track, and trace regime.

## References


1. J. Aston, Sky News 24 (2020), https://www.news.sky.com. Accessed 12 April 2020
2. D. Bird, The age of the SoC. Digital Leaders **1**(2), 22–23 (2016)
3. D. Bird, *Distributed Test, Track and Trace System as a Contributor to Epidemic Containment Management*. in Proceedings of 4th World Conference on Smart Trends in Systems, Security and Sustainability (IEEE Xplore, 27–28 July 2020)
4. British Broadcasting Corporation, BBC News (2020a), https://www.bbc.co.uk/news. Accessed 12 June 2020, 18 July 2020, 20 July 2020, 24 July 2020, 25 July 2020, and 15 August 2020
5. British Broadcasting Corporation, Coronavirus: Public Health England 'to be replaced (2020b), https://www.bbc.co.uk/news/health-53799854. Accessed 17 Aug 2020
6. British Broadcasting Corporation, New public health body 'vigilant for viral threat (2020c), https://www.bbc.co.uk/news/health-53813480. Accessed 18 Aug 2020
7. British Broadcasting Corporation, Coronavirus Councils in England get new powers for local shutdowns (2020d), https://www.bbc.co.uk/news/uk-53453720. Accessed 18 July 2020
8. British Broadcasting Corporation, Coronavirus will be with us forever, Sage scientist warns (2020e), https://www.bbc.co.uk/news/uk-53875189. Accessed 22 August 2020
9. J. Calvert, G. Arbuthnott, L. Leake, Coronavirus: 38 days when Britain sleepwalked into disaster (2020), https://www.thetimes.co.uk/article/coronavirus-38-days-when-britain-sleepwalked-into-disaster-hq3b9tlgh. Accessed 25 April 2020
10. S. Coates, Sky News 24 (2020), https://www.news.sky.com. Accessed 12 Oct 2020
11. E. Conway, Sky News 24 (2020), https://www.news.sky.com. Accessed 15 Oct 2020
12. N. Davies, Joint Biosecurity Centre (2020), https://www.instituteforgovernment.org.uk/explainers/joint-biosecurity-centre. Accessed 22 Aug 2020
13. J. Fitzsimmons, M. Fitzsimmons, S. Bordoloi, *Service Management: Operations Strategy and Technology* (McGraw-Hill, New York, 2014).
14. Georgia Department of Public Health, *Function and Form of the Command Center: Advanced Healthcare Emergency Management Course* (Department of Public Health, Georgia, n.d.)
15. G. Guru-Murthy, A. Costello, H. Lee, The country that beat the coronavirus (2020), https://www.channel4.com/programmes/the-country-that-beat-the-virus. Accessed 20 May 2020
16. J. Hardy, New Covid-19 mutation 'helping outbreaks spread faster around the world (2020) https://www.telegraph.co.uk/news/2020/07/22/infectious-covid-19-mutation-dominant-globally-fueling-faster/. Accessed 22 Aug 2020.
17. J. Hunt, Sky News 24 (2020a), www.news.sky.com. Accessed 17 April 2020
18. J. Hunt, Sky News 24 (2020b), www.news.sky.com. Accessed 29 March 2020
19. C. Hymas, At least 20,000 people with coronavirus arrived in UK before lockdown amid lack of restrictions (2020), https://www.telegraph.co.uk/news/2020/05/05/least-20000-people-infected-coronavirus-arrived-uk-lockdown/. Accessed 20 May 2020
20. P. Ingram, COVID-19 an Intelligence assessment (2020), https://greyharemedia.com/covid-19-an-intelligence-assessment/. Accessed 25 April 2020
21. S. Johnson, Confidential Computing (2020), https://www.cheltenhamfestivals.com/science-/. Accessed 3 June 2020.
22. R. Joshi, Sky News (2020), https://news.sky.com. Accessed 29 July 2020
23. D. King, BBC News (2020a), www.bbc.co.uk/news. Accessed 30 April 2020.

24. D. King, (2020b) Sky News. www.news.sky.com. Accessed 6 June 2020.
25. LogRhythm, LogRhythm NextGen SIEM Platform 7.4.9 (2020), https://aws.amazon.com/marketplace/pp/B085CGKMNR?ref_=srh_res_product_title. Accessed 23 August 2020.
26. A. Luft, The OODA loop and the half-beat (2020), https://thestrategybridge.org/the-bridge/2020/3/17/the-ooda-and-the-half-beat. Accessed 22 August 2020.
27. A. McLaine, Coronavirus: Daily Briefing (2020), https://www.bbc.co.uk/news. Accessed 19 May 2020
28. Microsoft, ArcSight ArcMC 2.90 (2020), https://azuremarketplace.microsoft.com/en-us/marketplace/apps/micro-focus.arcsight-arcmc?tab=Overview. Accessed 23 August 2020.
29. J. Muniz, G. McIntyre, N. Alfardan, *Security Operations Center: Building Operating and Maintaining Your SOC* (Cisco Press, Indianapolis, 2016).
30. J. Newton, Coronavirus: Daily Briefing (2020) https://www.bbc.co.uk/news. Accessed 4 May 2020
31. C. O'Grady, The U.K. backed off on herd immunity. To beat COVID-19, we'll ultimately need it (2020), https://www.nationalgeographic.com/science/2020/03/uk-backed-off-on-herd-immunity-to-beat-coronavirus-we-need-it/. Accessed 25 April 2020
32. Public Health England, New surveillance system for early detection of COVID-19, https://www.gov.uk/government/news/new-surveillance-system-for-early-detection-of-covid-19. Accessed 25 April 2020
33. T. Pueyo, Coronavirus: The Hammer and the Dance (2020), https://medium.com/@tomaspueyo/coronavirus-the-hammer-and-the-dance-be9337092b56. Accessed 25 April 2020
34. G. Rasche, *Guidelines for Planning an Integrated Security Operations Center* (Electronic Power Research Institute, California, 2013).
35. Reuters, More Infectious Coronavirus Mutation May Be 'a Good Thing', Says Disease Expert (2020) https://www.nytimes.com/reuters/2020/08/17/world/asia/17reuters-health-coronavirus-mutation.html. Accessed 17 August 2020.
36. K. Rushton, D. Foggo, Neil Ferguson, The scientist who convinced Boris Johnson of UK coronavirus lockdown, criticised in past for flawed research (2020) https://www.telegraph.co.uk/news/2020/03/28/neil-ferguson-scientist-convinced-boris-johnson-uk-coronavirus-lockdown-criticised/. Accessed 25 April 2020
37. D. Sabbagh, I. Sample, Senior counter-terror official put in charge of new UK biosecurity centre (2020) https://www.theguardian.com/world/2020/may/12/former-counter-terror-official-tom-hurd-put-in-charge-new-uk-biosecurity-centre-coronavirus. Accessed 12 May 2020
38. S. Schinagle, K. Schoon, R. Paans, *A framework for Designing a Security Operations Centre*. in Proceedings of the 48th Hawaii International Conference on System Sciences. Kauai, Hawaii 5–8 January 2015. https://doi.org/10.1109/HICSS.2015.270
39. G. Shapps, The Andrew Marr Show (2020) https://www.bbc.co.uk/iplayer/episode/m000hyq2/the-andrew-marr-show-03052020. Accessed 20 May 2020
40. D. Shridhar, BBC News (2020), www.bbc.co.uk/news. Accessed 4 June 2020
41. F. Smith, Malware and Disease: Lessons from Cyber Intelligence for Public Health Surveillance. In: Health Security Journal (October 2016), https://www.ncbi.nlm.nih.gov/pmc/articles/PMC5041502. Accessed 22 Aug 2020
42. S. SolarWinds, Core Cloud Monitor Cloud (2020), https://documentation.solarwinds.com/en/success_center/orionplatform/Content/Core-Cloud-Monitor-Cloud.htm. Accessed 23 Aug 2020
43. Splunk, Topology of Splunk on Azure (n.d.) https://www.splunk.com/content/dam/splunk-blogs/images/2016/02/Topology-of-Splunk-on-Azure.png. Accessed 23 Aug 2020
44. Z. Tidman, Government accused of 'passing the buck' with reported plan to scrap Public Health England (2020), https://www.independent.co.uk/news/health/public-health-england-scrap-phe-matt-hancock-coronavirus-a9672891.html/. Accessed 22 Aug 2020
45. P. Valance, BBC News (2020), www.bbc.co.uk/news. Accessed 5 May 2020

46. Worldometers, COVID-19 Coronavirus Pandemic (2020), https://www.worldometers.info/coronavirus/. Accessed 23 Aug 2020
47. C. Zimmerman, *Ten Strategies of a World-Class Cybersecurity Operations Center* (The MITRE Corporation, Bedford, 2014).

# Multimodal Routing for Connecting People and Events

**Marvy Badr Monir Mansour and Abdelrahman Said**

**Abstract** There are obviously many complex cases of multimodal route planning on a daily basis. This paper is intended to find an efficient and feasible solution to the optimal pathfinding algorithm and multimodal network modeling problem. The proposed solution can support the next generation of multimodal navigation service. To reach this goal, a navigation dataset which contains information about the multimodal route planning is required to be ready to use. Also, this paper tackles the following points: representation of multimodal networks for transportation, formalization of the problem of multimodal route planning, analysis and design of multimodal route planning algorithm. Finally, this paper does not cover only all the previous points but also presents for an efficient, effective and general approach that can help individuals to find optimal routes in transportation networks with reasonable mode combinations and accessible modes. We provide the simulation of the proposed approach where results show the applicability of our approach.

**Keywords** Multimodal navigation service · Multimodal network · Optimal pathfinding algorithm · Road networks · Route planning · Transportation networks

## 1 Introduction

Multimodal route planning will become a popular service very soon due to the expansion of joining accessibility of different types of road networks. This is because multimodal route planning is responsible for showing the user all the ways that can be taken to reach the desired destination through different options based on user needs. Nowadays, networks of transportation and routes that connects and links locations together have become very complicated due to the development of construction in

M. B. M. Mansour (✉) · A. Said
Department of Electrical Engineering, Faculty of Engineering, The British University in Egypt, Cairo, Egypt
e-mail: marvy.badr@bue.edu.eg

A. Said
e-mail: Abdelrahman49153@bue.edu.eg



A. Joshi et al. (eds.), *Sustainable Intelligent Systems*, Advances in Sustainability Science and Technology, https://doi.org/10.1007/978-981-33-4901-8_16

the infrastructure of communities for mobility facilitation and the increased rate of urbanization. However, high density complex multilayer and multimodal networks of transportation might not consequently be suitable to users as they might confuse the user when planning to travel from one place to another. Many users find it complex to determine the optimal paths because of the great information found in the highly developed transportation networks. Thus, for helping users efficiently to locate the best routes through the transportation infrastructures, route planning is gaining a very high importance in the fields of transportation and logistics.

Recently, personal navigation has spread worldwide, where its benefits are gained from the development of technologies, which include openness and standardization of navigation data, high accuracy positioning, powerful mobile computation and high-speed data communication. Although the personal navigation application has a very complex underlying workflow and contains multiple navigation-related functions, it only answers three questions that are: (a) where the destination is located?, (b) where I am? and (c) how I can reach my destination? Based on the results of the first and second question, route planning function is liable for half of the third question. While the route guidance function answers the other half of the third question that is the given guidelines turn by turn. This kind of service is available in the form of navigation function installed in smartphones and hand-held GPS systems. Other systems that provide online route planning are Google Maps and Map Quest which work on commercial navigation dataset, Bing Maps and open route service. These services are domain specific, which means that they run efficiently but in their respective domain only [1].

Moreover, people face some issues when planning routes in a complicated network of transportation over numerous distinctive vehicles. For example, car navigator cannot work with public transit system. At the same time, the route planner of the public transit cannot inform the user which transportation means is available on departure. Shared mobility options are enabled due to technological innovations, where they are heavily used by commuters. Companies are highly investing on services of shared mobility that are promptly accessible upon request of the user. For example, ride sharing applications like Lyft and Uber are very common. Such applications might be considered suitable than taking a public transportation, because of the nature of the on-request [2].

Furthermore, some cities in the USA, for example, the city of Nashville, uses the Travel Demand Management (TDM) programs [3] for complete trips. The objective of TDM programs is to improve the quality of air in a certain region by reducing the number of trips made by vehicles and extend the effectiveness of transportation system. While these systems are mainly concentrating on ad hoc networks and introducing new options, there is an inherent issue that is not being dealt with. This issue relates to how people decide which mode of travel to use, when to travel and what routes to use. The hypothesis is that the present transportation decision support systems available in market focus on short-term planning of people, where each user gets route suggestions or information based on that user's stated preferences instead of analyzing the system-level impacts of routing decisions across people where this analysis is used in distributing the mobility demand across the dimensions of modes of transport, time and space [4].

To tackle the previous problems, online route planning systems are used to provide a variety of options where the user can select the closest aligned option according to the chosen preferences. For example, it allows users to specify the desired mode of transportation means, such as bus, wheelchair, private car, bicycle, tram, suburban train, underground train, ship or aircraft, where mode could be public or private. Other options include departure/arrival time and certain routes. More clearly, the modes of the transportation are the means by which mobility is achieved for people [5], it may also refer to the type of the physical transportation network, such as bus line, pedestrian way, motorized road, inter/inner-city railway, sea line or airline based on the context specified [1]. Also, online route planning systems allow the user to see the estimated real-time congestion levels through a traffic heat map, either by data obtained from real-time samples for different users using the service or predicted from historical data.

In fact, when planning routes, if two or more modes are taken into consideration, much better solutions would be yielded. Despite that some routing problems seem to be unsolvable in mono-modal case, they do have solutions when using multimodal scenario. For instance, if a person travels from one place to a certain destination such as pedestrian area, and in between there is no direct motor way, walking route is possible though it is a time-consuming. However, a much faster route could be found easily by driving a vehicle or any other transportation means to get as close as possible to the destination, then park the vehicle in a parking lot and finally walk to the destination [1]. In this context, multimodal route planning deals with the route planning problems as it considers a variety of transportation means. It is sometimes called intermodal route planning in some of the transportation research field. Moreover, the term multimodal is adopted for consistency whenever trip or network are referred to. However, the algorithm design for multimodal transportation has its main bottleneck in solving the problem of route planning, where the different aspects should be considered. One aspect is the optimal pathfinding and the other is the multimodal network modeling approach used. Even though raw navigation dataset is generated via the process of data integration, it cannot be used directly for route planning. It needs a procedure of graph construction, in other words network modeling, which extracts the required data and builds suitable data structures to perform the algorithm of pathfinding. The pathfinding algorithm design and network modeling method aspects are closely related to each other [1].

Thus, the main motivation of this work is to find an efficient and feasible solution to the optimal pathfinding algorithm and multimodal network modeling problem. Our solution can support the next generation of multimodal navigation service via a navigation dataset which contains information about the multimodal route planning. Also, our proposed approach helps individuals to find optimal routes in transportation networks with reasonable mode combinations and accessible modes. The rest of this paper is organized as follows. In Sect. 2, the literature review is provided. Then, in Sect. 3, the network modeling used in the proposed approach is discussed. While in

Sect. 4, we present the proposed approach, and then in Sect. 5, we provide the results of our simulations. Finally, in Sect. 6, the paper is concluded.

## 2 Literature Review

Numerous applications in transportation field involve the design of solutions that can meet the specified goals or simply optimization. These goals could be decreasing the travel time, distance or any costs that can accumulate along the travel journey. One of the significant ways of optimization is concerned with choosing the optimal path or route planning that people who have mobility requirements follow. The term mobility, as seen from the transportation geography perspective means the ability to travel along various activity sites. It also has several levels related to efficiency of movement, capacity and speed [5].

Intelligent transportation systems (ITS) with navigation services are developed to assist the mobility of people in transportation network. People need this assistance to modify or plan routes according to updates in travel route. The process of controlling and reading the motion of the vehicles is called navigation, obviously, this indicates that navigation is related to vehicle trips. Navigation systems compared to mobile phones' navigation services provide better user experience because of the underlying technologies adopted by the former. However, mobile phones navigation services are excellent solution for usage outside the car. Gradually pedestrian navigation is being raised to include multimodal navigation and improved data map. This helps user to plan routes while considering all the available modes of transportation. Most of the navigation services available for a route planning are mono-modal. However, some public transportation systems are visualized as a layout of multimodal routing systems because they provide an overview map to the user which shows the routes that should be walked from user's current location to different stations [1]. People need this service during their trip, particularly, when they are in a new place.

The navigation services must work intermodal, since the mobility of human in urban areas often occurs in multimodal transportation network. One of the complexities that navigation services systems should consider is the multimodal usage requirement. In network analysis, route planning is one of the conventional problems that is widely studied in operational research, computer science and geographic information systems (GIS). Many approaches have been developed to address these objectives in practice and theory. For example, finding the shortest path on a weighted directed graph, where this problem could be modeled. Though, these tools, methodologies and theories have not considered yet the integration of multiple transportation modes [1].

## 2.1 Graph Theory and Shortest Path Problem

Transportation networks are modeled using graphs, where the concept of shortest path is behind route planning problem. Path(s) finding on network that minimizes the distance or time between two points is one of the problems in graph theory. Hence, an introduction to the basic concepts of graph theory and the shortest path problem is given below since these shape the methodological foundation of this research.

### 2.1.1 Graph Theory

A graph is representation of a set of discrete objects where some pairs of the objects are connected by links, which is donated as $G = (V, E)$. It consists of a finite set of vertices (nodes) $V$ such that $v \in V$, and a set of edges (arc) $E$, such that $(i, j) \in E$, and $i, j \in V$, where $i, j$ are adjacent vertices and the edge $(i, j)$ is incident with vertices $i$ and $j$ [6]. If any edge direction is discarded, $G$ is called undirected graph. While if the edges are directed, the graph is called directed graph or digraph $G$. An edge $(i, j)$ is said to be directed from $i$ (head) to $j$ (tail) of the edge. *A* graph $G$ is said to be multigraph if there are multiple arcs between two nodes [6].

Furthermore, a graph $G$ is called weighted graph if it is edges are given a weighting number. Moving from one vertex to its neighbors might involve a cost, where the weighting number assigned to the corresponding edge represents the cost. To map edges to real weight value, the weight is donated as cost function $c: E \rightarrow R$. Thus, in this research, vertices represent the public transport stations or road intersection, while the edges represent the roads and connections between them, and cost represents the weighting number [6].

### 2.1.2 Shortest Path Problem

A path with length $n$ from vertex $V_0$ to vertex $V_n$ in graph G is the alternating sequence of vertices and edges in the form $V_0, E_1, V_1, E_2, V_2, \ldots, V_{n-1}, E_n, V_n$ such that $E_1 = (V_0, V_1), E_2 = (V_1, V_2), \ldots, E_n = (V_{n-1}, V_n)$ are the edges of graph $G$. The path length is calculated by number of edges in the path [6]. The shortest path problem is defined as discovering the path between two edges on a graph where the cost $C(P)$ has minimal value. The first algorithms for finding the shortest path were suggested by Dijkstra, Bellman and Ford that are based on labeling method. Although these algorithms find shortest path, they take very long time in applications like road network of a county, where they are based on real-world dataset. In recent years, research has focused on improving the running time of Dijkstra's algorithm by introducing data structures which help reducing the running time of the worst-case scenario.

While algorithmic graph theory has two approaches: label setting and label correcting. They are both iterative approaches, they allocate temporary labels for the distance at each step for vertices. The labels of distance refer to the estimate of the shortest distance path. Both approaches differ in the way that how labels of distance update on each step and how they approach toward the shortest distance path [7, 8]. The labeling approach used such that the shortest path is found from one source node $z$ to every other node in the graph using the following method. Each node $v$ has a parent node $p(V)$, a temporary distance label $d(V)$ and a status $s(V)$. The status has three options: settled, explored or unreached. Each node $v$ sets $p(V) = \text{Nil}$, $d(V) = \infty$, $s(V) = \text{unreached}$. The algorithm then begins by initializing $d(V) = 0$ and $s(V) = \textit{explored}$. After each consecutive iteration, the node with explored status is selected and all its outgoing edges are relaxed, while its status is set to settled. Relaxing the edge $(v, x)$ means checking if $d(X) > d(V) + CVX$, then setting $p(X) = V$, $d(X) = d(V) + CVX$ and $s(X) = \textit{explored}$. The algorithm terminates when nodes with explored status are not found and if the graph is acyclic with its cost negative where it generates a tree for shortest path as shown in Fig. 1 [9].

The label setting method: The efficiency of the labeling method is influenced by determining the order in which a node is selected to be scanned next. Dijkstra algorithm was first presented in 1959 for the problem of shortest path, where it is currently classified as an algorithm of label setting. This is because the shortest path is calculated for only one node, which is the parent node, in each iteration, where *d(V)* is equal to shortest path length for this node, and hence *d(V)* is constant. The

**Fig. 1** Generic shortest path algorithm

**Initially,**

$$Q := \{s\}, \pi_s := -1, d_s := 0, d_v := \infty, \forall v \neq s.$$

**The typical iteration is as follows:**

STEP 1: *Node selection*

Remove a node $v$ from the candidate list $Q$.
Node $v$ is called the pivot node.

STEP 2: *Label update*

**for** each outgoing arc $(v, w) \in \mathcal{A}$ **do**
  **if** $d_w > d_v + c_{vw}$ **then**
    $d_w := d_v + c_{vw}$
    $\pi_w := v$
    **if** $w \notin Q$ **then**
      add $w$ to $Q$.
  **end if**
**end for**

STEP 3: *Stop condition*

**if** $Q \neq \emptyset$ **then**
  goto Step 1
**else**
  the algorithm terminates.

node with minimum label is the one removed from the queue, where each node is scanned only once. Also, nodes are scanned according to how far the node from the source node in a non-decreasing order. Finally, when the target node status is settled the algorithm terminates. The algorithm has a running time of $O(|V|2)$ [8] that can be improved by using binary heap that provides a running time of $O(|E|\log|V|)$, or by using Fibonacci heap which leads to a running time of $O(|E| + |V| \log |V|)$. Complexity can be improved theoretically if the weights of the edges are restricted to integers [1].

The label correcting method: the algorithm for label setting is time-consuming in terms of finding the parent node with smallest $d(V)$, hence, to avoid this, other nodes are chosen to be the parent in each iteration. However, the shortest path will be unknown until the algorithm terminates. The algorithm maintains a queue where the first node is the parent node in the queue. So, choosing node to be scanned is very easy and does not require much overhead as label setting algorithm, where the node with status settled can be inserted again in the queue and changed from settled to be explored several times. This method can be used in graphs with negative edges for shortest pathfinding, which is called Bellman–Ford algorithm. This algorithm is almost same as Dijkstra algorithm but all its edges are relaxed instead of choosing the minimum weight node to be relaxed. Also, it is performed $|V|-1$ times and its running time is $O(|V||E|)$. This method is more flexible and does not have input restriction, but when it comes to running time, it is less efficient compared to label setting method [1, 8, 9].

On the other hand, there are techniques used to speed up the previous methods are classified into three types, which are:

(1) Goal directed search: it is the shortest path algorithm; where the shortest path is found from a source node to all other nodes and does not take into consideration the location of the target. Nodes are checked according to their closeness to the source and does not depend on the target direction. When determining the shortest path between two locations only, the algorithm could be accelerated by considering the cost as heuristic estimation between the target and a location which is less than or equal to the real cost. The searching scope can be reduced if the algorithm did more searches in the target direction, which is called A* algorithm [1].
(2) Bidirectional search: it starts with two simultaneous searches, the first search is at the source node and the second one is at the target node, where both searches operate on the same graph. The number of the nodes scanned will be reduced almost to half. Algorithm terminates when both searches intersect where series produced from searches of partial paths form the shortest path [9].
(3) Network partitioning: billions of nodes and edges are contained in the transportation datasets used in route calculation inside the underlying networks. If a standard shortest path algorithm is used, it might take a very long time in optimum route planning on a real data network. Partitioning is developed to minimize the network size that has the following types: (1) Horizontal partitioning is when the network is divided into many disjoint sub-graphs. (2) Vertical partitioning,

also known as hierarchical organization, which is used in the preprocessing of car navigation systems as the resources of the hardware is very limited [1]. (3) Hybrid partitioning: a combination of the previous two partitioning types in order to gain more computation efficiency. Depending on the type of the graph, improved performance is yielded from different combinations [1].

## 2.2 State of the Art Work

Finding the minimum cost route in networks of transportation is one of application areas of shortest path algorithm, before routing algorithms are applied, modeling the transportation network as a graph (node/edge structure) is mandatory. This process is not trivial because the real transportation networks are very complex. Routing in public transit network is very complicated than routing in road networks because scheduling is performed using the edges of the network the same way that it is used in personal vehicles, where uncertainties are introduced in the timing plan as a result. Different models are presented for different application scenarios.

To handle the inherent time dependency of a public transit timetable, there are two ways: (1) Time dependent which has a faster query time as its graph size is small. (2) Time expanded which is more flexible in constraints and intuitive [10]. In networks of public transit, it is important to use multicriteria optimization for pathfinding because of the possible conflicting objects which may exist such as minimizing cost, number of transfers and travel time. The modeling methods used for road network in public transit networks need adaptation to the scheduled system and fixed route. For multilevel graphs used in an application scenario of a railway transport, where a hierarchical decomposition approach is adopted [11], results show an improvement in the computation performance for the shortest path.

While multimodal routing provides the following: information about how to reach the destination using the best path, consideration of all transportation modes available, shortest path, minimization of the travel time, some user preferences and constraints such as: users may add or exclude some transportation modes when using different transportation modes since they may want to limit the number of transfers between them, where feasibility should be assured that is a certain transportation mode is chosen only when available. This customization is a very important characteristic of multimodal routing. Regarding multimodal context, there are some basic functions and concepts, such as the navigation system based on multimodal which includes route presentation module, multimodal route data, routing module, cost determination module and an external interface that is optional [12]. The multimodal network of transportation was modeled by a multimodal graph, where every mono-modal subgraph represents one mode network [13]. The requirements for multimodal routing transportation system were described in detail in [14]. Multimodal routing is basically established on modeling methods of the multimodal network and algorithms for optimal pathfinding.

Currently, the industry is aware of the increasing growth of the multimodal navigation systems in the market that combine the public transit network and pedestrian. NAVTEQ launched in 2008 the Discover Cities TM project, which enables multimodal routing by combining the public transit network with the pedestrian. NAVITIME from Japan provides solutions for multimodal routing for mobile and Web-based devices [15]. OpenTripPlanner and Graphserver are active open-source projects for multimodal routing. Yet, all these solutions have a common drawback that is in transportation networks the intermodal facilities lack detailed modeling, which in turn limits the ability for issuing the multimodal route planning service.

## 3 Network Modeling

In this section, we elaborate on how the network modeling is performed in the proposed approach. For consistency, the use of the terms graph, network and their components are featured. Graph refers to edge/vertex model in the theory of graph, where it can be constructed for route planning by extracting it from the network. While the network refers to the model of data that consists of lines and nodes with their shapes and geographical positions as well as the topology of network [16]. Figure 2 illustrates the difference between graph and network. For network modeling, the following activities are carried out in this work:

- Development of a map that is a graph like which is made up of nodes and edges.
- Using different edges (colors) to represent different networks.
- Specifying the origin and destination.
- Choosing the preferable path and transportation mean.
- Determining the path using Dijkstra algorithm, where the path is shown using colored nodes.

In the following subsections, we discuss the single network that has different types of graphs where we will explain two types of them, which are walking network and public transit network (that includes railways and buses). Also, we discuss how these networks are combined to have a multimodal transportation network.

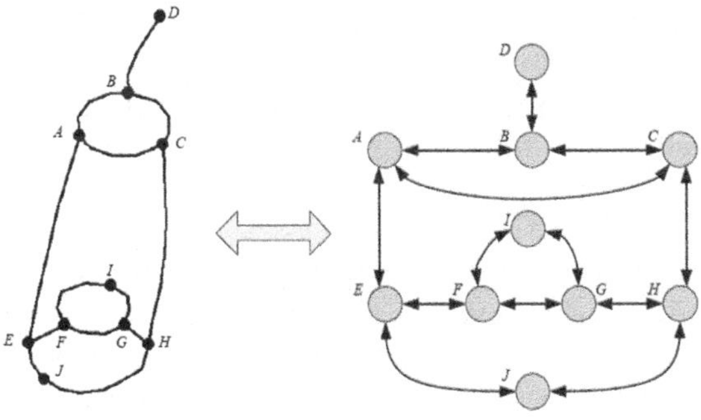


**Fig. 2** Sample of a network of 13 lines and 10 nodes on the right, and its corresponding graph of 13 edges (double way), 10 vertices on left

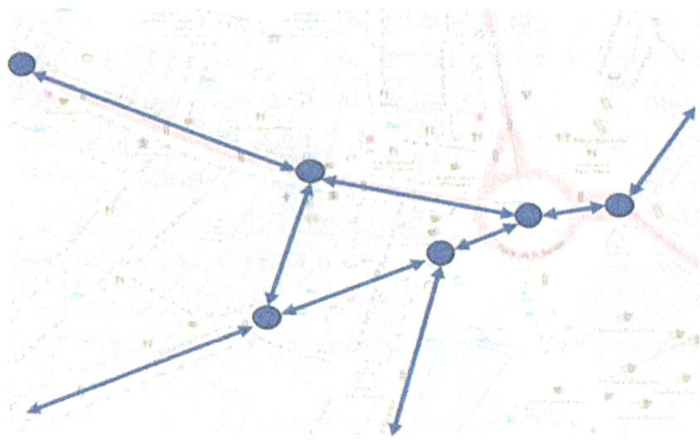

**Fig. 3** Walking network with vertices as intersections and edges as lines connecting vertices

## 3.1 *Single Networks*

### 3.1.1 Walking Network

Walking network construction is straightforward, and the intersection in the network model is represented by vertices. If a walking path is found between two or more intersections, an edge is found between these two vertices. The edge cost is determined by the average walking speed and the geographical length [9]. Figure 3 shows an example of a walking network model.

### 3.1.2 Public Transit Network

Public transit network includes several means such as railways and buses, where many approaches exist for modeling of this network. Most of the approaches are built on timetable concept, so we will introduce this concept in coming paragraphs. Also, we will provide a discussion of the following models: condensed model, time-expanded model and time-dependent model [9].

- Timetable Concept

This concept consists of the following tuple ($C, Z, B, \Pi$), where $C$ is an elementary connection, $Z$ is the public transit means, $B$ is the stop station of the transportation means and $\Pi$ is the timetable periodicity. Specifically, $C$ can be defined *as* $C = (Z, S1, S2, t1, t2)$, where $Z$ denotes a public transportation mean traveling from station $S1 \in B$ at time $t1 \in \Pi$ and arriving $S2 \in B$ at $t2 \in \Pi$, where $t1, t2 < \Pi$ and the vehicle goes from $S1$ to $S2$ directly with no stops in between [9].

- Condensed Model

The condensed model is the simplest approach for public transit model, because it is a time-independent approach where the network is represented in a structured way without including any scheduling. Maps of buses and railways are examples of this model because each map indicates only the lines, connections and the stations without considering the arrival or departure times. Each vertex on the graph $v \in V$ represents exactly one station $S \in B$, while an edge $(S_i, S_j)$ is only found if the timetable has an elementary connection that travels from $S_i$ to $S_j$. The cost of the edges can be omitted or described as shortest travel time among all the elementary connections that travel from station $S_i$ to station $S_j$. Thus, this model does not provide exact calculations for the shortest path, yet it only gives an overview of the structure of public transit network [9].

- Time-expanded Model

This model provides a complete representation of the public transit network where the arrival and departure times are addressed in the timetable and time-independent approach is still used. It has two types: the simple version and realistic version where the realistic version includes the realistic transfer times at stations between transportation means [9].

- Time-dependent Model

In this model, the travel times are represented as functions. This model overcomes all the problems that exist in the time-expanded model, but with introducing time dependency in the graph. Also, this model has two types: the simple version and realistic version, where the realistic version includes the realistic transfer times between transportation means [9].

## *3.2 Multimodal Network*

This part shows how different networks' types are combined to provide a multimodal network. First, locations, where different networks intersect, are identified and chosen, then networks are linked using transfer edges between vertices of different networks.

Given a set of a single network, graphs G1, ..., $G_n$, where each graph has a vertex, edge and label $(V_i, A_i, \sum_i)$, the addition of these graphs results in a multimodal graph of $G_{mm} = (V_{mm}, A_{mm}, \sum_{mm})$. The label and the vertex are the union of the labels and vertices of the input graphs, while the edge is the union of the edges of the input graphs and the transfer edges. The transfer edges are used to connect the graphs of single networks together in a way that it becomes possible to transfer from one single network graph to another. Figure 4 shows how to transfer from one layer to another with the help of transfer edges, such as from foot network to public transportation network, where the transfer between walking network to public transit network is only at the stations, so a transfer edge is inserted from a station vertex to nearest walking vertex on the network [9, 17, 18].

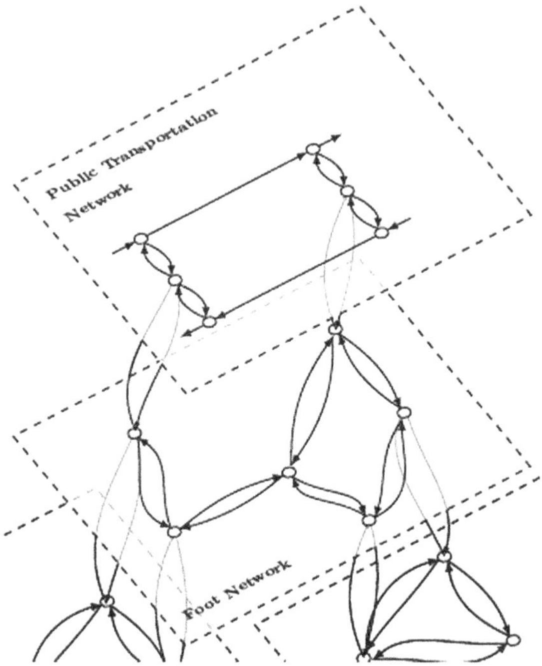


**Fig. 4** Graph of two layers that are linked together by transfer edges

# 4 Proposed Approach

In this section, we present the proposed approach. First, a graph-like map is developed, which is made up of nodes and edges. Different edges (colors) represent different networks. Then, "origin" and "destination" are entered. After that the "preferable path" is chosen, where the preferred transportation mean is chosen. Accordingly, the "path" is calculated using modified Dijkstra's algorithm. Finally, the path is shown with colored nodes as discussed in the next section.

A graph $G = (V, E)$ is used, where $V$ are vertices that represent the connections between $E$ or edges. These edges represent the streets, bus stations and metro connections, which have an associated weight that represents the length of the edge from its starting vertex to its ending vertex, a time that indicates the average time it takes to travel from starting vertex to ending vertex, and an object of the information class that contains additional information such as ID and name. The main aim is to have simple application user interface in order to facilitate choosing the best route considering the various criteria specified. As an output of the navigation system, a graphical resolution of the problem is presented with the origin, the traveled path until reaching the destination in different colors. Figure 5 presents the flowchart of the proposed approach.

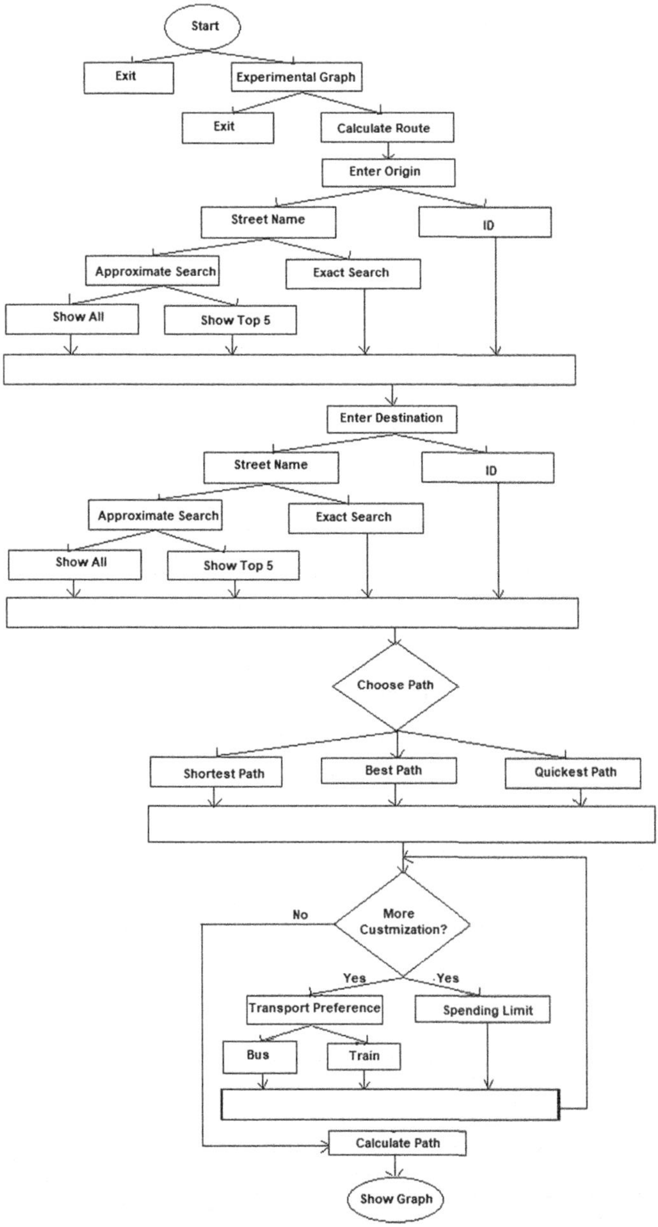


**Fig. 5** Flowchart of the proposed approach

## 5 Simulation and Results

In this section, the results of simulation of the proposed approach are presented. The simulation is performed using a Dell PC device with processor Intel i7-4500 CPU, RAM of 8 GB and × 64-based processor on Windows 10. Visual Studio program is used to simulate the algorithm and the results were satisfying. However, the screenshots are taken from an online.

GraphViewer. In the window of GraphViewer, every node on the graph is represented as a dot with yellow color. Also, every edge, which connects two nodes together and acts as roads, is represented by a line with different color in GraphViewer. There are three different colors for edges where each color represents different transportation means as shown in Fig. 6. These colors are:

- Blue: represents the public bus line route edge.
- Green: represents the public subway lines route edge.
- Red: represents the pedestrian walking path or road edge.

The results of our simulation, as shown in Fig. 7, are displayed on window of GraphViewer, where the node of origin is represented in blue color, and the nodes along the path are represented in green color and destination is represented in black color.

Despite that there are some issues faced during the implementation of the algorithm and the application, where the graph could not be implemented on the real map as it needs a higher computing capabilities. Also, choosing a city that has a different transportation means available online and extracting needed data was a challenging task that we have successfully performed.

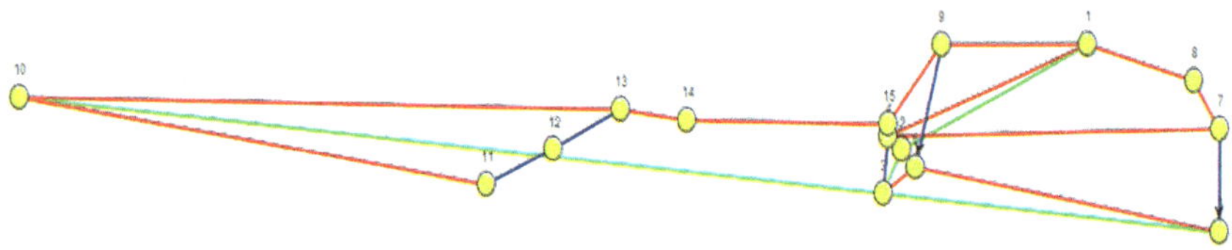

**Fig. 6** GraphViewer window for visualizing results

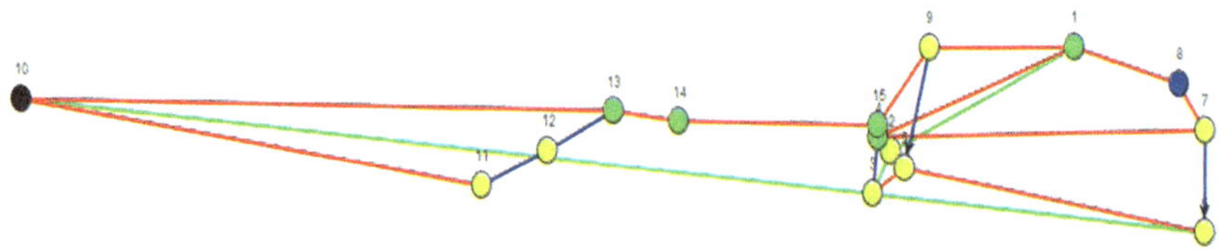

**Fig. 7** Result displayed after choosing the shortest path choice

## 6 Conclusion and Future Work

Multimodal routing with more focus on the public transit networks such as railways, bus lines and pedestrian ways was discussed in this paper. With the evolving demand on multimodal intelligent navigation services, the issue of the planning for multimodal routing has gain a great importance. The objective of these services is to discover an optimal and reasonable path with regards to transportation multimodal networks including pedestrian ways, bus lines and metro lines.

Moreover, in this paper, an efficient and feasible solution to the optimal pathfinding algorithm and multimodal network modeling problem was presented. Discussions in this paper showed that the proposed solution can support the next generation of multimodal navigation service. Also, this paper discussed the representation of multimodal networks for transportation and the formalization of the multimodal route planning problem. Besides that, we analyzed the design of multimodal route planning algorithm. This paper provided an efficient, effective and general approach that can help individuals to find the optimal routes in transportation networks with reasonable mode combinations and accessible modes. Simulation and result presented in this paper showed the applicability of the proposed approach.

Finally, there are some limitation in our work that we intend to work them out in the future, such as the time taken to travel from one node to next one on the path is not provided on the on the online GraphViewer, the total time of the trip and the total cost spent are not displayed too on GraphViewer. Also, it is worth to note some of the topics that are promising and so need more exploration in the future, such as visualization of results of multimodal routing, dynamic planning of multimodal route and planning of multimodal route for long trip distances.

**Acknowledgements** Our thanks to all reviewers for their dedication and valuable comments that are an added value and contributed much in the enhancement of this work.

## Reference

1. G.A. Klunder, H.N. Post, *The Shortest Path Problem on Large-Scale Real-Road Networks* (2006)
2. M. Muller-Hannemann, M. Schnee, *Finding All Attractive Train Connections by Multicriteria Pareto Search*. in Algorithmic Methods for Railway Optimization (2007)
3. F. Schulz, D. Wagner, C. Zaroliagis, *Using Multi-Level Graphs for Timetable Information in Railway Systems. Algorithm Engineering and Experiments* (2002)
4. J.-P. Rodrigue, C. Comtois, B. Slack, *The Geography of Transport Systems*, 2nd edn. (Routledge, New York, 2009).
5. T. Pajor, *Multi-Modal Route Planning*. Universitat Karlsruhe (2009)
6. W. Elhanafy, *Discrete Mathematics for Electrical Engineers*. 17SCIB05I. (The British University in Egypt, BUE, 2017), Lecture 8
7. R.K. Ahuja, T.L. Magnanti, J.B. Orlin, *Network Flows: Theory, Algorithms, and Applications* (1993)
8. M.F. Goodchild, *GIS and Transportation: Status and Challenges* (2000)

9. L. Filipov, B. Davidkov, *Development of Network Data Model for Urban Transportation System in GIS Environment for the City of Plovdiv, Bulgaria.* (International Conference on Cartography and GIS, Borovets, Bulgaria, 2006)
10. M. Muller-Hannemann, K. Weihe, *Pareto Shortest Paths is Often Feasible in Practice. Lecture Notes in Computer Science* (2001)
11. C. Samal, L. Zheng, F. Sun, J.R. Lillian, A. Dubey, *Towards a Socially Optimal Multi-Modal Routing Platform* (2018)
12. Y. Disser, M. Muller-Hannemann, M. Schnee, *Multi-Criteria Shortest Paths in Time-Dependent Train Networks* (Springer LNCS, 2008)
13. L. Lui, *Data Model and Algorithms for Multimodal Route Planning with Transportation Networks* (Technische Universitit Munchen, 2011)
14. NACTO Ride-Hailing Services: Opportunities and Challenges for Cities (2016) Retrieved from: https://nacto.org/policy-2016/ride-hailing-services-opportunities-and-challenges/
15. R. Bauer, D. Delling, D. Wagner, *Experimental Study of Speed up Techniques for Time-table Information Systems* (2011)
16. D. Kirchler, *Efficient Routing on Multimodal Transportation Networks* (Ecole Polytechniquue ParisTech, 2013)
17. Tennessee Department of Transportation. (2017). Transportation Demand Management Program. Retrieved from: https://www.tn.gov/tdot/article/transportation-demand-management-tdm
18. W. Zeng, X.-J. Chang, J.J. LV, Design of data model for urban transport. GIS. J. Geograp. Inf. Syst. (2010)